Farm Policy Analysis

Farm Policy Analysis

Luther Tweeten

Westview Press
BOULDER, SAN FRANCISCO, & LONDON

Westview Special Studies in Agriculture Science and Policy

This Westview softcover edition is printed on acid-free paper and bound in softcovers that carry the highest rating of the National Association of State Textbook Administrators, in consultation with the Association of American Publishers and the Book Manufacturers' Institute.

Published in 1989 in the United States of America by Westview Press, Inc., 5500 Central Avenue, Boulder, Colorado 80301, and in the United Kingdom by Westview Press, Inc., 13 Brunswick Centre, London WC1N 1AF, England

Library of Congress Cataloging-in-Publication Data
Tweeten, Luther G.
Farm policy analysis/by Luther Tweeten.
p. cm.—(Westview special studies in agriculture science and policy)
Includes index.
ISBN 0-8133-7745-5
1. Agriculture and state—United States. 2. Farms—Government policy—United States. I. Title. II. Series.
HD1761.T84 1989
338.1'873—dc20 89-9149
CIP

Printed and bound in the United States of America

The paper used in this publication meets the requirements of the American National Standard for Permanence of Paper for Printed Library Materials Z39.48-1984.

10 9 8 7 6 5 4 3 2 1

Contents

Preface

Farm Policy Analysis provides the foundation needed to understand, interpret, and analyze farm policy. It can be used as a text, a reference, or for selected readings. It is especially suited for upper-division undergraduate and graduate courses. The book was originally intended as the third edition of *Foundations of Farm Policy*. However, revisions were sufficiently extensive to require a new title.

Farm Policy Analysis does not require calculus but makes extensive use of classical welfare analysis to show the impact of farm policies on producers, consumers, taxpayers, and the *economy as a whole*. Analysis common in undergraduate texts which omits the latter and accentuates gains or losses only to producers is biased implicitly to favor producers. On the other hand, some texts are highly mathematical and rigorous but must be restricted to narrow policy issues because many of the principal dilemmas of farm policy do not lend themselves to formulas and quantification. *Analysis* attempts to avoid these pitfalls while maintaining rigor and comprehensiveness.

The book is intended to provide ready access to the fundamentals of farm policy but needs to be supplemented in places. Some readers will wish to consult U.S. Department of Agriculture sources for the latest details of farm commodity programs. Other readers will wish to supplement sections on technology with readings regarding the economic history of agriculture or supplement the section on the political process in Chapter 3 with data on the current composition of congressional agricultural committees.

The book rests on the proposition that farm policy can be studied properly only when it is placed within its social, economic, and political setting. It will challenge the upper-division undergraduate, who is now demanding a richer analytical diet, as his background is upgraded.

The term "farm policy" as used here refers to policies that affect groups and not just individuals. It deals with the ways farmers influence and are influenced by government policies. The table of contents further clarifies what I define as farm policy. The government policies that will be demanded (and also that will be accepted) by farmers depend heavily on economic characteristics and problems of farming discussed in Chapter 1. The conceptual framework in welfare economics helping to determine whether government interaction is warranted to deal with farm problems is laid out in Chapter 2. Farmers' goals, values, and the political process heavily influencing policy are discussed in Chapter 3. Farmers do not play the political game in isolation — government farm policies are subject to what urban America will accept.

The following three chapters analyze economic problems of low resource returns (Chapter 4), instability (Chapter 5), and macroeconomic linkages (Chapter 6). Monetary-fiscal policies outlined in Chapter 6 cannot be separated from international trade covered in Chapter 7. Agribusiness structure, conduct, and performance are found to contribute less to farm economic problems than commonly believed by laypersons (Chapter 8). Very real issues for farm policy are posed by problems of environmental and natural resources in Chapter 9 and by poverty and human and rural development in Chapter 10. Chapters 11 and 12 present the history and economic evaluation of current and proposed commodity programs.

To my colleagues, who have encouraged and stimulated my work; to the reviewers, especially James Bonnen, who have been so diligent and patient; to the Departments of Agricultural Economics at Oklahoma State University and Ohio State University and to the Food and Resource Economics Department at the University of Florida, all of which provided facilities and a stimulating academic environment; and to Brenda Jordan for competent typing — to each of these institutions and individuals I am very deeply grateful. None of them, needless to say, is responsible for any errors or views presented herein.

Luther Tweeten
The Ohio State University

CHAPTER ONE

Farm Characteristics and Problems

The United States food and fiber system includes (1) producers of agricultural commodities or the farm sector, (2) suppliers of farm inputs, and (3) marketers of farm production. The system is large and growing. Value added in the food and fiber sectors increased from $326 billion to $701 billion between 1975 and 1985 (Table 1.1). In real volume, the food and fiber system grew 14 percent in the decade. During the same period it increased from 20.1 million workers to 21.4 million workers. Of the major components, only the farm sector lost ground with employment falling from 2.8 million workers in 1975 to 2.5 million in 1985.

TABLE 1.1. Value-Added and Employment in the Food and Fiber System, 1975 and 1985

Item	1975		1985	
	($ billion)	(million workers)	($ billion)	(million workers)
Value-added:				
Farm sector	43.3	2.8	71.6	2.5
Nonfarm sectors	282.4	17.3	629.1	18.9
Food processing	38.7	1.5	83.0	1.6
Manufacturing	57.0	3.2	103.3	3.0
Transportation, trade, and retailing	96.8	5.7	220.4	6.6
Restaurants	25.7	3.1	58.3	3.6
All other	64.2	3.7	164.2	4.1
Total food and fiber	325.7	20.1	700.8	21.4
Total domestic economy	1,598.0	93.8	3,998.0	115.5
	(percent)			
Value-added:				
Farm sector	2.7	3.0	1.8	2.1
Nonfarm sectors	17.7	18.4	15.7	16.4
Total food and fiber	20.4	21.4	17.5	18.5
Total domestic economy	100.0	100.0	100.0	100.0

SOURCE: U.S. Department of Agriculture (January 1987, pp. 17, 18).

All components declined in *share* of the nation's total value-added and employment. The total food and fiber system dropped from about 21 percent of value added and employment in the U.S. economy in 1975 to about 18 percent in 1985. The farm sector share of the economy fell from 3 percent to 2 percent (Table 1.1).

For each worker or unit of output in the farm sector there was one unit in the input supply sector and there were eight units in the product marketing sector including wholesale, processing, storage, transportation, and retailing. Many input supply and marketing activities once provided by farmers are now provided by others.

The above numbers do not mean that an additional dollar of farm output adds $1 to input supply and $8 to product marketing output. A more typical output multiplier is $4. Each dollar of farm output adds about $1 to input supply and $2 to product marketing. In a typical situation each $1 of farm output adds $1 to the county but off the farm, another $1 to the state outside the county, and another $1 to the nation outside the state. Counting the original $1 of farm output, the multiplier of $4 is well below the ratio, 10, of farm output to total farm and food system output because much food system activity would remain even if all food were imported. The multiplier hints at what will be more apparent in Chapter 12 — a supply control program reduces not only farm output but food sector and national output and real income. Multipliers do not measure economic efficiency. If more output is not desired by consumers, an expansion of output will be economically inefficient even though the multiplier is large.

TERMS OF TRADE, INCOME, AND WEALTH

Farming is a heterogeneous industry. Some of the diversity and changes in structure and characteristics are apparent from aggregate income and balance sheet accounts. But first it is well to examine another measure of farm economic health, prices.

FARM PRICES

Farm prices are a useful measure of short-term changes in the economic health of the farming industry but are a poor measure of the economic position of farmers over time. The widely used price standard is the parity ratio, defined as the ratio of the index of prices received by farmers for crops and livestock to the index of prices paid by farmers for production inputs including interest, taxes, and wage rates. The ratio is often expressed as a percent of that which prevailed in 1910-14, a favorable economic period for farmers.

The parity price ratio needed to obtain a specified net farm income covering all costs of production (including a reasonable profit and a rate of return on farm resources comparable to what they could receive elsewhere) changes over time. With competitive equilibrium and constant returns to scale, all costs equal all returns so that

$$PQ - P'X = 0$$

where

P is prices received by farmers,
Q is aggregate output,
P' is prices paid by farmers, and
X is aggregate input.

Rearranging terms

$$\frac{Q}{X} = \frac{P'}{P}.$$

It is apparent that in equilibrium the output-input ratio (productivity rate) equals the inverse parity price ratio. All costs of production are covered by receipts. Productivity Q/X increased from an index of 100 in 1910-14 to 315 in 1986. This implies that the parity ratio could drop to an index of 32 while providing resource returns comparable to those in the 1910-14 period. The parity ratio fell from an index of 100 in 1910-14 to an index of 51 in 1986. Each unit of real farm production input producing one unit of output in 1910-14 produced 3.15 units of output in 1986 but at a real output price 51 percent of that in 1910-14. That means that the index of real prices received per unit of resources (*factor terms of trade*) was 3.15(51) = 161 in 1986 compared to 100 in 1910-14, a gain of 61 percent! *Commodity terms of trade* (the parity ratio) which fell nearly by half since 1910-14 is much inferior as a measure of farm economic health to the *factor terms of trade* which *increased* by over half since 1910-14 (Figure 1.1).

Figure 1.1 dramatically illustrates the importance of productivity advances to farmers' real prices as appropriately measured by factor terms of trade. Commodity terms of trade and factor terms of trade were close together until the late 1930s when productivity began to accelerate. Factor terms of trade initially rose sharply but advanced very slowly since the 1940s. Factor terms of trade vary widely from year to year because of the influence of weather on the productivity index. The slowing of productivity growth since the 1950s is a matter of considerable concern.

Today's farmers can cover all costs of production at a price well below parity standards of 1910-14. We will observe in Chapter 4 that adequate-size, well-managed farms covered all costs in 1986 at just over 50 percent of parity on average. Farmers prefer 100 percent of parity prices to parity income or rates of return because prices set at 1910-14 levels would provide massive windfall gains with today's productivity. Gains would be bid into land values so that rates of return on assets would remain unchanged.

In a dynamic industry, historic prices are a meaningless measure of "parity" defined as fairness. For example, productivity of computers has increased 1,000 times since 1960 and that benefit has been passed to consumers. It would make no sense to define the 1960 price as parity and pay computer producers the same real price per unit of computer capacity as in 1960 — 1,000 times today's prices! Income and rate of return on resources are far better measures of the economic health of the farming industry than parity prices.

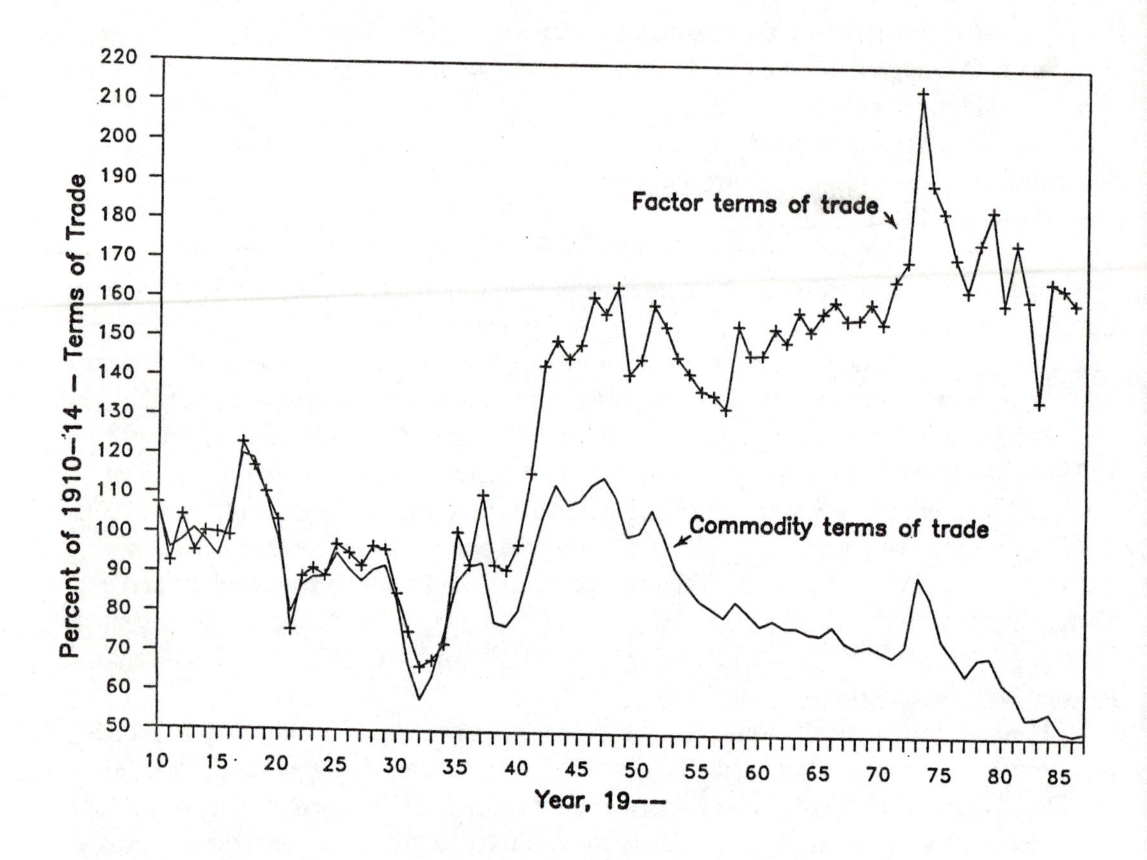

FIGURE 1.1. Commodity and Multifactor Terms of Trade

INCOME AND EXPENSES

Several characteristics of the farm sector are apparent from income and expense accounts by economic class of farms shown in Table 1.2:

1. Farm production is highly concentrated on larger farms defined as those with annual sales over $250,000. Large operations comprising less than 5 percent of all farms accounted for nearly half of farm output (gross income) in 1986.

2. *Large farms* accounted for a larger proportion of net farm income (66 percent) than of receipts (47 percent) in 1986. *Rural residences* (sales under $5,000) that accounted for 40 percent of all farms lost money from farming on the average in 1986 and in prior years of the 1980s.

3. Off-farm income grew faster than farm income from World War II to 1980, helping to level total income from all sources among farm classes. Whereas income from farming became more concentrated on larger farms, income from off-farm sources became more concentrated on the smallest farms. In fact, off-farm income was so high relative to net farm income for noncommercial farms (sales under $40,000) that the "farm operators"

TABLE 1.2. Farm Numbers and Income from Farm and Off-Farm Sources by Size of Farm (Including Operator Dwelling), 1980 and 1986, United States

	Farm Size by Sales Class					
	$250,000 and Over	$100,000 to $249,999	$40,000 to $99,999	$5,000 to $39,999	Less than $5,000	Total
Number of farms (1,000)						
1980	105	166	354	880	928	2,433
1986	95	210	294	725	887	2,211
Percent of all farms						
1980	4.3	6.8	14.5	36.1	38.3	100.0
1986	4.3	9.5	13.3	32.8	40.1	100.0
	(dollars per farm before inventory adjustment)					
Gross farm income						
1980	734,181	171,124	78,957	23,728	7,217	63,956
1986	811,463	179,086	80,476	24,066	7,851	73,605
Percent of all gross income						
1980	46.0	18.3	18.0	13.4	4.3	100.0
1986	47.3	23.1	14.5	10.7	4.4	100.0
Net farm income						
1980	169,552	25,451	6,016	-698	-1,419	9,223
1986	282,811	44,795	14,455	1,566	-1,028	18,426
Off-farm income						
1980	12,914	9,204	8,812	14,268	17,383	14,263
1986	17,508	12,602	13,780	17,439	26,704	20,212
Total net income						
1980	182,466	34,655	14,828	13,570	15,964	23,486
1986	300,319	57,397	28,235	19,005	25,676	38,638

SOURCE: USDA (December 1987).

might more appropriately be called machinists, mechanics, physicians, or whatever nonfarm occupation from which they made the vast majority of their income. The rise of off-farm income reduced poverty rates in farming from 50 percent in 1960 to near rates in the nonfarm sector by the late 1970s. The farm rate increased to 20 percent compared to 14 percent for nonfarm families in 1985 as farms experienced financial stress in the 1980s. The farm rate dropped to 13 percent in 1987, one percentage point *below* the nation's rate. These rates would be much lower if adjusted for in-kind welfare payments and net worth.

4. Among farm classes, total income from all sources was lowest on farms with sales of $5,000 to $40,000 in 1986, a major turnaround from 1970 when rural residences had the lowest incomes among farms. The full-time small farm is rapidly vanishing. An

economic unit, defined as a farm large enough to realize economies of size and provide a family operator income comparable to income elsewhere, required sales of approximately $150,000 and assets of $1.5 million in 1986. The figure varied by types of farms, of course. Operators of smaller farms tended to expand their operation or obtain off-farm income to achieve a satisfactory standard of living. Ever larger economic units are required to realize economies of farm size, that is, produce at low cost per unit. The "marginalization" process caused by lower real farm product prices and scale-increasing technology has now extended to medium-size farms. Farm labor and management demands on operators and their families make adjustments to part-time farming (off-farm employment) less feasible on medium-size farms than on small farms. Many medium-size farms are not large enough to achieve economies of size in production and marketing but too demanding of time to allow for much off-farm earnings.

5. Disposable personal income per capita of persons on the quintessential family farm with sales of $100,000 to $250,000 in 1986 averaged nearly double that of nonfarm persons although farm prices were only 51 percent of 1910-14 parity. On smaller farms, income below that of nonfarmers does not necessarily imply economic disequilibrium. Numbers of part-time small farmers increased in recent years despite lower incomes than elsewhere apparently because amenities of the farm way of life and tax advantages compensated for lower money income. Government payments of $12 billion, over one-fourth of net farm income, helped to maintain farm income in 1985. Rates of return on investment in farming averaged at least as high as rates on major alternatives from 1965 to 1980 (USDA, January 1981, p. 51). This impressive movement toward economic equilibrium has not been without adjustment costs, however.

6. Prices received by farmers declined 8 percent between 1980 and 1986, moving some farms of a given real scale to lower nominal sales classes. Adjusted for this price change, farms in smaller sizes and the lower end of the middle-size range declined in numbers and in the share of all farms and all output. Larger farms grew in numbers and shares of farm output. Agricultural census data show that the number of farms with 1 to 50 acres increased from 542,787 in 1978 to 636,917 in 1982, or by 17 percent. Farms with over 2,000 acres also increased in numbers while all acreage classes between 50 and 2,000 declined in numbers. The conclusion is that the position of middle-size farms is eroding but at a very slow pace.

BALANCE SHEET

Farm balance sheet data reveal several insights regarding farm structure.

1. The aggregate real volume of assets in the farming industry has remained virtually constant for several years. Real assets were the same level in 1986 as in 1970. Real *net* investment requirements for the farming industry as a whole were modest indeed and likely will continue to be modest.

2. The nominal value of farm assets *per farm* decreased from 1980 to 1986 (Table 1.3). Nonetheless, gross financial investments to maintain the individual economic

TABLE 1.3. Balance Sheet of the U.S. Farming Sector per Farm (Including Households) by Sales Class, January 1, 1980 and 1986

	Farm Size by Sales Class					
	$250,000 and Over	$100,000 to $249,999	$40,000 to $99,999	$5,000 to $39,999	Less than $5,000	Total
	(dollars per farm)					
Assets						
Real Estate						
1980	1,624,381	875,956	479,175	231,027	97,836	347,965
1986	1,525,600	613,465	343,198	151,898	77,908	250,565
Other						
1980	576,905	250,136	143,234	67,498	29,048	107,604
1986	618,589	247,356	146,910	65,170	38,465	106,468
Total assets						
1980	2,201,286	1,126,092	622,409	298,525	126,884	455,569
1986	2,144,189	860,821	490,108	217,068	116,373	357,033
Claims						
Liabilities						
Real estate debt						
1980	252,438	104,184	52,677	19,776	7,459	39,375
1986	346,745	116,066	58,685	19,098	8,277	43,329
Other						
1980	343,752	75,273	44,439	14,197	4,147	35,594
1986	319,655	80,025	38,810	12,303	3,960	32,112
Total						
1980	596,190	179,457	97,116	33,973	11,606	74,969
1986	666,400	196,091	97,495	31,401	12,237	75,441
Proprietors' equities						
1980	1,567,000	946,634	525,293	264,552	115,278	380,600
1986	1,447,789	664,731	392,613	185,666	104,136	281,592
Total claims						
1980	2,163,190	1,126,091	622,409	298,525	126,884	455,569
1986	2,114,189	860,822	490,108	217,067	116,373	357,033
Debt to asset ratio (%)						
1980	27.6	15.9	16.8	11.4	9.1	16.5
1986	31.5	22.8	19.9	14.5	10.5	21.1

SOURCE: USDA (December 1987 and earlier issues).

farming unit are large. They come from inflating asset values, increasing farm scale, replacing assets, and life-cycle entry-exit patterns of traditional family farms.

3. The debt-asset ratio for the farming industry is low in the aggregate but is higher for larger farms than for smaller farms (Table 1.3). Debt values rose faster than asset values from 1980 to 1986, leaving the debt-asset ratio higher in 1986 than in 1980.

4. Farmers have more wealth than nonfarmers. Although proprietors' equities declined by nearly $100,000 from 1980 to 1986, average wealth of $281,592 per farm was well above that, $136,000, of the average American taxpayer in 1985 (Oster, 1986, p. 24). Wealth of commercial farmers averaged several times that of nonfarmers. Farmers use twice as much capital per worker as other industries on the average.

The U.S. Bureau of the Census (1986) reported median net worth in 1984:

$32,670 for all U.S. households,
$60,420 for householders with a college education,
$50,120 for married-couple households, and
$73,660 for householders age 55 to 64.

These measures of net worth were below those of average farm households. Unfortunately, median measures of farm net worth are not available for more reliable comparisons.

5. Farmers are net debtors. They owe others about $3 for each $1 others owe them. Debtors gain and creditors lose when interest rates fall. This means that an unanticipated decrease in real interest rate (market rate less inflation) as occurred in the 1970s increases farmers' real wealth. An unanticipated increase in real interest rate as occurred in the 1980s decreases farmers' real wealth.

6. Farm ownership and output display similar degrees of concentration. Total income from all sources and wealth are much more equally distributed than are proprietors' equities or landholdings, however, because much of farmers' wealth is human capital. For that reason concentration of landholdings is much less of a concern in the U.S. than in Latin America where the distribution of human as well as material capital is highly unequal.

7. Farmers operate with lower debt ratios than other industries on the average although the ratios increased in the 1980s.

On the whole, operators of smaller farms have unusually low debt-asset ratios. They appear not to be fully utilizing their equity base to expand assets and income. Their observed conservative financial position for the most part is not caused by discrimination in lending to small farm operators. Operators, including minority races, of small farms in Oklahoma in 1981 indicated credit was readily available (Sanford et al., 1984). Small farm operators did not aggressively use their substantial equity bases to obtain available credit because they lacked profitable uses of credit and they desired to avoid financial risk. The favorable equity position of small farms suggests opportunities for operators to expand scale by leveraging net worth if management and profitability impediments are overcome. Despite lack of favorable investment opportunities on small farms, a growing body of evidence from studies in Oklahoma, Missouri, and other states reveals that small farming operations are a preferred and intended long-term position for many operators, particularly for part-time farmers who indicate no desire to "get big or get out."

CAPITAL-LABOR SUBSTITUTION

Among the numerous elements identifying the structural transformation of American agriculture, none is more basic than the substitution of capital for labor. Labor accounted for approximately half of all farm inputs from 1910 to 1940 but for only 13 percent of all farm inputs in 1986 (Table 1.4). This substitution in turn is identified with (1) adoption incentives provided by the declining real price of capital relative to labor along with the availability of technologically improved inputs through science and industry, (2) increased reliance on inputs purchased from the nonfarm sector, (3) economies of size along with larger and fewer farms, (4) increased reliance on debt capital to finance assets, and (5) increased vulnerability of farms to economic setbacks in the face of cash demands to finance purchased inputs and pay family living expenses. Today, farmers have less "nonpurchased" real estate and operator labor on which returns can be deferred to absorb economic shocks.

TABLE 1.4. Input Shares, United States, Selected Years, 1910-86

	Input		
	Labor	Real Estate	Capital
		(percent)	
1910	53	20	27
1920	50	19	31
1930	46	18	36
1940	54	17	29
1950	38	17	45
1960	27	19	54
1970	19	23	58
1980	14	23	63
1986	13	20	67

SOURCE: Basic data from USDA (April 1987 and earlier issues).

The pace of substitution of capital for labor has slowed as noted in Table 1.4. With the decline in labor's share during the period 1980-86 slowing to the lowest rate since 1940, a leveling off is also expected in farm numbers. The number of farm migrants to cities will be low relative to the past because comparatively fewer people remain on farms. The farm population numbered 5 million in 1987 compared to over 30 million in the 1920s and 1930s.

Input shares differ widely among classes of farms (Tweeten and Huffman, 1980, p. 53). In 1970, the operator labor-management share was 5 percent and the operating capital share 61 percent of all inputs used on farms with sales of over $100,000. In the same year, on farms with sales of $2,500 to $5,000, the labor-management share was 31 percent and operating capital share 17 percent of all inputs. These differences help account for differences in economic productivity and profitability among farms.

FAMILY FARMS

A focal point in much of the structure debate is the *family farm*. No precise definition of the family farm is acceptable to all, but there is some agreement that it ideally should be a crop and/or livestock producing unit in which the operator and his/her family:

- control most of the decisions (a subset of this proposition is that the preferred legal organization is a sole proprietorship as opposed to a publicly held corporation, that the preferred market arrangement is the price system as opposed to vertical integration, and that the preferred tenure pattern for operators is ownership rather than tenancy);
- supply most of the labor;
- derive most of their income from farming; and
- receive family income and rates of return on resources comparable to those in the nonfarm sector.

In addition, society would like to have as many family farms as possible without compromising the goal of having ample supplies of quality food and fiber at reasonable cost. Society would like to have the farming occupation accessible to a broad range of persons, particularly to young persons having the ability and desire to farm.

Table 1.5 summarizes the position of family and other farms in 1978. Because agreement on what constitutes a family farm is not complete, several definitions are used to appraise the position of the family farm.

TABLE 1.5. Summary Estimates of Family and Nonfamily Farm Numbers and Sales, United States, 1978

Characteristic	Farms		Sales	
	(number)	(percent)	($mill.)	(percent)
Family Farms	744,581	30.0	48,581	44.9
Nonfamily Farms				
Small Farms				
(1) Sales of $20,000 or less	1,584,163	63.9	8,682	8.0
Large Farms				
(2) Corporations, nonfamily	5,852	0.2	7,042	6.5
(3) Partnerships, 3 + partners	31,266	1.3	3,986	3.7
(4) Vertically coordinated (other than in 1-3 above)	9,640	0.4	4,822	4.5
(5) Excess labor[a] (other than in 1-4 above)	103,140	4.2	35,001	32.4
Total Farms	2,478,642	100.0	108,114	100.0

SOURCE: Tweeten (1984, p. 8).

[a]Over 1.5 person-years of labor hired.

The first definition is the broadest, including as family farms all operations except "nonfamily" corporations, partnerships, and vertically integrated operations along with farms that hire over 1.5 person-years of labor. By this definition, 94 percent of all units were family farms in 1978 — a percentage that has remained almost constant for decades (Nikolitch, 1972). It is highly misleading, however, to view this stable percentage as evidence of the staying power of family farms. To do so would be to contend that the family farm has been preserved if only 20 farms remained in the United States and 19 of them were family farms. The nation went from 6.3 million farms in 1940 to 2.3 million farms in 1985, a loss of 4 million or about two-thirds of all farms. Most were family farms.

Farms that persistently receive the vast majority of their income from off-farm sources might also be excluded from the family farm designation. On this basis, excluding farms with sales of $20,000 or less per year, family farms numbered about three-fourths of a million and accounted for 30 percent of all farms and 45 percent of all farm output in 1978.

Farms with sales of $20,000 to $39,999 also typically receive most of their income from off-farm sources. Excluding those farms from the "family" category changed family farm numbers very little but reduced the proportion of all output attributed to family farms to 36 percent in 1978. The proportion continued to fall in the 1980s but at a slow rate.

Farm numbers dropped at an annual rate of 3.2 percent in the 1950s, 2.8 percent in the 1960s, 1.9 percent in the 1970s, and 1.3 percent from 1980 to 1986. More farms were lost each year on the average in the 1950s than were lost over the entire first half of the 1980s despite distressed economic conditions.

Approximately 50,000 farm operators retire or die each year. Without replacements, farm operators and farm numbers would drop 2.4 percent annually from retirements alone. The fact that farm numbers have fallen at a slower rate since 1970 means that entrants have offset nonretirement attrition among established farmers. The nation will not run short of operators in the foreseeable future. Farm population fell 2.4 percent annually in the first half of the 1980s, a rate approximately half that experienced in the 1950s and 1960s.

The number of aged operators has held rather steady, a trend expected to continue. The number of full-time small-farm operators has declined sharply as the operators expanded to a larger farm, left the farm, or obtained part-time off-farm employment. Meanwhile, the number of part-time farms has grown markedly and probably will continue to do so. After dropping sharply for decades, the number of small farms will tend to stabilize with an expected leveling off of the past sharp decline in full-time small farmers, with increasing numbers of part-time small farmers, and with fairly stable numbers of aged farmers. Middle-size farms have relatively more full-time operators and have not yet completed the adjustment process. Hence, future numbers of middle-size farms may fall relatively faster than that of smaller farms whose economic viability is increasingly assured by off-farm income. Compared with the 1950s and 1960s, however, relatively little excess labor remains in farming.

FARMS BY TYPE

Table 1.6 highlights the diverse characteristics of farms by type, defined by the primary enterprise on the farm. (It is important to note that other enterprises in addition to the primary one are likely to be present.) Wheat farms average largest in acreage (958 acres), but sugar farms average largest in sales ($335,000 in 1982), land and building assets ($1.6 million), and machinery and equipment assets ($172,000). Horticulture specialty farms average the fewest acres (53) but beef cattle "other" (mainly cow-calf) farms average smallest in sales ($5,000), whereas tobacco farms average smallest in assets.

Cow-calf beef farms are far the most numerous, accounting for 28 percent of all farms. There are relatively few sugar, cotton, and rice farms (latter not shown in Table 1.6 but numbering 7,000 in 1982) but these farms receive much public recognition because of large commodity program payments per farm.

TABLE 1.6. Selected Characteristics of U.S. Farms by Type, 1982

Farm Type	Number of Farms		Acres per Farm	Value of Assets per Farm: Land and Buildings	Value of Assets per Farm: Machinery and Equipment	Value of Sales per Farm	
	(1,000)	(%)	(acres)	($1,000)	($1,000)	($1,000)	(%)
Wheat	96	4	958	508	69	51	4
Corn	177	8	332	513	64	62	8
Cotton	21	1	836	887	103	150	2
Tobacco	131	6	105	118	20	21	2
Sugar	3	.1	105	1,607	172	335	1
Vegetables & melons	31	1	107	440	48	128	3
Fruits & tree nuts	84	4	107	420	34	69	4
Horticulture specialty	29	1	53	274	38	130	3
Beef cattle feedlots	85	4	369	336	42	193	12
Beef cattle, other	619	28	636	269	20	5	9
Hogs	130	6	183	243	39	66	7
Dairy	165	7	303	333	72	110	14
Poultry and eggs	42	2	140	217	41	240	8

SOURCE: U.S. Bureau of Census (1984, p. 9).

LAND TENURE

Three prominent elements of land tenure are noted: the rise of the part-owner operator; the lack of significant inroads of foreign ownership; and the slow, almost imperceptible, movement toward nonfarm ownership.

TENURE OF OPERATORS

After farm numbers peaked at 6.8 million in 1935 (and declined every year through 1980), the number of full-tenant operators decreased sharply (upper panel, Figure 1.2). The number of full-owner operators also dropped while that of part-owner operators increased slightly over the years. In 1982, 29 percent of operators were part owners, 59 percent were full owners, and 12 percent were full tenants.

The change in farmland operated by part owners is more dramatic than the change in their numbers (lower panel, Figure 1.2). In 1974, part owners accounted for 27 percent of all farms but 53 percent of all farmland, of which 28 percentage points were owned and 25 percentage points rented. Part-owner farms were about twice as large as other farms. Full tenants accounted for about one-tenth of all farms and all acres operated in 1974, hence their farms were of average size. The share of all farms operated by full owners has increased over time. But because full-owner farms were smaller than average size, their two-thirds share of all farms accounted for only about one-third of farmland operated in 1974.

The part-owner tenure arrangement is an adaptation to economic pressure. The part owner rents land to achieve economies of size while reducing instability, cash-flow, and asset acquisition problems. The part owner owns some land to gain the security of a home unit and an outlet for savings and investment.

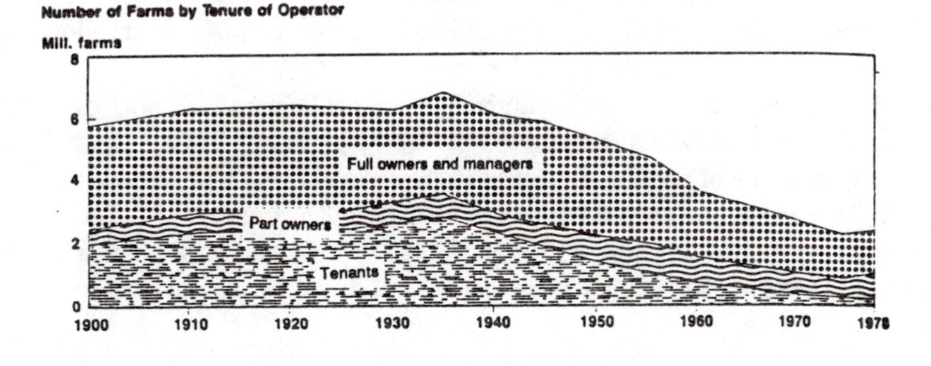

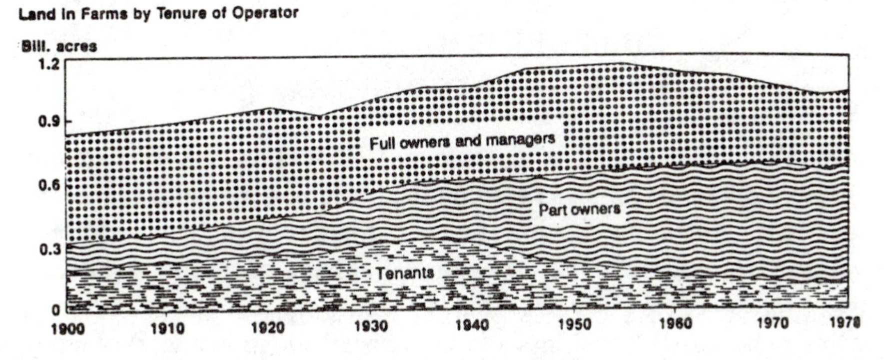

SOURCE: See Tweeten (1984, p. 11).

FIGURE 1.2. Number of Farms and Land in Farms by Tenure of Operator

FOREIGN OWNERSHIP

In 1986 foreigners owned about 12 million acres of U.S. farmland, or 0.6 percent of the total land. They owned less than 1 percent of all privately held cropland, pasture land, and forest land. Foreign individuals accounted for 8 percent of foreign owners, corporations for 79 percent, and partnerships for 11 percent. Another 2 percent was held by foreign estates, trusts, and institutions. Forest land accounted for 52 percent of foreign owned acreage, cropland for 17 percent, and other agricultural and nonagricultural land for 31 percent.

American citizens own approximately half of the capital in foreign corporations and partnerships holding U.S. land. Excluding this portion, fully foreign owned farmland accounts for less than 0.5 percent of all U.S. farmland. This amount is inconsequential in the aggregate but of some importance in selected local areas where foreign holdings are sizable. Also foreign ownership is prominent in some agribusiness industries.

NONFARM OWNERS OF FARMLAND

The family farm ideal holds that families should own the land they operate. A departure from that pattern might be regarded as a compromise of the family farm. On the other hand, it is unrealistic to expect a young family to begin farming by owning an economic farming unit and perhaps even to retire owning an economic unit. If some operators must rent to get a toehold in farming and realize economies of size, it follows that others must be nonoperator landlords.

According to the Census of Agriculture, 36 percent of farmland was owned by nonfarmers in 1978 and in 1982. The proportion was 32 percent in 1969. These figures exaggerate the active role of nonfarm investors in the farmland market. Data suggest that about half of the farmland owned by nonfarmers is in the hands of retired farmers or the spouses of retired and deceased farm operators. Thus, approximately 18 percent in 1982 and 16 percent in 1969 of all farmland was owned by other than present or former farm operators or their spouses.

ENVIRONMENTAL AND NATURAL RESOURCES

Of the nation's 1.4 billion acres of agricultural land, 413 million acres are cropland. Soil erosion is more severe on cropland than on the remaining agricultural land which is range and forest land. According to the National Resources Inventory conducted by the Soil Conservation Service (SCS, 1982), the national average of sheet and rill erosion on cropland was 4.4 tons per acre per year. This is roughly equal to the soil loss tolerance level (the maximum permissable soil loss allowing sustained productivity) of many soils. Data on *average* erosion rates across the nation conceal the fact that some regions have serious problems. About 44 percent of all cropland in the U.S. is eroding at levels greater than the soil loss tolerance. In 14 intensely cropped areas in the U.S. average erosion rates on cultivated cropland exceed 10 tons per acre (Lee, 1984).

The costs of these productivity losses are difficult to estimate. One set of estimates places the current costs of erosion-induced productivity losses for land in corn and soybeans at about $40 million per year, with present values over 100 years of $4.3 billion to $17 billion, depending on the discount rate (American Agricultural Economics Association Soil Conservation Policy Task Force, 1986). These estimates do not include the costs of offsetting management practices, the costs of erosion reduction measures such as terraces, and the costs of damage to growing crops from deposition of wind-blown soil. Another study of soil productivity losses concluded that if present levels of wind and sheet and rill erosion continued for another 100 years, productivity on soils with the greatest erosion problems nationwide might decline about 4 percent (Alt and Putman, 1987). Other studies provide similar estimates and indicate that improved crop varieties and other purchased inputs can offset yield losses due to erosion. Productivity declines in some regions will be much larger than the national average.

Estimates of the acreage lost to urban and built-up uses nationally have ranged from .9 to 1.1 million acres per year over the last several decades. Recent data from the 1982 National Resources Inventory suggest that between 1967 and 1982, urban, built-up, and rural transportation uses increased about .9 million acres a year (Lee, 1984). These estimates reflect the loss of all rural land, not just cropland or agricultural land. According to estimates from the 1975 Potential Cropland Study, less than 40 percent of land converted to urban and built-up uses between 1967 and 1975 was from prime farmland, considered by the Soil Conservation Service to be the best farmland. Not all of this land was in crops, but implies that 360,000 acres of prime farmland could be lost to other uses each year.

Groundwater pollution from farm pesticides and nitrogen fertilizers is of growing concern not just to nonfarmers but to farmers, many of whom rely on groundwater for drinking supplies. Too little is known about the extent of the problem and appropriate cures to reach definitive conclusions, but farm commodity program policy and environmental policy increasingly will intertwine.

LEGAL AND MARKETING ARRANGEMENTS

Individual or family proprietorship farms dominate farm numbers and, to a lesser extent, sales. Such units accounted for 87 percent of all farms and 70 percent of all farmland in 1982. Corporations, partnerships, and vertical coordination, compared to individual proprietorships coordinated by the price system, compromise the family farm ideal to the extent that they diminish control of farm operators over decisions. Sometimes these alternative legal and marketing forms give rise to units where the operator and family provide a small share of the farm equity capital, labor, and management.

CORPORATIONS

The Census of Agriculture reported 59,792 farm corporations in 1982, an increase of 11 percent since 1974. Most were family corporations. Corporations accounted for only 3 percent of all farm numbers but for 14 percent of farmland in 1982. The share is expected to expand slowly in the future. The share of nonfamily corporations in farm numbers and receipts was shown in Table 1.5.

PARTNERSHIPS

The 1982 U.S. Census of Agriculture (1984) identified 223,274 farms operating as partnerships. Partnership farms accounted for 10 percent of all farms but for 16 percent of all farmland. Partnership farms tended to be smaller than corporate farms but larger than individual proprietorship farms. Seventy-six percent of partnership farms had only two partners and another 22 percent had only three partners in 1974. Judging by the number of partners and the family relationship of partners and operators, most partnerships were family farms.

VERTICAL COORDINATION

Vertical coordination is of two types: (1) *production contracts* in which a farmer or group of farmers agrees to provide a commodity according to specifications to a food processor or other contractor, and (2) *vertical integration* in which two or more phases of the food supply chain are controlled or performed by one firm. Examples of the latter are production input supply firms or food processing firms also engaged in farming.

Production contracts in fruits, vegetables, and milk are frequently arranged through group bargaining between producers and processors under federal marketing orders. In such cases, the individual producer sacrifices little control over production and marketing decisions but may enhance economic rewards through group bargaining and efficiency gains passes by processors to producers. Vertical integration is a more serious threat to the family farm ideal than is production contracting.

The proportion of all crops and livestock produced under production and marketing contracts increased from 15.1 percent in 1960 to 22.9 percent in 1980, while the proportion of all crops and livestock under vertical integration increased from 3.9 percent to 7.4 percent during those years (Table 1.7). Vertical coordination is extensive in vegetable, fruit, sugar, and seed crops production. Major crops such as food and feed grains, soybeans, and tobacco have experienced comparatively little vertical coordination.

Vertical coordination is more extensive in livestock than in crops. And production contracts are used more extensively than vertical integration in both crops and livestock. Production and marketing contracts dominate fluid milk and broiler production. Vertical integration increased from 10 percent in 1960 to 37 percent in 1980 for eggs and from 4 percent to 28 percent in those years for turkeys. For the United States as a whole, vertical

TABLE 1.7. Crops, Livestock, and Livestock Products Output Under Production Contracts and Vertical Integration as a Percentage of Total Output of These Respective Commodities, 1960, 1970, and 1980

	Production Marketing Contracts			Vertical Integration		
Products	**1960**	**1970**	**1980**	**1960**	**1970**	**1980**
			(percent)			
Crops	8.6	9.5	14.3	4.3	4.8	5.3
Feed grains	0.1	0.1	7.0	0.4	0.5	0.5
Hay and forage	0.3	0.3	0.5	----	----	----
Food grains	1.0	2.0	8.0	0.3	0.5	0.5
Vegetables for fresh market	20.0	21.0	18.0	25.0	30.0	35.0
Vegetables for processing	67.0	85.0	85.0	8.0	10.0	15.0
Dry beans and peas	35.0	1.0	2.0	1.0	1.0	1.0
Potatoes	40.0	45.0	60.0	30.0	25.0	35.0
Citrus fruits	60.0	55.0	65.0	20.0	30.0	35.0
Other fruits and nuts	20.0	20.0	35.0	15.0	20.0	25.0
Sugarbeets	98.0	98.0	98.0	2.0	2.0	2.0
Sugarcane	40.0	40.0	40.0	60.0	60.0	60.0
Other sugar crops	5.0	5.0	5.0	2.0	2.0	2.0
Cotton	5.0	11.0	17.0	3.0	1.0	1.0
Tobacco	2.0	2.0	2.0	2.0	2.0	2.0
Oil-bearing crops	1.0	1.0	10.0	0.4	0.5	0.5
Seed crop	80.0	80.0	80.0	0.3	0.5	10.0
Miscellaneous crops	5.0	5.0	5.0	1.0	1.0	1.0
Livestock items	27.2	31.4	38.2	3.2	4.8	10.1
Fed cattle	10.0	18.0	10.0	3.0	4.0	6.0
Sheep and lambs	2.0	7.0	7.0	2.0	3.0	3.0
Hogs	0.7	1.0	1.5	0.7	1.0	1.5
Fluid grade milk	95.0	95.0	95.0	3.0	3.0	3.0
Manufacturing grade milk	25.0	25.0	25.0	2.0	1.0	1.0
Eggs	5.0	20.0	52.0	10.0	20.0	37.0
Broilers	93.0	90.0	89.0	5.0	7.0	10.0
Turkeys	30.0	42.0	62.0	4.0	12.0	28.0
Miscellaneous	3.0	3.0	3.0	1.0	1.0	1.0
Total farm output	15.1	17.2	22.9	3.9	4.8	7.4

SOURCE: Manchester (1983, p.7).

integration accounted for 3.2 percent of livestock production in 1960 and for 10.1 percent in 1980.

Overall, vertical coordination in crops and livestock, defined as production contracts and vertical integration combined, increased from 19 percent of farm output in 1960 to 30 percent of farm output in 1980. Farming is very slowly taking on the conglomerate, integrated economic form that has long characterized nonfarm industry.

HUMAN RESOURCES IN AGRICULTURE

FARM OPERATORS

The average age of farm operators dropped from 51.2 years in 1969 to 50.5 years in 1982 because many young operators began farming during the halcyon years of the 1970s. A total of 356,146 persons less than 35 years of age operated farms in 1982 compared to only 327,000 in 1969. Operators used credit on concessional terms, leasing arrangements, and, most importantly, assistance from parents to get a toehold in farming. Without technical, managerial, and financial assistance handed from parents to their children who farm, the family farm would nearly end in a generation. That assistance will continue, however, assuring a supply of new family farm operators in the foreseeable future. The ratio of retiring and deceased operators to farm youth is increasing. Hence, would-be young operators will have a higher chance of getting into farming during the next two decades than during the 1950s and 1960s unless the pace of farm consolidation accelerates.

Operators of black and "other" races numbered only 54,143 in 1982, or 2 percent of all farm operators. Female farm operators outnumbered black farm operators in 1982. In that year, female farm operators numbered 121,578, constituting 5 percent of all operators and 4 percent of all farmland operated. The proportion of operators underestimates the role of women on farming. Women increasingly share in the decisions made on farms although men by tradition continue to be listed as *the* operator.

OFF-FARM INCOME AND EMPLOYMENT OF OPERATORS

Before the integration of the farm and nonfarm economies, human resources tended to be employed either in the farm or nonfarm sectors but not in both at once. In recent decades, the farm and nonfarm labor economies have integrated, making it possible for small farm operators to stay on the farm and earn a living from nonfarm jobs. The extent of the integration is noted below.

1. An estimated 21 percent of all farmers resided off the land they operated in 1982, little change from 23 percent in 1969 (U.S. Bureau of the Census, 1984).

2. Over 90 percent of all farm families had at least one source of off-farm income (Carlin and Ghelfi, 1979).

3. Sixty percent of all income of farmers came from off-farm sources in 1985.

4. Forty-five percent of farm operators listed their principal occupation as other than farming in 1982.

5. In 1982, 58 percent of all farmers spent at least some days in off-farm employment (U.S. Bureau of the Census, 1984). The figure ranged as follows by size of farm in 1982:

	Percent of Farm Operators with Off-Farm Employment	
Farm Size by Sales	Some Days	200 Days or More
$250,000 and over	21	8
$40,000 to $249,999	32	11
$5,000 to $39,999	61	39
Less than $5,000	75	57
All farms	58	38

HIRED WORKERS

Increasing separation of farm labor from management could be apparent in a rising proportion of hired workers in farming. Hired workers comprised about one-quarter of all farm workers (hired and family) for several decades, but the ratio began to rise in the 1970s and was 40 percent in 1986. This ratio has drawbacks as a measure of separation of labor from management and hence of the industrialization of agriculture. The number of hired workers was nearly the same in 1970 and 1986 while total hours worked fell 29 percent. Part-time hired labor became more prominent and each worker put in fewer hours. Family worker numbers fell 51 percent in the same period. The share of hired labor costs in farm production expenses not only is low (6 percent in 1986) but was lower from 1980 to 1986 than from 1970 to 1980 (U.S. Department of Agriculture, December 1987, p. 33).

Oliveira and Cox (p. iii) list characteristics of the hired farm work force in 1985:

- In that year, 2.5 million persons were hired farm workers, 2.9 million persons were farm operators, and 3.8 million persons were unpaid farm workers, the latter often spouses and children of operators.
- Most hired workers were young. Just over half were under 25 years of age; another 24 percent were 25 to 34 years of age.
- In part because workers were young and because hired farm work is available to persons with little schooling, education attainment of hired farm workers was low. Over half of all hired farm workers had not completed high school and 40 percent of those age 25 and over had not completed high school in 1985.
- Nearly one-fourth of hired farm workers were Hispanic or black and other, compared to less than 5 percent of either operator or unpaid farm workers.

- Hired farm workers earned an average of only $3,247 from farm work and $2,579 from nonfarm work in 1985.
- Hired farm workers averaged 107 days of hired work on farms. Less than one-third worked 150 days or more on farms in 1985. Almost 40 percent, mostly students, of the hired farm workers were not in the labor force most of the year.
- Migrant laborers comprised only 6 percent of the total hired farm work force.

OTHER CHARACTERISTICS AND PROBLEMS OF THE FARMING INDUSTRY

Several farm problems have been alluded to in this chapter. Before providing a more complete list, it is well to review additional characteristics of farming that give rise to economic problems.

Several forces of markets and nature create unique features of the farm economy:

1. Biological processes of crop and animal production sometimes require long lags between production planning, planting or breeding, and marketing of the final product. The process, unlike a factory production line, is characterized by unpredictable and often uncontrollable influences of weather, insects, disease, and other pestilence as well as by changes in prices.

2. Demand and supply at the industry level are price inelastic for most farm products and for the industry as a whole in the short run. The elasticity of supply, the percentage change in quantity associated with a 1 percent change in price, varies among commodities but tends to average approximately .2 for individual crops and .3 for livestock and livestock production, and .1 for aggregate farm output in the short run (Tweeten, 1979, ch. 9; Henneberry, 1986). Long-run supply response is much larger, approximately .8 for major crops, over 1.0 for livestock and livestock products, and 1.0 for the aggregate farm output, but annual price variability is influenced most by short-term parameters. However, when stocks are included in supply and reserves are large, even short-run supply can be very price-elastic.

Short-run price elasticities of demand mostly are low, with domestic demand elasticities ranging from -.6 for beef and veal to -.5 for pork and chicken to -.2 or lower for feed grains, wheat, soybeans, and cotton at the farm level. Adjustment for domestic inventory demand raises elasticities (absolute value) in the short run and export demand raises elasticities especially in the long run. Still, aggregate price elasticity of demand for farm output is probably no higher than -.3 in the short run and -1.0 in the long run (see Tweeten, 1979, ch. 11). Price elasticity of meat demand may have fallen along with income elasticities in recent years as noted below.

3. Income elasticities of demand are low for most farm products. Demand for beef was long considered to be one of the most responsive of all farm products to consumers' income. Typical income elasticities of demand for beef were .5, for pork .4, and for poultry .2. More recent studies (Chavas, 1983) suggest income elasticities near .2 for beef, .3 for pork, and .3 for poultry meat. Some studies (see Hudson and Vertin, 1985)

show negative income elasticities of demand for pork and beef. The latter results could be spurious — caused by changes in tastes and preferences away from fat and cholesterol in diets which correlated with trends in income.

At any rate, price and income elasticities of demand for meat appear to have fallen. The good news for farmers is that demand for red meat is less sensitive to business cycles. The bad news is that in a mature economy such as the United States where the great nutrition problem is excessive food consumption people do not eat more pounds of food as incomes rise. They do improve quality of food so that the income elasticity of demand for farm products in aggregate is approximately .1. The income elasticity of demand for all goods and services is 1.0. It follows that as incomes rise most of the increase goes to purchase other goods and services such as recreation or computers and little is used to buy more food. With more single parents, with housewives working outside the home, and with more income, demand for food convenience in terms of additional processing of food and eating out rises. Hence the income elasticity of demand for marketing services, perhaps about .5, is much greater than the income elasticity of demand for farm ingredients. It follows that demand for farm food ingredients grows mainly from population and exports and it grows little from rising domestic consumer income. Demand grows much faster for food processing and other marketing services and for nonfood goods and services.

4. Demand faced by the individual farmer is highly price elastic. Farmers are price takers, not price makers. If an individual farmer raises his offer price above the market price, he has no market — implying an infinitely elastic (horizontal) demand curve at the farm level. Individual firms in many industries can reduce output to increase price or can raise their price and still sell merchandise. Of course, that does not assure them a profit because quantity may fall sharply as price rises — witness the high failure rates for firms even in concentrated industries in recent years. Farmers are not alone as price takers. They are not an island of pure competition in a sea of monopoly. They are not the only Americans who "buy retail and sell their products wholesale." Most Americans sell their goods and services as price takers in highly elastic demand markets — e.g., laborers, most small businesses, and large businesses exposed to international competition. Most Americans, like farmers, sell their goods and services wholesale and buy retail. Small nonfarm businesses have had higher failure rates than farm firms over the years.

5. Farmers depend heavily on precarious foreign markets. Agriculture accounts for only 2 percent of gross national product but accounts for nearly one-fifth of U.S. exports. The export market is subject to the vagaries abroad of weather, pestilence, economic conditions, and politics.

6. Farm resources are highly specialized to farming. A farm tractor, combine, hay baler, or skilled operator is not easily transferred to other occupations when demand fails or supply jumps due to a quantum improvement in technology. In contrast, fewer adjustment problems arise with demand failure for output of a nonfarm service firm utilizing mainly resources of office space, computers, secretaries, salespersons, lawyers, and accountants. Because these resources are not specialized to the industry and are well-suited for other industries, they are more easily transferred.

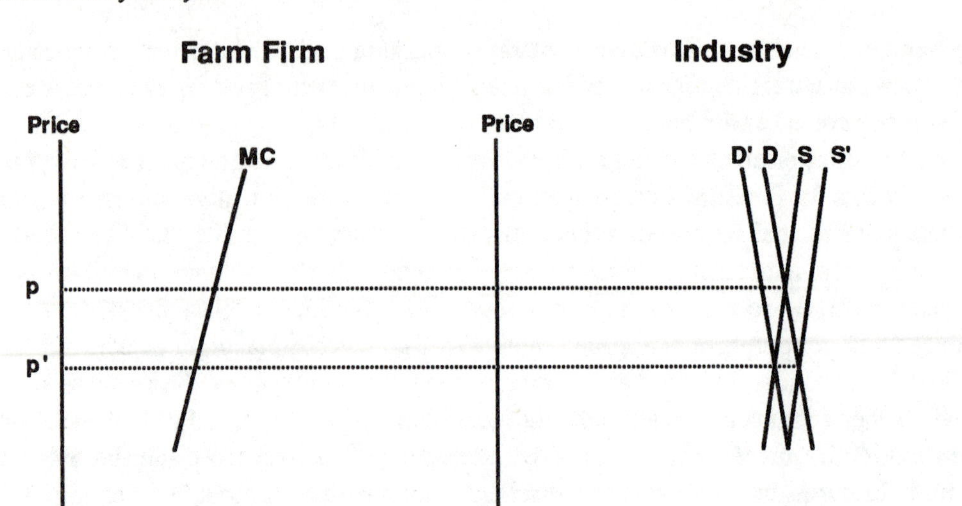

FIGURE 1.3. Industry and Farm Level Demand and Supply

Figure 1.3 illustrates the impact on farmers of characteristics of the agricultural industry noted above. Industry demand D and supply S are initially in equilibrium at quantity Q (million tons) and price p. The farm firm demand curve is the horizontal line at price p, marginal cost is MC, and the farm produces q at price p. (The industry supply curve is the horizontal summation of many individual firm marginal cost curves, only one of which is shown in Figure 1.3.)

Suppose that new technology unexpectedly shifts the supply curve forward to S' or favorable weather provides quantity Q_S rather than the normal Q. Because of the inelastic demand, the fall in price is much greater than the increase in quantity, revenue falls, and the farmer cuts back output to q'— perhaps a slow and difficult process.

Or suppose that supply is S but export demand drops from D to D' due to unforeseen considerations such as an export embargo or unusually favorable weather in Europe. Quantity falls to Q_d and price falls proportionately much farther from p to p', hence revenue falls. The farm firm can only accept the price change and make appropriate adjustment of output from q to q' to minimize loss if its average costs remain at initial equilibrium — equal to price p at q. Instability and other difficulties caused by structural characteristics of the farming industry and its markets are discussed below.

INSTABILITY

Sometimes it seems that the only "constant" in American agriculture is instability. Farmers gladly adjust to persistent high prices and with some difficulty adjust to persistent low prices but they cannot adequately adjust to unstable prices. As agriculture has been more commercialized and the economy has become more industrialized, man-made sources of instability have become more prominent relative to the uncertainties of nature which have traditionally troubled agriculture.

Given the inelastic short-run supply curve S in Figure 1.3, even small changes in demand from D to D' and shifting quantity from Q to Q_d cause a large reduction in price from p to p'. Principal sources of shifts in demand include:

— Macroeconomic policies, some as explained later influencing export demand through real interest rates, exchange rates, and foreign income growth. Inflation does not influence all prices proportionately, and can cause unpredicted real price changes.

— Population growth, a major but rather steady, predictable shifter of demand.

— Political decisions at home and abroad including export embargoes, decisions by the Soviet Union to buy from the U.S., price controls on food, commodity price support adjustments by the European Community, and petroleum price changes by OPEC.

— Other factors that influence export demand including weather, economic growth, and technological change in other countries.

Given the inelastic short-run demand D in Figure 1.3, even relatively small changes in supply from S to S' and shifting quantity from Q to Q_s cause a more than proportional decrease in price from p to p'. Principal sources of supply shifts include:

— Weather and pestilence such as insects and diseases. These are perennial problems not unique to modern agriculture. Modern technology has not removed yield instability as apparent in the 1988 drought.

— Technology, the major but rather steady and predictable supply shifter.

— Political or government administrative decisions such as a major acreage diversion and stock release under the payment-in-kind program in 1983.

— National economic growth and inflation which affect prices unequally, hence causing real price effects. For example economic growth raises the cost of labor relative to capital, causing substitution of capital for labor which may shift the supply curve.

— Commodity cycles which average about 4 years for hogs and 10 years for beef cattle. Commodity cycles arise from imperfect expectations coupled with the economic and biological characteristics of agriculture noted earlier.

The business cycle ravished agriculture approximately every 20 years in the 1800s. War induced booms frequently are followed by reduced demand, disinflation, and agricultural recession. In the past decade, the so-called inflation cycle and macroeconomic policy became a prominent source of instability and uncertainty to farmers. In the 1970s, instability was exacerbated by OPEC oil price increases and excessive expansion in credit and money supply at home and abroad. The result was inflation and cash-flow problems to farmers in the 1970s. This was followed by contractionary macroeconomic policies which brought recession in 1981 and 1982. Expectations that inflation might revive and high full-employment deficits brought high real interest rates from 1982 to 1987. The result was financial stress to agriculture in the 1980s.

The grain embargo imposed against the Soviet Union in 1980 had little short-run impact by induced uncertainty and eroded the image of the U.S. as a reliable food supplier. Foreign economic policies such as the decision of the Soviet Union to enter or leave the market, trade barriers, and trade skirmishes have created uncertainty for farmers increasingly dependent on export markets. Commodity programs have reduced but not eliminated uncertainty. Commodity programs introduce another uncertainty — the possibility that the programs will be terminated or changed for political reasons.

A highly inelastic demand gives rise to greater variation in prices and receipts than in quantity. If demand elasticity is -.5, the variance of receipts equals the variance in quantity (Tweeten, 1983). If demand elasticity is -1.0, changes in price and quantity exactly offset each other so that the variance of receipts is zero. A rising share of volatile exports in total demand increases variation in demand quantity and receipts directly but decreases variation in prices and receipts indirectly by increasing the price elasticity of demand. The net impact of these factors on volatility of the farming economy is an empirical question. Typical coefficients of variation (standard deviation divided by mean expressed in percent) for commodities are 20 percent for farm production and receipts at the industry level but up to 50 percent or more for production and net receipts at the farm firm level (see Tweeten, 1981; 1983). The coefficients show no distinct pattern among decades but showed a tendency to rise in the 1970s as U.S. agriculture became more internationalized.

Finally, some characteristics of agriculture make farmers less able to withstand instability and uncertainty. Fixed obligations to service debt and provide for family living have been growing. As farmers have become more commercial, they use a higher proportion of purchased to home-produced inputs, have a higher percentage of receipts going to pay for cash inputs, and rely more on debt capital. As a consequence, the impact of instability, a perennial problem to farmers, has become more severe.

Cash Flow and Financial Stress

The total real rate of return on farm equity capital, defined as the residual farm income (real capital gain plus gross income less production costs for cash inputs including operator labor and management) divided by net worth, averaged 10.8 percent from 1975 to 1979. Of the highly favorable total, 1.5 percentage points were current returns and 9.3 percentage points were real capital gain. Historically, a more normal real return on equity has been 4 percent from current earnings and real capital gain. If real return in the last half of the 1970s was more than double long-term normal past or equilibrium levels, why were farmers widely complaining and some were protesting in tractorcades and in the halls and malls of Washington, D.C.?

A reason is because farm mortgage interest rates were 12 percent. Real interest rate (the market interest rate less the inflation rate) was zero and hence real returns were far greater than real costs. But interest costs are immediate and capital gains are deferred and unrealized returns until assets (mostly land in the case of farmers) are sold. To be sure, a farmer could refinance and borrow on appreciated capital to service cash flow. But farmers did not like to incur new debt each year and lenders did not like to lend on unrealized and perhaps transitory capital gains. The result was a cash-flow problem of large proportions at a time of general farm prosperity.

The cash-flow problem transformed into a financial stress problem in the 1980s. Farm market interest rates held at about 12 percent to 1986. And the current return on farm equity averaged only a little less than in the late 1970s and was 1.5 percent, the 1975-79 average, in 1985. The principal change that transformed the cash-flow problem of the late 1970s into a financial-stress problem of the 1980s was the fall in inflation rate without a corresponding fall in market interest rates. With inflation reduced to 4 percent, the real interest rate of 12 - 4 = 8 percent was triple historic levels. Land value, which is mainly determined by expected land earnings divided by expected real interest rates (adjusted for risk, premium placed on owning land versus other assets, etc.), fell 50 percent or more in many states and capital losses totaled nearly one-half trillion dollars on farm assets including dwelling from 1981 to 1986 (U.S. Department of Agriculture, December 1987, p. 69).

Financial stress was acute among those with considerable debt. With rising real costs of debt, and with cash-flow shortfalls combined with declining collateral and rising debt-asset ratios, even many prudent farm operators were driven to bankruptcy. With land values down over 50 percent in many states, a prudent operator purchasing land while holding a conservative debt-asset ratio of 50 percent in 1980 could find himself technically insolvent in 1985. Later, financial stress was intensified as high real interest rates raised the value of the dollar. Resulting lower exports reduced commodity prices, receipts, and net income for many producers and raised costs of farm commodity programs for taxpayers.

Table 1.8 shows real cash income and cash flow before and after interest payments. Gross cash income and net cash income before interest payments were more favorable in the problem years of 1985 and 1986 than in 1970-71 years which were typical of the 1960s. Greater debt and higher interest rates doubled real interest cost in 1985-86, causing real net cash flow after interest payments to be lower in 1985-86 than in 1970-71. Residual income to assets was low in the 1980s but was in line with that in 1970-71 and the 1960s.

Aggregate data in Table 1.8 mask the seriousness of financial stress in the 1980s. We must turn to more disaggregated data. Based on a complex classification using debt-asset ratio, cash-flow, value of equity, and rate of return to equity and assets, Johnson, Morehart, and Erickson (1987, p. 12) estimated that 74 percent of farm operators with sales of over $40,000 were in good financial condition and 9 percent were vulnerable to financial failure in 1985. In the same year, 66 percent of the assets and 57 percent of the debt were judged to be in a good financial position while 8 percent of assets and 18 percent of debt were judged to be vulnerable to financial collapse.

Financial stress was somewhat evenly distributed among enterprises, but rates were highest among beef, hog, cotton, and rice farms and were lowest among poultry farms and among fruit, vegetable, and nursery farms. Stress was most severe in the Plains, Lake, Corn Belt, and Delta states and least severe in the Northeast and Western states. Areas least distressed were characterized by strong nonfarm demand for land and produced commodities mainly utilized in domestic rather than export markets. Small farms, because they had substantial off-farm income, and large farms, because they had high rates of return on average, had lower rates of financial stress than mid-size farms.

TABLE 1.8. Farm Sector Cash Flow, Assets, and Rate of Return on Assets

	Average for Period				Year	
Item	1970-71	1972-74	1975-79	1980-84	1985	1986[a]
	(billions 1982 dollars)					
Gross cash income[b] (including net CCC loans)	128.9	165.7	157.8	152.2	139.8	131.3
Less expenses excluding interest	75.1	87.4	94.5	90.1	78.2	75.7
Equals cash flow before interest payments	53.8	78.3	63.3	62.2	61.6	55.6
Less interest paid	7.6	8.9	12.8	19.6	16.0	13.4
Equals cash flow after interest payments	46.1	69.3	50.7	42.4	45.6	42.2
Total farm assets (end of previous year, Dec. 31)	637.6	705.8	867.6	994.1	766.4	673.7
Residual income to assets as percentage of asset value (%)	3.1	5.9	2.7	2.0	3.3	3.0

SOURCE: Johnson et al. (1987, p. 3).
[a]Forecast.
[b]Excludes inventory adjustment, gross imputed rental value of farm dwellings, and home consumption.

GOVERNMENT INVOLVEMENT

The United States has a long tradition of government provision of public goods such as defense, basic research, education, roads, and parks but a more recent tradition of large government interventions to stabilize income of private sectors. Government involvement in such activity would be normal in a socialist economy but value judgments in our society are that each farmer should be free to make production and marketing decisions and that income from the market is superior to income from the Treasury.

Federal costs of farm income stabilization increased sharply from $4 billion in 1981 to near $26 billion in 1986 (Table 1.9). This contrasts with a federal cost of $4 billion per year for farm price and income stabilization in the peak two years 1963-64 of previous decades. It is of interest that the relative cost of farm programs, defined as federal outlays for price and income stabilization as a percent of the total federal budget, was lower in 1986 than the 3.7 percent recorded in years 1963 and 1964. With expanding exports and domestic drought in 1988 commodity program costs fell to $13 billion.

Federal outlays for farm price and income supports never reached 50 percent of net farm income in the 1960s or 1970s but averaged over 50 in the 1981-88 period. By that measure, government support for the farming economy in the 1980s was unprecedented.

Some contend that the high Treasury outlays for farm income support is a substitute for farm receipts from the market. Thus the world "subsidy" which implies that the government is transferring funds to farmers who give nothing of benefit in return is not

TABLE 1.9. Federal Budget for Farm Price and Income Stabilization, Agricultural Research and Services, and Food and Nutrition, 1981-87

Fiscal Year	Food and Nutrition	Agricultural Research and Services	Farm Price and Income Stabilization		
	(billion dollars)			(% of federal budget)	(% of net farm income)
1981	16.2	1.5	4.0	.6	14.9
1982	15.6	1.6	11.7	1.6	49.8
1983	18.0	1.6	18.9	2.3	148.8
1984	18.1	1.7	7.3	.9	22.7
1985	18.5	1.8	17.7	1.9	54.8
1986	18.6	1.8	25.8	2.6	68.8
1987	19.4	1.9	22.4	2.2	48.4
1988[a]	---	---	13.1	1.2	32.8

SOURCE: Office of Management and Budget (1987) and USDA sources.
[a]Preliminary.

entirely appropriate. However, neither the total food bill nor the farm ingredients portion was down by anything near the rise in federal outlays, hence the real total cost of food and agriculture (domestic food and fiber consumer costs plus tax costs to support agriculture) was up substantially in the late 1980s. Gross farm income was the same nominal value in 1985 as in 1981. The sharp rise in federal income support cost for agriculture just offset lower export receipts. Taxpayers rather than domestic consumers and producers bore the major burden of the drop in farm exports in the 1980s.

Farmers sometimes resent inclusion of food and nutrition programs in the U.S. Department of Agriculture, feeling that these programs divert the Department's attention from serving farmers to serving consumers. Food and nutrition programs were larger than income support outlays in the late 1970s and early 1980s (Table 1.9). Including consumer food and nutrition programs with farm income stabilization programs strengthens support in Congress for both types of programs. Consumer-oriented members of Congress from urban areas join with farm-oriented members from more rural areas to more strongly support the overall USDA budget.

Excess Capacity. Federal outlays for agriculture are closely related to another dimension of farm problems, *excess capacity,* defined as the surplus of production capacity over market utilization with normal weather and at politically acceptable prices. It is usually expressed as a percentage of production capacity.

Excess or reserve capacity consists of three components: (1) net stock accumulations, (2) the portion of subsidized farm commodity exports judged to be for U.S. farm price and income support rather than foreign economic aid, and (3) production diverted by cropland withdrawal programs (Tyner and Tweeten, 1964). Because three diverted acres are required to reduce the harvest by two acres and because diverted acres on the average have yields only three-fourths that of the average in production, two acres must be diverted to reduce production by the equivalent to one acre in production. In the 1960s, 50-60 million acres per year were diverted from production. That plus subsidized exports and additions to stock totaled between 5 and 7 percent of farm production capacity at prices of that decade (Quance and Tweeten, 1972). Excess production capacity reached even higher levels in the mid-1980s and was 9 percent in 1986 after dropping in the 1970s (Figure 1.4). Because random elements such as weather influence excess capacity, a seven-year moving average was used in Figure 1.3 to even out chance elements. Weather-adjusted excess capacity dropped to 4 percent in 1987 and to 2 percent in 1988 as exports expanded under a lower exchange rate and lower commodity prices and higher incomes abroad.

Excess capacity exists at the will of government; at some price markets will clear. Political and other factors have caused excess capacity to be concentrated primarily in grains, cotton, and dairy. Excess capacity is greater in crops than in livestock. Dairy excess capacity was 8 percent in 1986 (Dvoskin, 1987, p. 36). Excess capacity of the seven major crops reached 13 percent of their total production during 1979-85 compared to 6 percent for U.S. agriculture as a whole. Wheat excess capacity was one-third in 1985. Of this one-third, 16 percentage points were production potential on diverted acres, 14 percentage points were excess supply as measured by the rise in ending stocks, and 6 percentage points were from exports that would not have moved commercially in the absence of programs to dispose of surpluses. Given a market price to be protected by

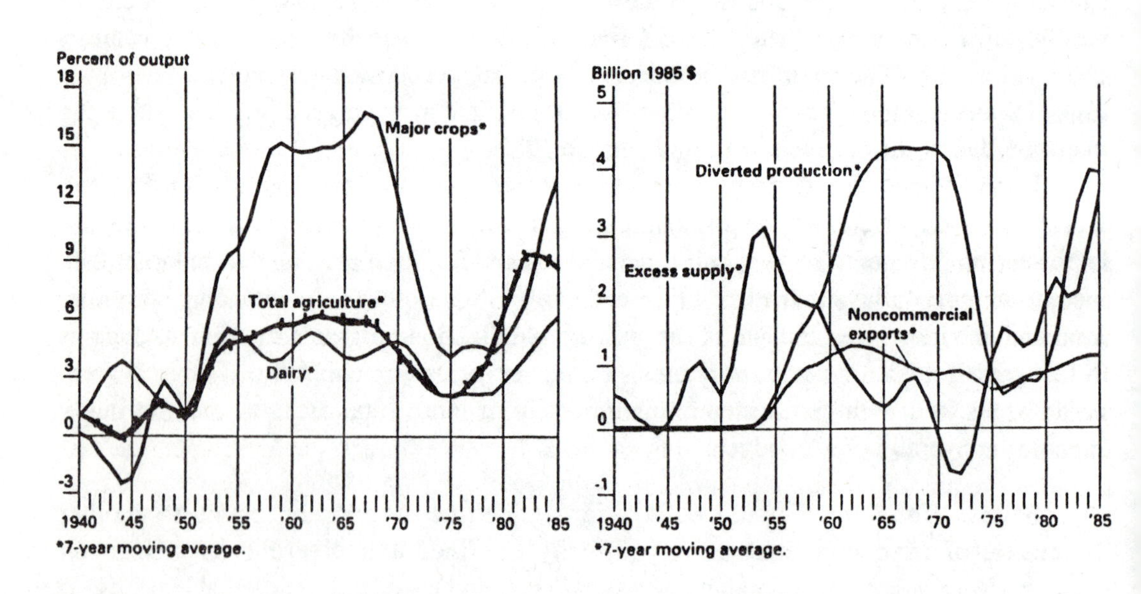

SOURCE: Dvoskin (1987, p. 21).

FIGURE 1.4. Excess Production Capacity in U.S. Agriculture

commodity programs, government failure to control production results in either accumulating of stocks or subsidizing of exports.

Removing excess capacity from markets raises farm prices and incomes in the short run although not necessarily in the long run. Some measure of the short-run benefit to farmers is apparent by observing the unfavorable consequences of releasing excess capacity to the market. The broad impact of releasing excess capacity on the market can be estimated in the short run from the aggregate price elasticity of demand. A key parameter is the short-run price elasticity of demand, assumed to be -.33, and its inverse, the price flexibility of demand F = -3. That is, each 1 percent increase in quantity placed on the market reduces real farm prices by 3 percent. The elasticity of receipts with respect to quantity is 1 + F = -2. If the length of run is short so costs do not have an opportunity to make adjustments, the short-run elasticity of net income with respect to excess capacity placed on the market is 1+F times the ratio of gross farm receipts GR to net income NI. If that ratio is 3, then the elasticity of net farm income with respect to release of excess capacity is -6.

Short-Run Elasticity or Flexibility of Variable I with Respect to Quantity	Percentage Change in Variable I with Release of 6 percent Excess Capacity
Price F = -3	-18
Gross receipts 1 + F = -2	-12
Net farm income $(1+F)\cdot\frac{GR}{NI} = -6$	-36

Multiplying excess capacity by the above elasticities, release of 6 percent excess capacity is estimated to reduce farm prices by 18 percent, gross receipts by 12 percent, and net farm income by 36 percent.

Termination of government payments would reduce net income even farther — to zero or even negative. Release of stocks is not considered above and would further depress the farm economy. These results clearly point to severe economic repercussions for the farming industry if commodity programs were terminated "overnight." But it does not say whether farm prices or incomes would be lower in 1990 if commodity programs would have been terminated and never revived again after excess capacity reached zero in 1973. And it does not necessarily mean that farm prices, gross receipts, or net income would be lower in the future with termination of commodity programs. It does imply that a transition program would be essential to ease trauma while shifting the farming industry to a market orientation. These issues will be discussed in depth in later chapters.

SUMMARY

Five major economic problems of the farming industry are apparent from the foregoing review of characteristics.

Cash Flow and Financial Stress

In the 1970s, *total* real rates of return on investment and farm income per capita favorable in relation to those elsewhere were of little consolation to farmers unable to service cash-flow requirements. The cash-flow problem is primarily associated with high inflation rates, commercial farming, and of leveraged operators. Inflation raises immediate costs and defers returns, creating the "live poor — die rich" problem.

The financial-stress problem is a cash-flow problem made far more serious by high *real* as well as high market interest rates. Sharply higher real interest rates in the 1980s brought capital losses, higher debt-asset ratios, and reduced collateral for farmers to support debt. Farm financial failure was widespread. Causes and consequences of cash-flow and financial stress problems will be analyzed in Chapter 4 providing an overview of farm problem causes and in Chapter 6 analyzing macroeconomic linkages.

Instability

Unstable production, prices, and incomes affects farms of all sizes and types. Instability has been seasonal, annual, and cyclical. Financial stress in the 1980s is a manifestation of cyclical instability because the farming economy is depressed for more than one year but not permanently. Variation in food quantity and price also troubles consumers at home and abroad. Annual and cyclical instability is the oldest and most serious commercial farm problem. It may be growing more severe. Rising cash costs and fixed obligations for debt service and family living have made farmers highly vulnerable to economic setbacks. Instability will be examined in depth in Chapter 5. Instability from monetary and fiscal policies is analyzed in Chapter 6 and from international trade is analyzed in Chapter 7. Chapter 8 on agribusiness structure makes the case that farm problems would be essentially unchanged if farm input supply and product marketing industries were perfectly competitive rather than imperfectly competitive.

Environment

Soil erosion and groundwater polution are principal environmental problems in agriculture. Efficient use of water supplies for irrigation and other purposes is a worthy but illusive goal. Chemical residues from fertilizers and pesticides in groundwater, surface water, and in foods is a major concern not only to farmers but also to consumers and American society. Long-term economic, social, and environmental sustainability of agriculture is a perennial concern. Chapter 9 will address problems of maintaining a sustainable food and agriculture through wise utilization and protection of air, water, soil resources, and safeguards against contamination of food and resources by agricultural chemicals.

POVERTY

Chronic low farm income, once a sector-wide problem, has become a case-poverty problem associated with certain high-risk groups within agriculture. A modern, efficient, commercial farming unit entails such large asset requirements that persons with limited resources cannot enter or remain. Hence poverty is either transitory or nonexistent among commercial farm operators and their families when ***net worth*** as well as income is considered. Hired workers on commercial farms may experience poverty, however. Poverty is higher in agriculture than in other sectors on the average and is particularly prominent on small farms. Small farms without off-farm income have a high probability of experiencing poverty. Low income is the most serious problem of small- to medium-size full-time farms. Poverty, human resources, and rural development will be the focus of Chapter 10.

FAMILY FARM DECLINE

Farming has always been a heterogeneous industry but has gradually evolved into three major subsectors. One is large industrial type farms with sales of over $250,000 per year which account for a major portion of farm output but for few farms. These farms on average enjoy relatively high rates of return on resources and are increasing in numbers. Another subsector is the small farm, increasingly dominated by part-time farms because full-time small farms cannot survive long in farming. Part-time farms increasingly dominate farm numbers but account for a small proportion of farm output. These farms are not efficient measured by economic returns in relationship to the opportunity cost of farm resources. But they provide a farm way of life and tax advantages. They grew in numbers from 1978 to 1982.

In between these two subclasses of farms is the traditional commercial family farm class with sales of $40,000 to $250,000 per year. These farms account for one-fourth of all farms and two-fifths of farm output. The family farm is increasingly squeezed between larger farms able to realize greater economies in production and marketing and the small part-time farms which thrive on off-farm income. The family farm is very slowly fading in numbers, proportion of all farms, and proportion of farm output. Of course, a well-managed medium-size family farm is more efficient than a poorly managed large farm. The family farmer provides unequalled operational management apparent in husbandry of crops and livestock and timeliness of operations and care of equipment. However, family farms often must compete against well-managed larger farms — not against poorly managed farms. The family farm realizes most cost economies but the 10 percent of potential economies it does not realize can be decisive.

The family farm has the advantage of expertise and capital passed from one generation to another. The family farm is in no immediate danger of extinction and will be around for generations to come, although in decreased numbers. But many young operators of mid-sized units selected to farm because their fathers were farmers find it difficult to cope with the high level of organizational and risk management decisions now required for survival in commercial agriculture. The family farm will continue to be a major source of political agitation because resource rewards will be judged inadequate to compensate for the demands placed on operators and their families for survival.

Society may have many reasons to preserve the family farm as discussed later in Chapter 3. However, the family farm need not be preserved to maintain farm output and food supplies. Output does not fall as family farmers exit because their operations are mostly taken over by other family farmers who produce more per acre than the operator who left.

Problems of maintaining the family farm and theories explaining the family farm exodus will be presented in Chapter 4 and in Chapter 8 on agribusiness structure. Efforts to maintain the family farm and address other farm problems with international trade will be presented in Chapter 7 and with commodity programs will be presented in Chapters 11 and 12.

An important principle of policy analysis is that the *presence of an economic problem is not sufficient justification for public intervention.* It is necessary to show that the social costs of public intervention are less than the social costs of the market failures or other problems the public programs attempt to address. The challenge of Chapter 2 is to develop a conceptual framework suitable for analyzing whether markets fail and whether government intervention is justified.

REFERENCES

Alt, Klaus and John Putman. April 1987. Soil erosion dramatic in places, but not a serious threat to productivity. *Agricultural Outlook* AO-129. Washington, D.C.: ERS, USDA.

American Agricultural Economics Association Soil Conservation Policy Task Force. 1986. Soil erosion and soil conservation policy in the United States. Occasional Paper No. 2. Ames: AAEA Office, Iowa State University.

Carlin, Thomas and Linda Ghelfi. November 1979. Off-farm employment and the farm sector. Pp. 270-73 in *Structural Issues of American Agriculture*. Agricultural Economics Report 438. Washington, D.C.: ESCS, USDA.

Chavas, Jean-Paul. 1983. Structural change in the demand for meat. *American Journal of Agricultural Economics* 65:148-153.

Council of Economic Advisers. 1987. Economic report of the President. Washington, D.C.: U.S. Government Printing Office.

Dvoskin, Dan. 1987. Excess capacity in U.S. agriculture. Staff Report No. AGES870618. Washington, D.C.: Resources and Technology Division, ERS, USDA.

Dvoskin, Dan. October 1986. Excess capacity and resource allocation in agriculture, 1940-1985. *Agricultural Outlook* pp. 31-33.

English, Burton, James Maetzold, Brian Holding, and Earl Heady, eds. 1984. *Future Agricultural Technology and Resource Conservation*. Ames: Iowa State University Press.

Henneberry, Shida. 1986. A review of agricultural supply responses for international policy models. Agricultural Policy Analysis Project Background Paper B-17. Stillwater: Department of Agricultural Economics, Oklahoma State University.

Hudson, M.A. and J.P. Vertin. 1985. Income elasticities for beef, pork, and poultry. *Journal of Food Distribution Research*. Urbana-Champaign: Department of Agricultural Economics, University of Illinois.

Johnson, James, Mitchell Morehart, and Kenneth Erickson. 1987. Financial conditions of the farm sector and farm operators. *Agricultural Finance Review* 47(Special Issue):1-18.

Lee, Chinkook, Gerald Schluter, William Edmondson, and Darryl Wills. 1987. Measuring the size of the U.S. food and fiber system. Agricultural Economic Report No. 566. Washington, D.C.: Economic Research Service, USDA.

Lee, Linda. 1984. Land use and soil loss: A 1982 update. *Journal of Soil and Water Conservation* 39:226-28.

Manchester, Alden. 1983. The farm and food system: Major characteristics and trends. Series 1 in J. Shaffer, V. Sorenson, and L. Libby, eds., *The Farm and Food System in Transition*. East Lansing: Cooperative Extension Service, Michigan State University.

Nikolitch, Radoje. 1972. Family-size farms in U.S. agriculture. ERS-499. Washington, D.C.: ERS, USDA.

Office of Management and Budget. 1987. The budget in brief. Washington, D.C.: Executive Office of the President.

Oliveira, Victor and E. Jane Cox. 1988. The agricultural work force of 1985. Agricultural Economic Report No. 582. Washington, D.C.: ERS, USDA.

Oster, Merrill. 1986. Do religious values suggest that "family farms" are more socially desirable than "commercial farms"? Proceedings of Conference on Religious Ethics and Technological Change. Ames: Religious Studies Program, Iowa State University.

Quance, Leroy and Luther Tweeten. 1972. Excess capacity and adjustment potential in U.S. agriculture. *Agricultural Economics Research* 24:57-66.

Sanford, Scott, Luther Tweeten, Cheryl Rogers, and Irving Russell. November 1984. Origins, current situation, and future plans of farmers in east central Oklahoma. Research Report P-861. Stillwater: Agricultural Experiment Station, Oklahoma State University.

Soil Conservation Service (SCS). 1982. Basic statistics: 1977 national resources inventory. Statistical Bulletin No. 686. Washington, D.C.: USDA.

Tweeten, Luther. December 1984. Diagnosing and treating farm problems. Pp. 95-118 in *Trade Policy Perspectives*. Washington, D.C.: Committee on Agriculture, Nutrition, and Forestry, 98th Congress, 2nd Session, United States Senate.

Tweeten, Luther. 1984. Causes and consequences of structural change in the farming industry. NPA Report No. 207. Washington, D.C.: National Planning Association.

Tweeten, Luther. December 1983. Economic instability in agriculture: The contributions of prices, government programs, and exports. *American Journal of Agricultural Economics* 26:922-931.

Tweeten, Luther. 1981. Prospective changes in U.S. agricultural structure. Pp. 113-146 in D. Gale Johnson, ed., *Food and Agricultural Policy for the 1980s*. Washington, D.C.: American Enterprise Institute.

Tweeten, Luther. 1979. *Foundations of Farm Policy*. Lincoln, Nebraska: University of Nebraska Press.

Tweeten, Luther and Wallace Huffman. 1980. Structural change. Part 1 in *Structure of Agriculture and Information Needs Regarding Small Farms*. Washington, D.C.: National Rural Center.

Tyner, Fred and Luther Tweeten. 1964. Excess capacity in U.S. agriculture. *Agricultural Economic Research* 16:23-31.

U.S. Bureau of the Census, U.S. Department of Commerce. September 1986. *Data User News*. Washington, D.C.: U.S. Government Printing Office.

U.S. Bureau of the Census, U.S. Department of Commerce. October 1984. *1982 Census of Agriculture*. Volume 1, Part 51, United States summary and state data. Washington, D.C.: U.S. Government Printing Office.

U.S. Department of Agriculture. April 1987. Economic indicators of the farm sector: Production and efficiency statistics. ECIFS 5-5. Washington, D.C.: Economic Research Service, USDA.

U.S. Department of Agriculture. January 1987. Economic indicators of the farm sector: Farm sector review, 1985. ECIFS 5-4. Washington, D.C.: Economic Research Service, USDA.

U.S. Department of Agriculture. December 1987. Economic indicators of the farm sector: National financial summary, 1986. ECIFS 6-2. Washington, D.C.: Economic Research Service, USDA.

U.S. Department of Agriculture. January 1981. A time to choose: A summary report on the structure of agriculture. Washington, D.C.: Economic Research Service, USDA.

U.S. Department of Agriculture. November 1979. *Structural Issues of American Agriculture*. Washington, D.C.: ESCS, USDA.

CHAPTER TWO

Public Welfare and Economic Science

Economics can be defined as the science of allocating scarce resources among competing ends to satisfy these ends as fully as possible. The definition raises several issues. The term *science* implies not only a meaningful classification of facts and a systematic body of knowledge but also a logical structure and ability to predict. Whether economics can meet these requirements for a science is conjectural.

Of great interest is the term *ends* in the definition. What ends or goals are to be achieved, and for whom? In farm management economics, the dilemma of alternative goals poses no serious philosophic problems. The economist can hold out a profit-maximizing allocation of resources computed by linear programming, and the farm manager can take or leave it depending on his personal goals and financial circumstances. An economist employed by a farm organization with well-defined goals also has no confusion over how and whom to serve.

The task is not easy for the policy economist whose salary is paid by taxpayers. Farm policies have a far-reaching impact not only on U.S. farmers and consumers but sometimes on the world. The individual farmer often must accept the policy that the majority of farmers or members of society voted to accept. The goal in broad terms may be defined as well-being, satisfaction, welfare, or utility in society. But there are problems in trying to maximize something as elusive as utility by a policy that affects many individuals, each with unique objectives in life. Welfare economics deals with such issues. It is a part of general economic theory that attempts to answer the question "what can the economist in his professional role of a scientist say about public policy?" This chapter outlines the origins of welfare economics and criteria for making economic prescriptions. The chapter deals with economic equity and economic efficiency and with factors that interfere with economic efficiency in society. Most importantly, this chapter establishes the conceptual framework for the analysis in subsequent chapters.

DESIRED ATTRIBUTES OF ECONOMIC SCIENCE

A *worthwhile science* has *internal validity* and *external worth.* A science is internally valid if it is *clear, logical,* and *reliable in prediction.*

Predictive ability: Internal validity of science depends on reliability as measured by ability of its propositions to predict. That predictive ability is apparent by how closely the propositions correspond to past experience and, more importantly, how accurately future or previously unknown phenomenon are predicted. It must be possible to state within some acceptable bounds of error that *if* this action is taken, *then* this outcome will result. This reliability separates economics from history which is not a science; history is only a meaningful classification of facts and systematic body of knowledge.

Logical structure: Internal validity requires a logical or coherent structure. Ideally, it is possible to make a case that A is caused by systematic structural forces B, C, and D but A may also be influenced by random elements. Conceptual validity and internal consistency enhance usefulness of science. The laws or principles of economics combine logical consistency with predictive capabilities.

Clarity: If economics can be unambiguously and easily communicated, its value to society is enhanced. The fewer the assumptions required, the more useful the science, other things equal. A science does not have to be simple, but it needs to have clarity in the sense of being no more complex in analysis nor demanding of assumptions than necessary.

External worth: A science with only internal validity may have value as a curiosity or arcane diversion. External worth is in turn a function of (1) the ability to turn human and material resources devoted to the science into output, an attribute closely related to internal validity, and (2) the value of that output. Tradeoffs are apparent. Meteorology is unable to turn its resources into a product with much reliability. But weather information is highly valued so the imprecision is tolerated. Economics is similar. Farm management economics and policy analysis can be equally worthwhile to society because the greater imprecision of policy analysis is compensated by the greater value of its output. Ideally, a science directly or indirectly helps meet real needs of society. Economics needs to be practical and nothing is more practical than good theory.

A science is *objective* if researchers are willing to have their work reviewed for clarity, logic, and predictiveness and abide by the results. This procedure helps to ensure that two researchers independently studying the same phenomenon reach the same conclusions.

The above demanding requirements are not met fully by economic science or by any science. But they remain a measure of performance and hence a foundation for improving economics. In continually striving to make economic science more objective, accurate in predicting, logical, clear, and worthwhile, it is useful to classify approaches to economics. Particular attention is paid to objectivity and predictive ability, two issues of much controversy.

Particular concern centers on whether economists are making unobjective value judgments regarding the ends or goals of society. If economists substitute their own values for the ends and values of society, economics will not be workable in meeting real needs. Advocacy of assumed ends or means to assumed ends makes economics unobjective and hence unscientific. In a search for intellectual integrity and respectability, economists have classified economics into various forms or approaches as a way of ruling out certain forms judged to be unscientific. Two broad classifications are normative economics which is shunned and positive economics which is embraced.[1]

NORMATIVE ECONOMICS

Normative economics is the economics of "what ought to be." In its extreme form, normative economics entails advocacy by economists of ends and means which they themselves have judged to be best for society rather than what society wants and would improve the well-being of people. An economist who advocates Marxism, *laissez faire,* or other policies making society worse off is practicing normative economics and is operating as a politician and not a scientist. There are times for an economist to step out of his role as professional economist and into a role of private citizen or politician free to practice free speech and advocacy, but that changing of hats from professional economist to private citizen needs to be recognized by all concerned.

POSITIVE ECONOMICS

There are several variants of positive economics, all of which are legitimate activities for professional economists. *Descriptive economics* is the purest form of the positive economics of "what is." It merely describes the economy without making judgments of "what ought to be" or predicting "what will be." Although it may be analytical, because it does not predict it is a primitive form of economics. It is useful for some purposes but does not fulfill the capability of worthwhile economic science.

Predictive economics is the economics of "what is likely to be." This approach to positive economics is mostly concerned with forecasting and need not confront issues of what goals are being met for whom. Instead, it focuses on the likely state of the economy or price of hogs next month or next year.

Prescriptive economics is the economics of "what could be" and is the principal basis for economic policy analysis. It does not advocate objectives invented by the economist but holds that if the economic decision maker wishes to pursue objective X, then the most efficient use of resources is allocation Y. Prescriptive economics may assume a norm or objective such as profit maximization or risk minimization for the firm or utility maximization for consumers and society. Linear programming or simulation used for determining allocations to increase income, employment, well-being, or other recognized objectives or norms of society is positive prescriptive economics. Such analysis does not

[1]Some broadly define normative economics as that dealing with values regarding good or bad (Johnson, 1987). Virtually all scientific economic analysis deals with values. For example, it is important to know how prices and economic theory relate to values and ethics — a topic of this chapter.

specify what ought to be and hence is not normative. Because prescriptive economics often analyzes potential implication of policies to increase profit or other norms, it frequently is *conditionally positive.* It is wise in keeping with the tradition of positive economics under such prescriptive economics to specify alternative means to achieve various objectives, letting the decision maker observe tradeoffs before choosing policy options.

Because major goals of society are well known, the publicly employed economist need not resort to normative economics of making up goals and advocating them along with the means to achieve them. The overriding objective of society is to increase the well-being of people, a concept expressed in the social welfare function. The social welfare function (utility function for society) expresses satisfaction of society as a function of the level, variability, and distribution of full income where "full income" ideally includes the value of leisure time and other nonmonetary goods and services which contribute to satisfactions. Economists cannot precisely specify the full social welfare function but know enough to make a useful beginning.

Regarding the *variability* of income, most people tend to be risk adverse but differences among persons are sizable. On the average, society is willing to sacrifice relatively little income to gain greater stability of income (see Tweeten and Mlay, 1986, p. 20).

The *level* of income relates to economic efficiency and growth. If resources and products are used efficiently, income will be generated, a necessary but not sufficient condition for economic growth defined as an increase in real income over time. Economic efficiency is important for well-being, but if the discount rate is high well-being may be greatest if all production is consumed rather than saved and invested in future production and economic growth. In a very poor country with a high premium on current consumption to survive, economic efficiency produces income that is consumed rather than saved and invested in human and material capital to provide future income streams.

In long-term equilibrium the discount rate will be the interest rate which is jointly determined by the time preference for consuming now rather than later and the rate of return on investment. In typical advanced economies, the real interest rate is low, about 3 percent, implying that people place little premium on current consumption. If people prefer more to less, the central assumption of economics, economic efficiency is important to well-being whether a country is experiencing economic growth or not.

The *distribution* of resources and income also influences utility and has been a troublesome issue for economists over time. Modern statistical techniques and socio-psychological attitudinal scales have greatly improved the objectivity of procedures used to calculate the contribution of income and other variables to utility. Much of the remainder of this chapter details how economics relates to the well-being of people.

In short, a most important objective of public policy economics as a positivistic prescriptive science is to show the impact of alternative policies on the well-being of producers, consumers, taxpayers, and society. Analysis herein relies heavily on the classical welfare economics model presented later in this chapter. Objectivity is promoted in translating income level, distribution, and variation into well-being by estimating the marginal utility of income from validated socio-psychological scales. The analysis does not advocate well-being as the objective of society, it recognizes that well-being of people *is* the most important objective. Analysis which fails to recognize this broader objective for economists employed by society and concentrates instead on income to producers inadvertently becomes normative advocacy economics.

GENESIS OF WELFARE ECONOMICS

The judgments on welfare economics of political economists in the classical era entailed the Benthamite notions of cardinal utility (cf. Scitovsky, 1951). The so-called hedonistic concepts of gratification and the calculus of pleasure and pain continued from the classical into the neoclassical era. There was no contention that utility could be measured like pounds of butter or bushels of corn, but it was assumed that economists could make sufficiently valid assessments of satisfactions to determine whether utility in society was increased by a public policy and hence whether that policy was desirable.

Virtually all economic policies, even those not explicitly intended to do so, entail some redistribution of income. And it is clear that the change in total utility that attends economic policies redistributing income depends on the nature of individual utility functions. Because interpersonal comparisons required cardinal measures of utility and were not judged to be reliable, it follows that the economist would not know precisely whether a public policy adds to or detracts from total utility in society.

TOWARD A POSITIVE ECONOMICS

These considerations, coupled with the new formulations of economic theory based on the indifference curve analysis of Pareto (which permitted derivation of economic theories with use of ordinal rather than cardinal utility measures), led to rejection of interpersonal comparisons of utility. Lionel Robbins' *An Essay on the Nature and Significance of Economic Science* was a clear statement of the new philosophy in welfare economics. Robbins (1935, p. 30) says:

> Economics is not concerned at all with any ends *as such*. It is concerned with ends in so far as they affect the disposition of means. It takes the ends as given in scales of relative valuation, and enquires what consequences follow in regard to certain aspects of behavior.

If the economist is not to be concerned with judging ends such as utility, income, or freedom, not only is he to be neutral among ends but also he is to make no interpersonal comparisons of utility. Presumably Robbins' assertion would be motivated by the fact that economists cannot make reliable appraisals of utility. But he goes on to conclude (1935, p. 142) that, even if economists *could* measure utility, and even if, "proceeding on this basis, we had succeeded in showing that certain policies *had the effect* of increasing social utility, even so it would be totally illegitimate to argue that such a conclusion by itself warranted the inference that these policies *ought* to be carried out."

Robbins' statement of positive economics leaves the economist basically concerned about specifying alternate means to reach ends *given* to him by Congress, the President, or his employer. He is to be a "social" engineer concerned with specifying alternative ways of reaching given goals. He is to be concerned with "what is," not with the normative economics of "what ought to be." For economics to be an objective science, the economist cannot take sides in policies that make some worse off, others better off. He can be a technician but not an advocate. In the role of economist, he can be an adviser but not a politician. He can maintain a political dialogue only as long as politicians are asking the questions.

TOWARD A NEW WELFARE ECONOMICS

Because this philosophy of positive economics removed the profession of economics from the indeterminacy of utility measurement and moved it closer to the objectivity of a science, positive economics was widely accepted by the profession. But all was not well. If the economist were to have nothing to say about the merits of programs and policies that entailed interpersonal comparisons of utility, his education and policy role was virtually ruled out. Were not economists, because of their backgrounds, the appropriate persons to articulate value judgments on economic issues when lack of time, economic education, or imperfections in the political process precluded public involvement? Furthermore, the Great Depression was calling for economic prescriptions which positive economics was unable to voice (Scitovsky, 1951). And was not ruling out interpersonal utility comparisons in deference to the recognized indeterminacy of individual utility functions really just tantamount to assuming naïvely that marginal utility of income is equal and constant for all persons and all income levels? And was waiting for politicians to specify ends pandering to special interest groups represented by those politicians while ignoring the public interest?

Many economists were not bound by the philosophy of positive economics, and Keynesian prescriptions were loudly proclaimed. What the profession sought was a new welfare economics to legitimatize intellectually their activity. Four criteria have been successively proposed to provide a rationale for policy recommendations when interpersonal comparisons of utility are ruled out. The first is the Pareto criterion: *An economic policy is desirable when it makes one or more individuals better off without making anyone worse off.*

The use of the Pareto criterion is illustrated in Figure 2.1 by an *Edgeworth box* for two individuals, I and II (cf. Reder, 1947). The initial distribution of two commodities A and B is $0_I a_I$ and $0_I b_I$ to individual I, and $0_{II} a_{II}$ and $0_{II} b_{II}$ to individual II. Indifference curves showing successively greater satisfactions to I are indicated by the solid lines 1, 2, and 3. The origin of the indifference curve mapping for II is in the upper right-hand corner, and successively higher levels of satisfaction for him are designated by broken line indifference curves 1, 2, and 3.

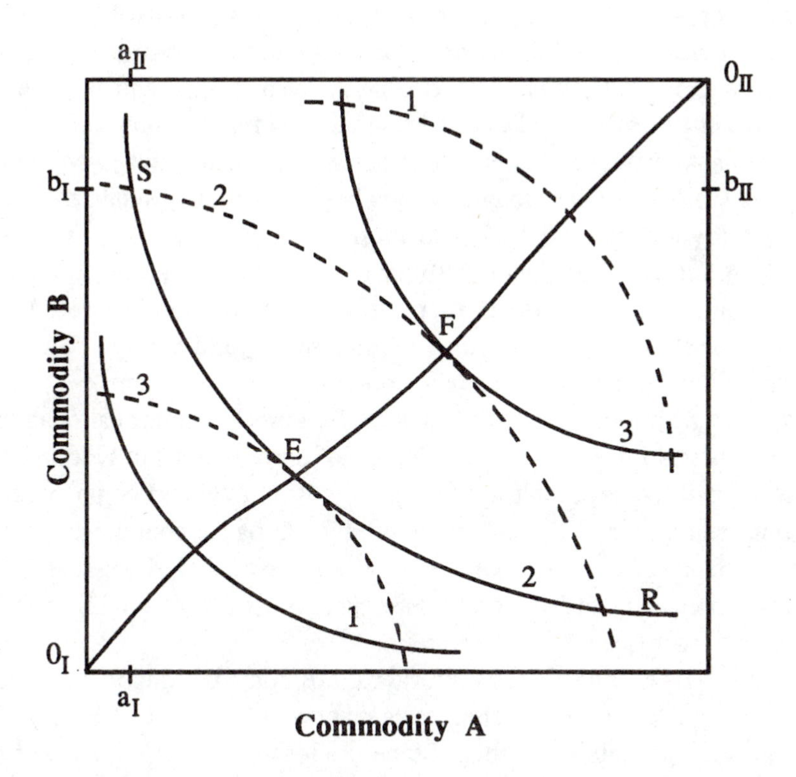

FIGURE 2.1. An Edgeworth Box, Showing Indifference Curves for Individuals I and II

It is apparent that given the initial distribution of commodities at point S, I can be made better off while II is made no worse off by moving to point F — that is, II remains on the same indifference curve while I moves to curves denoting higher satisfaction levels. Point F is a Pareto optimum, defined as a position from which it is not possible to make someone better off without making someone else worse off. Also from point S, by moving along I's indifference curve to E, II can be made better off without making I worse off. The *contract curve* $0_I EF 0_{II}$ is the locus of points of equal marginal rates of substitution between A and B for I and II — represented by tangency of an infinite number of indifference curves.

Given the initial distribution of commodities (resources) at S, any point on the contract curve between E and F is a Pareto optimum in the sense that it is not possible to make one party better off while the other is made no worse off. Any redistribution of A and B that falls in the area ESF or ERF is a *Pareto better*, defined as a position increasing utility for one individual without a decreasing utility for the other. But utility can be increased for one without decreasing that of the other by moving fully to the contract curve. Given the initial point S, the actual ending location along EF depends on the relative bargaining power of the two individuals.

Any position on the contract curve from 0_I to 0_{II} is possible with some distribution of the total quantity of A and B designated by the dimensions of the "box" in Figure 2.1. A

Pareto optimum represented by any given point along EF is not likely to maximize total utility. There is a third, vertical dimension not shown in the two-dimensional, horizontal figure which measures total utility to I and II from consuming A and B. The total utility surface may have one or more local maximums that lie along the contract curve. If there is one point that is the highest point on the utility surface, it is called the *global maximum.*

If the global maximum lies along 0_IE, then beginning at, say, F and moving along the contract curve toward the global maximum the disutility of I from sacrificing simultaneously A and B is less than the utility gained by II. Rejection of cardinal utility means that we cannot determine the third, or utility, dimension in Figure 2.1. Any given Pareto optimum will rarely be the point of maximum total utility. But the point of maximum total utility will be *one* of the many possible Pareto optimums.

It is apparent from Figure 2.1 that the specific Pareto optimum allocation on which society settles depends on the initial distribution of commodities or resources. This is a major shortcoming of the criterion for public policy. If individual I begins with a very low amount of commodities A and B, the Pareto optimum along the contract curve may mean a very low level of utility for him alone and perhaps for I and II together. The Pareto optimum is also an inadequate criterion because there is virtually no public policy that does not require some to sacrifice while others are made better off.

The Kaldor, Hicks, and Scitovsky criteria are contributions to the so-called *new welfare economics*. The Kaldor criterion states that a given policy is desirable if those who gain from it can compensate the losers. Since the losers are made no worse off and the gainers are made better off, this criterion is essentially the Pareto criterion with a compensation principle added. Because it is stipulated that compensation be made payable only potentially (it would *not* be paid in practice), the Pareto criterion is not in fact met. The condition that compensation need not be made was dictated by practical considerations and justified by the contention that gains and losses are likely to be distributed somewhat randomly over time among members of society (Scitovsky, 1951).

The Hicks criterion is a modification of the Kaldor criterion and states that if those who lose from a public policy cannot bribe those who gain into not accepting the policy, then the policy is desirable. The Scitovsky criterion goes further and states that a policy is socially desirable if gainers can potentially bribe the losers into accepting the change and if, in turn, the losers cannot bribe the gainers into not making the change.

EQUITY AND EFFICIENCY IN WELFARE ECONOMICS

The fundamental breakthrough of the new welfare economics was to separate the equity and efficiency components of policies. *Equity* here refers to the distribution of gains and losses, and *efficiency* refers to the real quantity of goods and services per unit of input. Public policies increasingly entail compensation, but for many policies compensation for loss is neither attempted nor made. And since there is an imperfect economic and political bargaining process between gainers and losers, the new welfare economics condenses to the issue that a policy is to be recommended if it either increases the real value of goods and services produced with given resources or reduces the resources required to produce a given output. Thus if social benefit-cost ratios are favorable, the policy meets the efficiency criterion and can be recommended.

The new welfare economics has failed. It has resulted in economists dividing the world into efficiency and equity dimensions and throwing away equity as unscientific. Other social scientists also divide the world into efficiency and equity dimensions and throw away efficiency because their training was too superficial to understand it. Neither world view suffices.

Economists clearly need to distance themselves from the narrow positive economics of responding only to questions posed by politicians and examining only questions of economic efficiency. To do so, three conditions are helpful.

1. Economists need to recognize that improving the well-being of people is a well-recognized objective of society and hence is consistent with sound, positive economists of responding to real objectives rather than those made up by economists. That recognition is not a venture into the forbidden normative economics of what ought to be.

2. Economists need to specify a social welfare function including how the stability and distribution as well as the level of income contributes to the well-being of people. This social welfare function should be estimated as objectively and comprehensively as possible. It needs to be continually refined and updated as multidisciplinary conceptual tools become more refined.

3. Economists need to avoid advocacy and continue to present policy options, recognizing that economists inform whereas the political process makes policy decisions. Economists educate the public and decision makers on options to increase employment, income, well-being of society, and to serve other objectives — many of which may conflict but which inform the decision process.

Progress is being made on constructing a social welfare (utility) function relating well-being of people to income level, distribution, and variability, and to other variables such as education, occupation, and health (see Tweeten and Mlay, 1986). In many instances, the *weak form* of the social welfare function suffices:

1. More income is preferred to less income.
2. A more equal distribution is preferred to a less equal distribution of income and wealth.
3. More stability of income is preferred to less stability of income.
4. More freedom is preferred to less freedom in making choices and decisions.

These do not exhaust the arguments in the social welfare function. Obviously, each proposition implies "other things equal." The *strong form* of the utility function quantifies these tradeoffs.

Happily, more freedom often attends systems that provide higher income. But unhappily, total national income often is sacrificed by policies to provide a more stable and equitably distributed income. These tradeoffs are the heart of economic analysis presented later which provides procedures to estimate the national income lost (net social cost or deadweight loss) by policies to increase farm income or utility gained by transfers from those with high income to these with low income. Hence later analysis reports initial work on a strong form of the social welfare function which indicates the impact on well-being of people from the income level (economic efficiency and growth) and its distribution among taxpayers, consumers, producers, and society. Less attention is given to risk but empirical data indicated risk aversion is not decisive in the utility function (Tweeten and Mlay, 1986, p. 20). No more than a 10 percent risk premium was estimated to be required on the average. Thus a "normal" risk-free rate of return of 10 percent on farm assets would be raised to 11 percent by the risks normally encountered in the farming industry.

PERFECT COMPETITION AS AN EFFICIENCY NORM

Before turning to classical welfare analysis of efficiency and equity, we here review briefly the Pareto optimum concept of perfect competition as a measure of economic efficiency. Two types of efficiency are involved, technical and economic. Technical efficiency is involved in choices of production practices and mechanical input-output ratios. One chooses the practice that requires the least input to produce a given output or that produces the most output from a given input. Technical efficiency entails operating on the highest production possibility curve with given resources or the lowest isoquant to produce a given output. However, economic efficiency is involved in optimal allocation when more than one resource or product is introduced.

Perfect competition is widely used by economists as a norm for judging economic efficiency in the market. This section demonstrates that perfect competition is a Pareto optimum — it is the most efficient allocation economists can prescribe if they are unable to make valid judgments about the marginal utility of money.

ASSUMPTIONS OF PERFECT MARKET

1. The market is atomistic. There are many buyers and sellers so that actions of any one have no perceptible influence on price. Each market participant does not influence the market but the market influences him.

2. Resources and products are divisible. The assumption of divisibility is closely related to problems of scale, because a costly, lumpy input often gives rise to asset fixity that in turn may lead to a decreasing average cost curve for a firm.

3. External economies and diseconomies of scale are absent in production and consumption, and private and social costs (or benefits) do not diverge.

4. Knowledge of markets and production possibilities is complete, including the production function for developing new techniques. Individuals must be aware not only of the wages but also of the satisfactions associated with alternative employment opportunities.

5. Mobility of commodities and resources is perfect; markets are not restricted by institutional impediments. Firms have freedom of entry and exit.

PARETO OPTIMALITY IN A PERFECT MARKET

The physical Pareto optimum concept is illustrated in Figure 2.2 for two firms I and II producing two commodities A and B. The production possibility curves for the firms are designated respectively I and II. By producing at point E, the bundle of resources used by firm I produce a_I and b_I of the commodities. The origin of the production possibility curve for firm II is in the upper right-hand corner. Producing at E, its output is a_{II} and b_{II}. The firms at E together produce A and B in quantities designated by the solid line dimensions of Figure 2.2. The same output would also be produced at H.

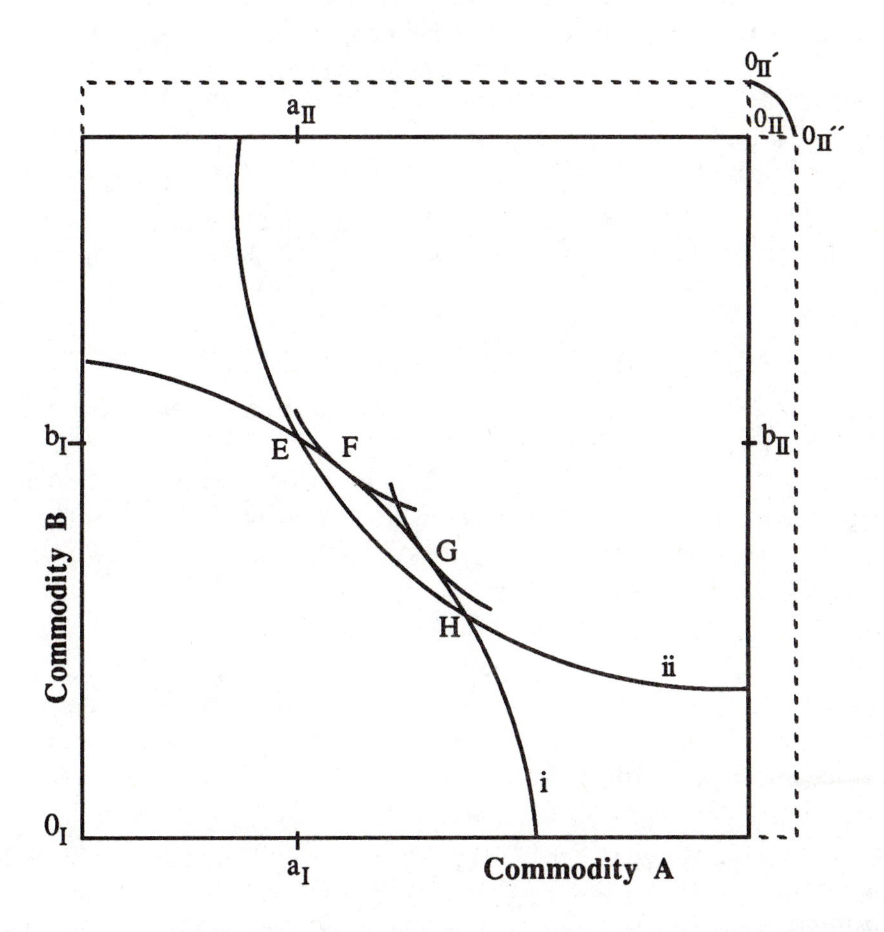

FIGURE 2.2. Production Possibility Curves — I for Firm I and II for Firm II

Neither E nor H is a Pareto optimum. More of B can be produced with the same resources without sacrificing A by shifting the production possibility curve II straight up until it is tangent to I at F. The new origin for II is $0_{II}'$. Or more of A can be produced without sacrificing B by shifting II directly to the right until II is tangent to I at G. The new position of the origin for firm II is now $0_{II}''$. Thus points F and G, or any tangency point in between along I, represent Pareto optimums. And the locus of these points is a physical contract curve FG = $0_{II}'0_{II}''$. There is no unique Pareto optimum — the actual outcome depends on the worth of having more A or B, on the production possibility curve, and on the initial distribution of resources. But the most efficient point, which is one, but probably not the only, Pareto optimum, will be at a tangency of the two production possibility curves. At that point the marginal rate of substitution in production of the two commodities for one firm is equal to the marginal rate of substitution in production of the two commodities for the other firm.

The analyses in Figures 2.1 and 2.2 provide the basis for proof that a perfectly competitive price economy is a Pareto optimum. From economic theory, it is known that to maximize utility under competitive market conditions, a consumer i equates the marginal rate of substitution of A for B to the price ratio as in the following equation:

$$-\frac{\Delta B_i}{\Delta A_i} = \frac{P_a}{P_b} \quad \text{(Consumption).} \qquad (2.1)$$

If equation 2.1 holds true for consumers i and j and each faces the same price ratio, then it is also true that the marginal rate of substitution of A and B is the same for all consumers:

$$\frac{\Delta B_i}{\Delta A_i} = \frac{\Delta B_j}{\Delta A_j} \quad \text{(Consumption).} \qquad (2.2)$$

Hence, Equation 2.2 meets the conditions for a consumption Pareto optimum along the contract curve in Figure 2.1.

It is also well known that under the assumptions of perfect competition any firm i equates the marginal rate of substitution between products to the price ratio as in:

$$-\frac{\Delta B_i}{\Delta A_i} = \frac{P_a}{P_b} \quad \text{(Production).} \qquad (2.3)$$

Because equation 2.3 holds for firm j as well as firm i, and because both are confronted by the same price ratio, then

$$\frac{\Delta B_i}{\Delta A_i} = \frac{\Delta B_j}{\Delta A_j} \quad \text{(Production)} \qquad (2.4)$$

holds for all firms, and the conditions for a production Pareto optimum in Figure 2.2 have been met.

Finally, because marginal rates of substitution in consumption and production are equal to the same price ratio, they must be equal to each other.

$$\frac{\Delta B_i}{\Delta A_i} \text{ (Consumption)} = \frac{\Delta B_i}{\Delta A_i} \text{ (Production)} \qquad (2.5)$$

The result is illustrated in Figure 2.3. A portion of the "societal" production possibility curve T, the indifference curve U, and the price ratio P_a/P_b are shown.

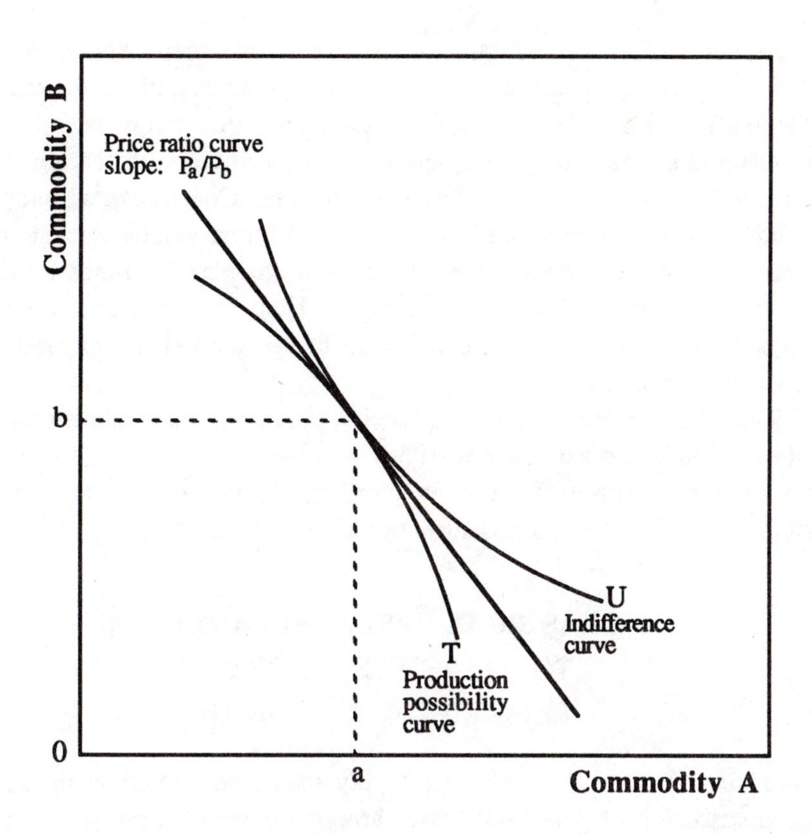

FIGURE 2.3. Societal Indifference Curve and Production Possibility Curve

It is clear from the graph that production and consumption of any combination of commodities other than a and b mean attaining an indifference curve and utility level below U. Tangency of T and U to the price ratio line means a Pareto optimum for society under perfect competition. Any deviation from this point, where the marginal rates of substitution between two commodities in production and consumption are equal, would mean someone is made worse off. Thus if we do not make judgments about the distribution of income attending the competitive equilibrium, the competitive equilibrium is as close to full economic efficiency as the economy can get.

It is necessary to add further marginal conditions for an equilibrium that is optimal. In addition to the conditions stated above (that the marginal rate of substitution between any two commodities be equal in production for all firms and in consumption for all consumers), it is necessary that:

1. The value of marginal product for any resource must be equal to its factor price in all uses; hence the value of marginal product must be equal in all uses of a given resource. If this conditions holds, the marginal rate of substitution between any two factors will be equal in all production uses, and the marginal rate of substitution between any two products will be equal, as in equation 2.4.

2. An individual free to allocate personal resources between production and consumption, including leisure, will employ labor resources until, at the margin, the value of satisfaction from leisure just equals the monetary reward from work.

3. Assuming free access to a perfectly competitive bond market, producers and consumers can adjust their income streams over time. Consumers will adjust borrowing and lending until their marginal rate of discount on future versus present consumption is equal to the interest rate. And entrepreneurs will compete for funds until the marginal efficiency of capital is equal to the interest rate. Hence the marginal rate of discount on consumption between two periods is equal to the marginal efficiency of capital. And capital grows at the interest rate.

4. Second-order conditions of convexity, concavity, and stability must also be met (Reder, 1947; Henderson and Quandt, 1958).

It is again emphasized that the level and distribution of income under competitive equilibrium is subject to the initial distribution of resources.

CLASSICAL WELFARE ANALYSIS TO MEASURE ECONOMIC EFFICIENCY AND EQUITY

Classical welfare analysis employs supply and demand relationships to determine the level and distribution of gains and losses among consumers, producers, taxpayers, and society from changes in economic policy. The technique is useful for policy analysts determining an economically efficient allocation of resources and determining whether it is appropriate or not to intervene in markets. The technique is useful for analysts to estimate who gains and who loses from market failure and from government distortions of markets. With appropriate modification, classical welfare analysis is suited to analyze equity (distribution) issues as well as economic efficiency. Equity and efficiency are joined in analysis of overall well-being of society.

In contrast to conventional marginal analysis which shows economic efficiency only at the margin, classical welfare analysis shows whether inefficiency entails loss of $1 or $1 billion. The strong assumptions (some listed later) required for classical welfare analysis caused it to fall into disuse for decades. It has been revived and has largely replaced the foregoing Pareto analysis because it is so well suited to empirical applications. The technique applies not just to market economies. Diminishing returns, scarce resources, and foregone opportunities expressed by the supply curve and diminishing incremental satisfaction from greater consumption of a commodity as expressed by the demand curve characterize centrally planned economies and market economies alike. Classical welfare analysis requires the supply curve to measure opportunity costs of resources and the demand curve to measure marginal utility but does not require perfect knowledge, perfect mobility of resources, or other consumptions of perfect competition.

UNDERSTANDING THE CONCEPTS OF CLASSICAL WELFARE ANALYSIS

Classical welfare analysis with industry supply S and demand D curves in Figure 2.4 provides an alternative proof of the efficiency of competitive market output. If a perfectly discriminatory monopolist charges consumers the maximum demand price d_1 for the first unit(s) q_1 of a commodity, the revenue from the first unit(s) is d_1q_1. The maximum price for the second unit is d_2, hence revenue from it is d_2q_2. Revenue from the third unit is d_3q_3. Continuing this procedure until price is zero (demand curve D intersects the horizontal axis at q_{14}), it is apparent that the demand curve is a marginal revenue curve. The area beneath it is total or gross revenue, and total revenue is greatest with quantity q_{14}.

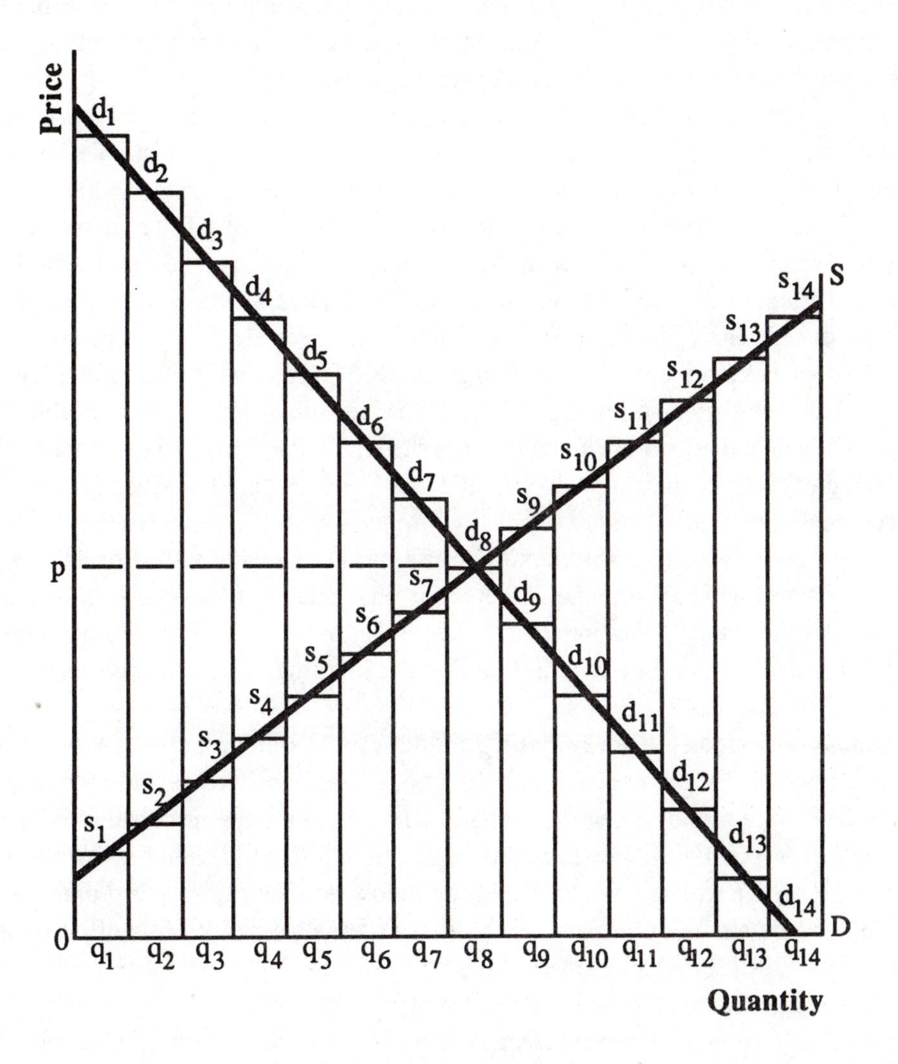

FIGURE 2.4. Economic Costs and Benefits Illustrated by Supply and Demand Curves

The marginal cost of producing the first unit(s) q_1 is s_1, hence the total variable cost for the first unit(s) is s_1q_1. The marginal cost of the second unit is s_2 and of the third unit is s_3, hence total variable costs for these units are s_2q_2 and s_3q_3, respectively. The sum of these marginal variable costs, the total variable cost, is the area beneath S in Figure 2.4.

Net incremental revenue, the difference between the demand price d and the marginal cost s for each quantity, is positive proceeding from 0 to q_8. Net revenue for incremental quantities to the right of q_8 are negative because the marginal cost exceeds marginal revenue. It follows that for a perfectly discriminating monopolist, maximum net revenue is triangle $d_1d_8s_1$ forthcoming by producing and consuming q_8. The demand curve shows the maximum price consumers would pay for various quantities of a commodity. The demand curve is a measure of the incremental benefit derived by consumers from another unit of the commodity. Moving to the right along the demand curve, price falls because less and less satisfaction is derived from consuming more units of q.

If producers are rational and supply additional output if the supply price covers marginal costs, then the incremental costs $s_1, s_2, \ldots, s_{14}$ trace out the supply curve. The supply price may be viewed as the opportunity cost or incremental value to society foregone by use of resources for producing and consuming commodity q rather than other goods and services. The supply price (marginal cost) rises because the first units of a farm commodity, for example, can be produced at low cost on very productive land. As more units are produced, less productive land and other resources are brought into production. As additional variable resources are shifted from producing other goods and services to q, incremental costs rise and less goods and services other than q are available. Fixed costs are irrelevant and are not included because they are committed to q and cannot be used to produce other commodities; they do not influence the decision to produce q.

Because a competitive market results in output q_8 where D and S intersect, it follows that such a competitive allocation maximizes net revenue or net social benefit which is loosely termed full national income. If the marginal utility of income is constant, each dollar of net revenue can be interpreted as utility. It follows that the competitive output also maximizes net utility under the assumption that the marginal utility of income is constant and equal among people.

Whereas a perfectly discriminating monopolist would receive all net revenue, competitive markets divide net social benefit among many consumers and producers. The area above the price p and below the demand curve D, the amount consumers would be willing to pay for q in excess of what they are required to pay, is called *consumer surplus* and is triangle d_1d_8p in Figure 2.4. The area below equilibrium price p and above the supply curve S, the amount producers are paid in excess of variable costs of production, is called *producer surplus* and the is area of triangle ps_8s_1 in Figure 2.4. Producer surplus is gross revenue to producers less total variable cost.

Consumer surplus plus producer surplus equal net social benefit. Net social benefit is maximized at the equilibrium quantity under workable competition. It is important to note that the equilibrium price and quantity under classical welfare analysis depends on the initial distribution of resources. The allocation at q_8 is the efficiency of a *Pareto optimum.* There are many Pareto optimums, depending on the initial distribution of resources before markets work. The global maximum net social gain which maximizes the common good or utility of society is also a Pareto optimum. Although the allocation in Figure 2.4 is described as a *market* equilibrium, the same principles apply for a socialist, barter, or other economy equilibrium. The final distribution of net social benefits will be most equitable if the initial distribution of resources and access to opportunity are most equitable — a subject to be examined later.

Given time, all resources become variable and values are bid up to remove pure profit. With producer surplus zero in the long run, it follows that consumers receive all net social benefits. This outcome is called *consumer sovereignty* in economics.

Classical welfare analysis is a powerful tool of economic policy analysis. It is sufficiently flexible to encompass not only social costs from an inefficient output (foregone net social benefit or deadweight loss) as shown in Figure 2.4 but also from an inefficient resource mix (x-inefficiency), from excessive spoilage and administrative cost, and from lost output in resources "wasted" in political-economic seeking of transfers (PEST) activities.[2] This cardinal approach requires stronger assumptions than Pareto-partial analysis which reveals inefficiency only as disequilibrium at the margin. Pareto-partial marginal analysis is capable of indicating presence of inefficiency but not its magnitude. Classical welfare analysis also shows disequilibrium at the margin but adds marginal units to indicate impacts on national income.

Classical welfare analysis is sufficiently flexible to estimate the net social benefits from *workable competition,* a term which recognizes that perfect competition is an abstraction and that economic efficiency must be measured in an environment of imperfect information, varying degrees of mobility of resources, and sometimes of an industry able to achieve economies of size with only one or a few firms. If costs are sufficiently lower with monopoly than with atomistic (many buyers and sellers) competition, then a well-performing monopoly may be preferred over inefficient atomistic competition — based on classical welfare analysis. Classical welfare analysis recognizes that adjustments are made and information is obtained only if it pays at the margin. In short, classical welfare analysis can be used to measure efficiency even when the assumptions of perfect competition are not met.

The Marshallian concepts of producer surplus and consumer surplus have been controversial as well as useful. Major criticisms are listed below.

1. Consumer surplus must be estimated from a demand curve which ordinarily is estimated from time series data. The historic price-quantity data used to estimate the demand curve are likely to be narrow in range. The shape of the demand curve outside the range of historic price-quantity data is unknown. Consumer surplus outside the historic price-quantity data range can be measured imperfectly at best. The appropriate means to

---

[2]The term "x-inefficiency" is often used to refer to inexplicable cost increases characteristic of absence of a competitive environment but here may also be interpreted as deadweight losses in input markets which shift the commodity supply price upward at any given quantity.

PEST activities include lobbying and other means to obtain favors from government. Such activities require resources which have value in other uses, hence entail real costs and are not merely transfers.

alleviate the shortcoming is to estimate consumer surplus only within the historic price-quantity data range and only as deviations from equilibrium as shown in the illustrations to follow.

2. A second major criticism of consumer surplus is that price as a measure of marginal utility and the area beneath the demand curve as a measure of total utility requires constant marginal utility of income. Ideally, consumer surplus would be estimated from an income-compensated demand curve. Such a curve is not easily estimated but an approximation was formulated by Willig (p. 594) to adjust conventional measures of consumer surplus for changing marginal utility of income. If income elasticity of demand is low and the change in income is small from changes in consumer and producer surpluses being considered, error introduced by failure to adjust demand for changing marginal utility is small.

3. Marginal utilities differ among individuals, creating problems in using aggregate demand to measure utility. Although differences within groups such as consumers or producers may average out, systematic differences are apparent between groups by income level. Classical welfare analysis applies to group rather than to individual behavior and hence is well suited for policy analysis. Method of adjusting for marginal utility among groups and income levels are shown later.

4. Producer surplus is a return to fixed resources — a net return after subtracting variable costs from gross returns. Supply curves are behavioral relationships. They are imprecise measures of economic rent in the long and short run. Often the best solution to this problem is to estimate producer surplus as gross receipts less all operating costs defined as costs which are variable in the time period under consideration. Estimates of operating costs may be available for commodities from farm management surveys or other sources.

5. Classical welfare analysis often is but need not be confined to a partial, single commodity analysis. Equilibrium prices and quantities ideally would account for interactions among commodities in a simultaneous system of demand and supply equations.

6. Long-run as well as short-run elasticities often are not but can be recognized in analysis.

7. Finally, classical welfare economics primarily relates to static economic efficiency and current income rather than to long-term efficiency and economic growth which requires savings and investment in high-return durable capital. Net social benefit may be consumed today or invested to produce future social benefits. If net social benefit is consumed today, it provides satisfaction today. If it is saved and invested, it provides a flow of future output benefiting later generations. The later will be apparent in dynamic efficiency. Dynamic efficiency can best be measured by *trends* in resource productivity or in national income. Historical data indicate that countries with the least amount of government interventions distorting markets (lowest net social cost) have had the fastest rates of economic growth (Agarwala, 1983).

NET BENEFITS AND REDISTRIBUTIONS FROM PUBLIC INTERVENTIONS IMPROVING THE ECONOMY

Individuals pursuing self-interests without a coordinating framework would not necessarily increase well-being of society. Some coordinating system is essential, either the private market sector or the public sector. The latter includes the family and all other non-market allocators but here refers mainly to allocations by government. Classical welfare analysis can help to identify situations of market failure or public sector failure which reduce net social benefits. Figure 2.5 is used to illustrate market failure — where

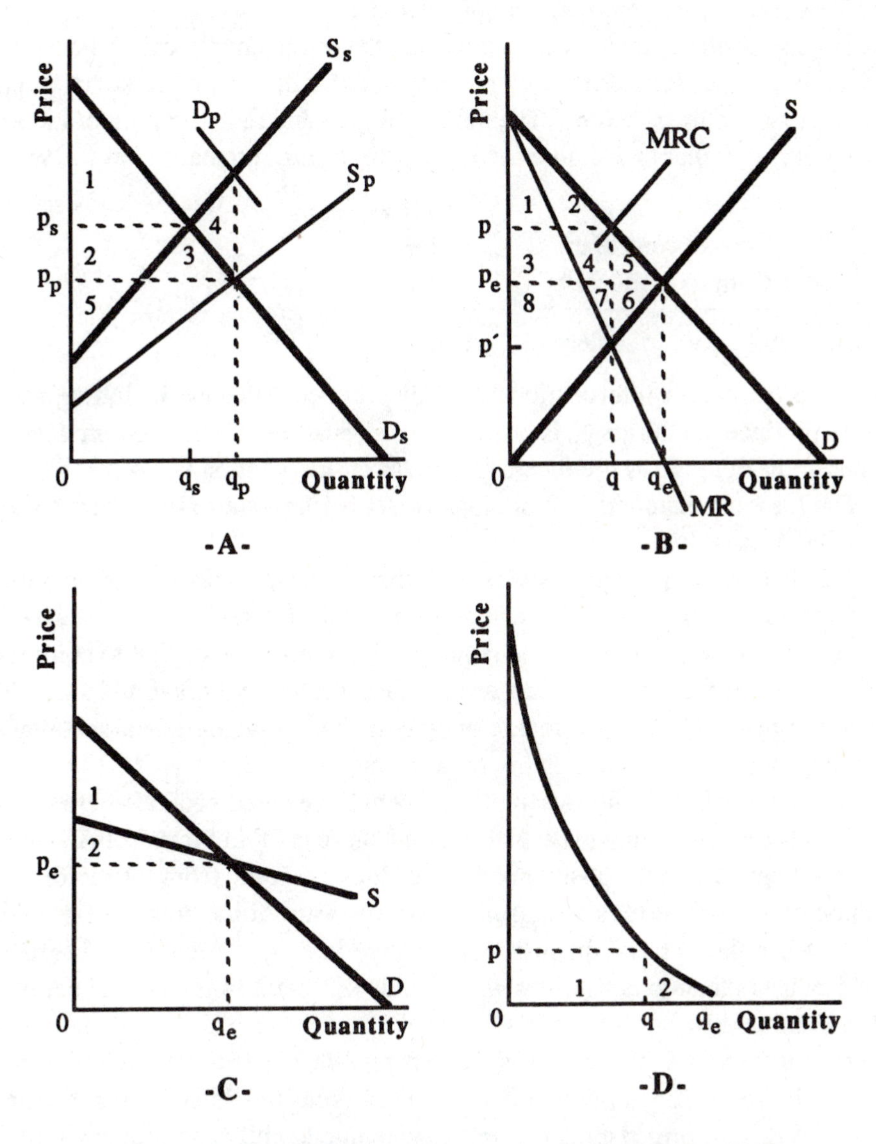

FIGURE 2.5. Redistributions and Costs of Market Failure

markets alone do not minimize social cost and where the public sector potentially can intervene to increase well-being.

1. Externalities. Panel A in Figure 2.5 shows the impact of disassociation of private and social costs — an *externality* such as costs incurred by downstream farmers when an upstream farmer allows high rates of soil erosion. The private supply curve S_p reflects costs incurred by and entering the accounts only of the upstream producer but does not reflect costs incurred by downstream producers when sediment destroys their crops. S_s is the social supply curve which includes costs borne by all producers. The social demand curve for the good being produced is D_s. *Net social costs are appropriately measured from social rather than private supply and demand curves.*

The firm responds to the private rather than the social supply curve, hence output is q_p and price is p_p. Efficient pricing and output determined by the social supply and demand curves would be q_s and p_s. The net social gain and redistribution of income from efficient output at q_s rather than at the inefficient level q_p are summarized as follows:

	Area
Loss to consumers	2 + 3
Gain to producers	2 + 3 + 4
Net gain to society	4.

In the absence of market intervention, the full producer surplus, including the cost to downstream producers, at price p_p is 5 - 3 - 4. After market intervention producer surplus is 2 + 5, hence the net gain to producers from market intervention is 2 + 5 - (5 - 3 - 4) or 2 + 3 + 4. The loss to consumers is more than offset by the gain to producers so society is better off with intervention.

In other instances, private costs exceed social costs, reversing the position of the supply curves in Figure 2.5A. An example is costs of storing buffer stocks of farm commodities. The private storage trade requires a high return on capital to compensate for high risk. For society as a whole, risks are less because many unfavorable and favorable outcomes average out for a large storage operation. The private trade alone may provide too little storage to bring optimal stability to farm and food prices.

Another externality occurs when private benefits exceed social benefits so that the private demand curve is D_p and the social demand curve is D_s in Figure 2.5A. An example is health problems caused consumers by residues in food from pesticides used by producers on crops or livestock. Consumers are unaware of the source of problems and purchase q_p when the optimal quantity would have been q_s. Benefits of pesticides are greater to producers than to consumers when residues remain in foods. The net social cost is area 4 in Figure 2.5A.

In other instances social benefits exceed private benefits. D_s and D_p then trade positions in Figure 2.5A. An example of such externalities is general schooling. The benefits of a literate, informed citizenry able to vote intelligently and provide military skills exceed the sum of private benefits of schooling. Hence, reliance on the market alone results in an inefficient, low quantity of schooling and a net social cost or lost real income measured by the value in area 4 if D_s and D_p are parallel.

Considerable evidence indicates that producers and consumers are rational in responding to private incentives. Thus a system of subsidies and taxes correcting externalities so that private costs (benefits) align with social costs (benefits) give signals to producers and consumers causing them to act in the public interest. In other instances regulations are required to bring socially efficient output q_s. These corrections to improve the economy will raise *real* national product but will reduce the *money value* of national income. The former rather than the latter is the appropriate measure of well-being and the appropriate target for expansion.

2. Imperfect Competition. Imperfect competition is another source of inefficiency in markets. The social gain from intervention to bring about a competitive market is shown in Figure 2.5B. A monopolist produces where marginal revenue MR equals marginal cost S at output q and price p. Such pricing and output maximizes net revenue to the firm, but not to society. Competitive pricing at p_e and output at q_e increases consumer surplus by area 3 + 4 + 5 and reduces producer surplus by area 3 + 4 - 6. The net gain or addition to real national income is 5 + 6 from competitive markets. On the other hand if economies of size are substantial so that an atomized industry of many sellers would have a supply curve intersecting D above price p in Figure 2.5B, then monopoly pricing and output would have less social cost than would competitive pricing and output.

Now turning from the case of monopoly (single seller) to monopsony (single buyer), the monopsonist utilizes a resource q to the point where marginal revenue D equals marginal resource cost MRC. Quantity is q and the price paid suppliers is p′. Net social cost is area 5 + 6.

Market efficiency is relatively robust with respect to number of firms. A market with only one domestic firm may perform efficiently if the firm is not protected from foreign competition or entry into the industry by other firms. Antitrust action or termination of government regulations that shield imperfect competition may be appropriate responses.

A characteristic which gives rise to imperfect competition is a decreasing cost industry depicted in Figure 2.5C. Competitive pricing and output result in price p_e and quantity q_e. Consumer surplus is area 1 + 2 and producer surplus is area 2, a loss. Because producers lose money, they will choose not to produce. The benefits from producing and consuming at competitive price p_e and output q_e rather than not producing are summarized as follows:

	Area
Gain to consumers	1 + 2
Loss to producers	2
Net gain to society	1.

Intervention to bring output q_e increases national income by area 1.

An industry or geographic area in which only one firm can operate in an industry at lowest cost per unit is called a *natural monopoly*. Frequently, natural monopolies also are characterized by decreasing marginal cost illustrated by the curve S in Figure 2.5C. If one private firm operates, however, it is likely to practice monopoly pricing such as in Figure 2.5B to avoid losses.

Given a situation of natural monopoly, several options are available to society. Cooperatives, public utilities, or regulated private firms are sometimes the answer where only one firm can serve the industry at low cost. A private firm may be regulated and subsidized out of taxes or the industry may be served by a public utility owned by government. In many rural areas, cooperatives such as grain elevators operate to avoid the onerous consequences of natural monopolies. Atomization of an industry (breaking up a single dominant firm into so many firms that no one can influence price) may raise production costs so high that the social loss from higher costs are larger than from monopoly pricing and output. In reality, social costs of inefficiently administered regulations or of poorly managed cooperatives or public firms frequently exceed the social cost of private monopoly the interventions were attempting to eliminate.

3. Public Goods. Markets function best where goods are rival and exclusionary. In the case of a pure public good, consumption by one person does not reduce consumption available to another, hence the good is not rival. For example the marginal cost is zero as depicted in Figure 2.5D for providing another person national defense or another car access to an underutilized highway or bridge. Because marginal cost is zero, the social supply curve lies flat on the horizontal axis. To arbitrarily charge a price p to each individual who uses a highway or bridge reduces quantity from optimal amount q_e to q. Termination of the toll p increases use at no cost to society. The net gain to society from marginal cost pricing in an efficient market has components in Figure 2.5D as follows:

	Area
Gain to consumers	1 + 2
Loss to producers	1
Net gain to society	2.

Benefits of some goods cannot be appropriated by private firms. For the market to function efficiently, goods must be exclusionary; that is, if the good is made available to one person X it must be possible to exclude it from other persons Y and Z. If the latter can obtain the good without charge if X purchases it, the supplier will be unable to appropriate benefits. Everyone will want to be a free rider. No one will pay to produce it and the good will be under-produced by the market. For goods such as the technology in a seed variety easily reproduced by farmers, it is not possible for a private firm to appropriate enough benefits from users to cover costs of developing the technology however large the social benefits. In some cases of public goods such as a sidewalk or lighthouse, it is possible to collect for use but not practical because the cost of collection is large relative to benefits.

Some goods such as agricultural research, extension, and statistical (data) information have some but not necessarily all properties of public goods. Extension of patent rights can in some instances make nonrival biological technology appropriable and hence provide incentives to private firms but private firms are likely to price higher than the social marginal cost specified above as economically efficient. Marginal cost pricing requires public sector funding for developing public goods technology. Such funding is frequently not forthcoming especially in developing countries despite high social rates of return on investment. The resulting economic inefficiency of under-investment implies that use of the private sector to develop applied technology may be more efficient even if the sector prices in excess of marginal cost. Basic research, however, is likely to be under-funded if left solely to private firms because even with patent rights the firm may not be able to appropriate receipts to pay for research or may view outcomes as too risky. Because reliance on the public sector alone or the private sector alone for public goods such as agricultural technology has advantages and disadvantages, mixed public-private systems are often preferable and are widely used.

NET COSTS AND REDISTRIBUTIONS FROM PUBLIC INTERVENTIONS DISTORTING AND DEPRESSING THE ECONOMY

We now turn to public interventions in markets which reduce economic efficiency and national income as illustrated in Figure 2.6. Removing these interventions will raise national income but not necessarily income of agriculture or consumers. The private supply and demand curves are assumed to be the social curves.

1. Commodity Program Interventions. Panel A is used to show distributions and costs of mandatory supply control and of price supports without supply control. Deviations are examined from open market equilibrium at price p_e and quantity q_e. Suppose mandatory supply control reduces output to q_1 and raises price to p_1. Loss of consumer surplus compared to open market equilibrium is area 2 + 3. The increase in producer surplus is area 2 - 8, a value likely to be positive if demand is inelastic. Results for mandatory production controls are summarized as follows:

	Area
Loss to consumers	2 + 3
Gain to producers	2 - 8
Net loss to society	3 + 8.

The net loss to society, the excess of consumer losses over producer gains, is area 3 + 8. *Transfer inefficiency* as measured by net social loss divided by the gain to producers is (3 + 8)/(2 - 8). Inefficiency is zero for an efficient transfer.

Next consider costs and redistributions of income from price supports without production controls. Price supported at p_1 brings production q_2. At least two options are available to clear the market which will not clear at price p_2. One is to dispose of $q_2 - q_1$ outside the market. The second is to subsidize the difference between supply price p_1 and the demand price p_2 which clears the market.

Consider the latter first. In the case of transfer (subsidies) from taxpayers to hold producers' price at p_1 and consumers' price at p_2, the price p_1 requires a transfer from

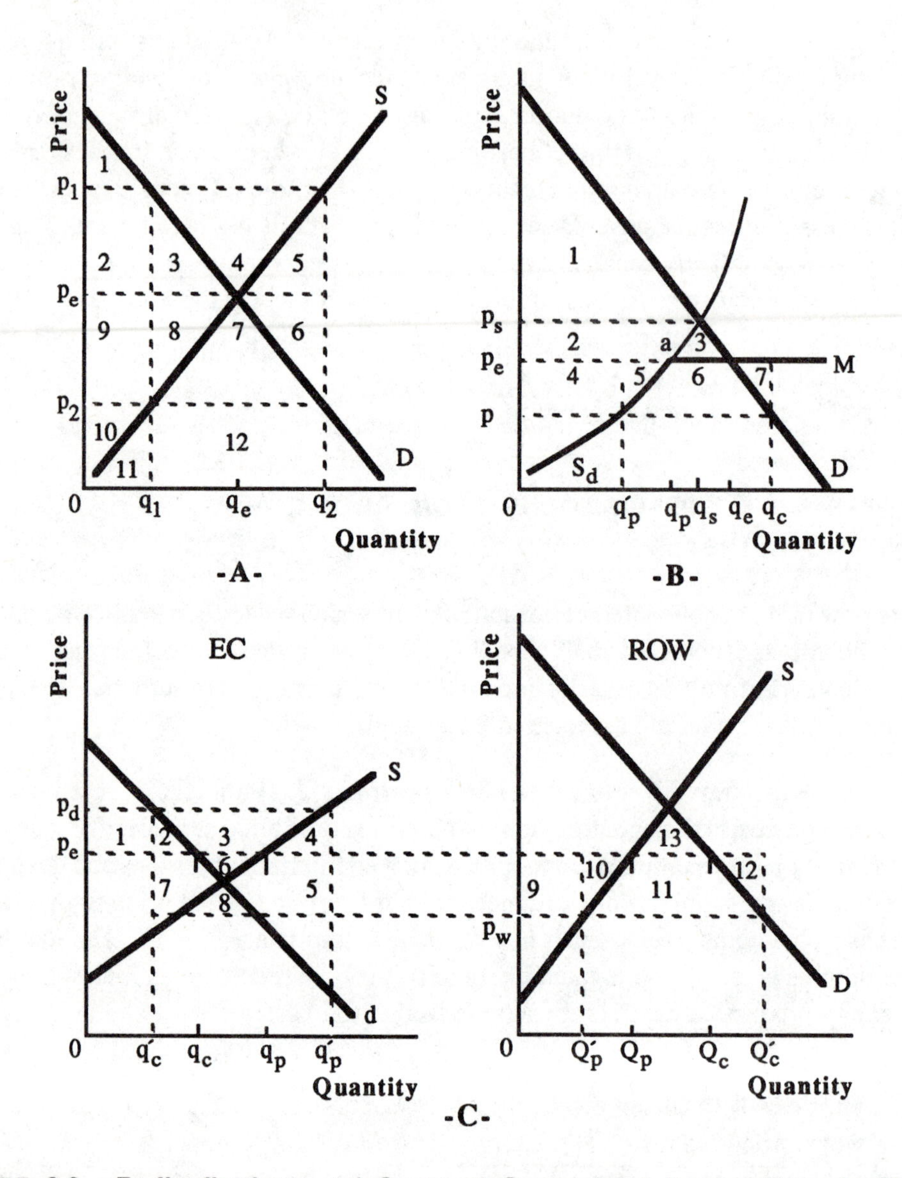

FIGURE 2.6. Redistributions and Costs of Government Interventions in Markets

taxpayers of 2 + 3 + 4 + 5 to producers, increasing their surplus by 2 + 3 + 4. Hence, net social cost is 5 and transfer inefficiency is 5/(2 + 3 + 4). For consumers, holding price at p_2 instead of p_e requires a tax transfer equal to area 6 + 7 + 8 + 9 which increases consumer surplus by area 7 + 8 + 9. Taxpayers' cost exceeds consumers' gain by area 6, the net social cost. Transfer inefficiency is 6/(7 + 8 + 9) and *transfer efficiency* is (7 + 8 + 9)/(6 + 7 + 8 + 9). Compared to a well-functioning market equilibrium, the resulting combined welfare implications are summarized as follows:

	Area
Gain to consumers	7 + 8 + 9
Gain to producers	2 + 3 + 4
Loss to taxpayers	2 + 3 + 4 + 5 + 6 + 7 + 8 + 9
Net loss to society	5 + 6.

The overall transfer inefficiency is (5 + 6)/(2 + 3 + 4 + 7 + 8 + 9).

All countries experience a *food price dilemma* which may be defined as a free market equilibrium price deemed too low to avoid hardship and poverty among producers and too high to avoid hardship and poverty among consumers. Subsidies to producers and consumers to address the food price dilemma tend to leave one or both groups worse off. That is because taxpayers are either producers or consumers. The subsidies are a negative sum game; that is, consumers and producers lose more as taxpayers than they gain as consumers and producers. Despite the shortcomings of this game, many countries play it.

An alternative to subsidies is to dispose of surpluses by spoilage or by allocating them to uses with little or no value. Disposing of quantity $q_2 - q_1$ in such manner provides the following welfare implications from Figure 2.6A:

	Area
Loss to consumers	2 + 3
Gain to producers	2 + 3 + 4
Loss to taxpayers	3 + 4 + 5 + 6 + 7 + 8 + 12
Net loss to society	3 + 5 + 6 + 7 + 8 + 12.

The welfare loss is large relative to transfer gain to the beneficiary, producers, from the policy. Transfer inefficiency is (3 + 5 + 6 + 7 + 8 + 12)/(2 + 3 + 4).

Figure 2.6B illustrates redistributions and social costs from government interventions in markets to provide self-sufficiency and to reduce food cost through a ceiling on food price. Domestic demand is D and domestic supply is S_d. The horizontal, perfectly elastic import supply curve M is for a small country which can import all it wants without changing world price from p_e. Total supply is the curve of lowest cost sources of q, or S_daM. Open market equilibrium quantity is q_e which is equal to consumption. Supply quantity is q_p from the domestic market and $q_e - q_p$ from imports.

Raising price to p_s attains self-sufficiency by increasing domestic quantity supplied by $q_s - q_p$ and decreasing quantity demanded by $q_e - q_s$. The welfare impact is as follows:

	Area
Loss to consumers	2 + 3
Gain to producers	2
Net loss to society	3.

It is apparent that the loss to consumers offsets the gain to producers, hence society is worse off from self-sufficiency. As with other welfare measures in this section, only economic efficiency (national income) is considered and other objectives such as equity and stability are ignored. Instability increases danger of food shortages if self-sufficiency is achieved at the expense of lost economic means to purchase supplies from abroad when domestic production is short, if self-sufficiency is achieved by use of fertilizers and pesticides purchased from abroad more subject than food imports to interruption, and if domestic production is variable from year to year. Equity may not be served by self-sufficiency because high food prices especially oppress the poor, who spend a high proportion of income for food.

Food prices may be held down by a price ceiling at p in Figure 2.6B. This reduces domestic production to q_p' and increases consumption to q_c. An import subsidy of 5 + 6 + 7 is required to avoid a massive food shortage. Welfare impacts are summarized as follows:

	Area
Gain to consumers	4 + 5 + 6
Loss to producers	4 + 5
Loss to taxpayers	5 + 6 + 7
Net loss to society	5 + 7.

Transfer inefficiency (5 + 7)/(4 + 5 + 6) per dollar of gain to consumers tends to be low if $p_e - p$ is small and domestic supply and demand are highly inelastic.

2. Interventions Influencing World Markets. Many less developed countries "tax" producers by holding down food prices as shown above; developed countries often subsidize agriculture. The example in Figure 6.3C is hypothetical but broadly resembles the European Community which generates excess production with high price supports in the absence of supply controls and subsidizes export of surpluses. With free trade, welfare is maximized at equilibrium world price (no transport costs assumed) p_e, domestic quantity supplied q_p, and quantity consumed q_c in the European Community (EC), and Q_p and Q_c respectively in the rest of world (ROW). Small initial exports from the EC, $q_p - q_c$, equal small imports $Q_c - Q_p$ by ROW. Price supports in EC bring prices to p_d, raising production to q_p' and lowering consumption to q_c'. In the absence of supply control, the EC elects to export the surplus $q_p' - q_c'$ to ROW which lowers world price to p_w. The subsidy per unit is $p_d - p_w$ and the total subsidy is area 2 + 3 + 4 + 5 + 6 + 7 + 8. The area 2 + 3 + 4 of this subsidy raising price from p_e to p_d is a transfer to producers, but area 5 + 6 + 7 + 8 is a subsidy to ROW, equal to area 10 + 11 + 12.

The welfare impact compared to open market equilibrium is summarized as follows:

EC	***Area***
Loss to consumers	1 + 2
Gain to producers	1 + 2 + 3
Loss to taxpayers	2 + 3 + 4 + 5 + 6 + 7 + 8
Net loss to society	2 + 4 + 5 + 6 + 7 + 8.

The loss to society of area 2 + 4 can be charged to holding domestic price above p_e; the loss of 5 + 6 + 7 + 8 can be traced to forcing world price below p_e. The impact on ROW is:

ROW	***Area***
Gain to consumers	9 + 10 + 11
Loss to producers	9 + 10
Net gain to society	11.

Consumers gain more than producers lose in ROW, leaving a net gain to ROW of area 11 from the policies of the EC. Of the net loss 2 + 4 + 5 + 6 + 7 + 8 to the EC, 2 + 4 was deadweight loss and 5 + 6 + 7 + 8 = 10 + 11 + 12 was a transfer to ROW. Of the transfer, only 11 was a real gain to ROW; 10 + 12 was deadweight loss. Hence the world deadweight loss from the EC support programs without supply control is 2 + 4 + 10 + 12. It is apparent that losses to consumers and taxpayers in the EC fall well short of gains to producers, thus the EC would appear to benefit from ending the Common Agricultural Policy (CAP). On the other hand, consumers gain more than producers lose in the ROW, hence the ROW would appear to benefit from the EC continuing the CAP. That the EC in fact favors continuation of the CAP while the ROW opposes it probably is not testimony to the inadequacy of the analysis in Figure 2.6C but rather testimony to the political power of producers in the EC.

In concluding this section it is well to note that governments can go too far or not far enough to bring about efficient allocations as noted in Figures 2.5 and 2.6. Frequent government policy errors include an overvalued exchange rate, price controls holding traded goods prices below world price levels, excessive reliance on parastatal corporations or government agencies to produce and market, neglect of infrastructure and social overhead in general, provision of credit at excessively subsidized rates, too rapid increase in money supply and inflation, federal budget deficits, import substitution, excessive central planning, and, finally but not least important, unstable policies that thwart sound long-term decision-making.

In some cases, the decision whether the government should intervene or let markets work for efficiency is clear-cut. However, in many instances the case for government interventions in markets is not clear-cut. Experience with problems actually encountered in developing countries suggests that when in doubt the appropriate policy is to rely on the private market. That is because government interventions often entail higher social costs than the private market failure the policies attempt to correct.

EQUITY

Economic efficiency entails allocating resources to uses contributing the most to output. Economic growth results from economic efficiency over time, including savings and investment allocated to earn high returns in building human and material capital. As indicated earlier, the supply price is a schedule of incremental cost to firms and society of production whereas the demand price is the incremental benefit to society from consuming

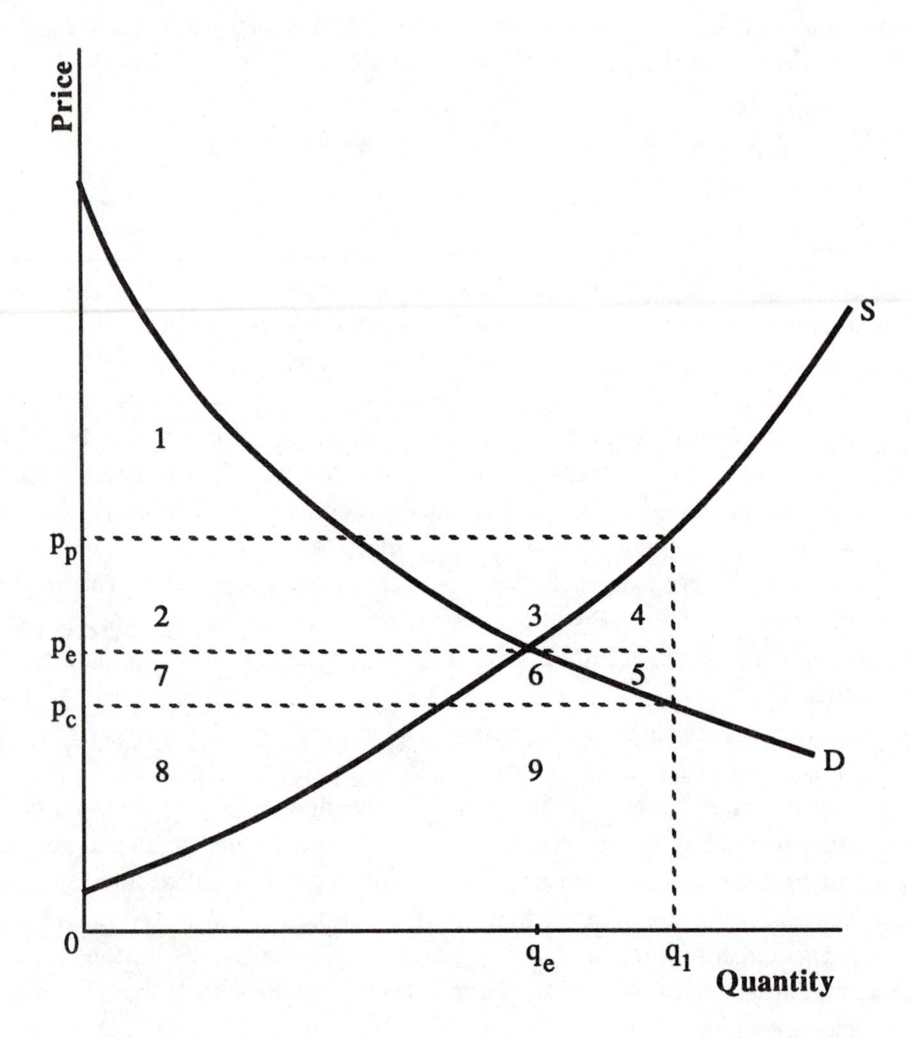

FIGURE 2.7. Economic Efficiency Illustrated with Demand and Supply Curves

another unit of the commodity. But an efficient allocation is not necessarily an equitable allocation. Markets can be in equilibrium while some persons are starving and others chronically obese.

The food price dilemma with producers wanting higher prices than p_e and consumers wanting lower prices than p_e to promote equity is illustrated in Figure 2.7. Suppose taxpayers provide a subsidy $p_p - p_e$ to producers, bringing the supply price to p_p and inducing output to expand to q_1. Suppose also that a subsidy of $p_e - p_c$ is provided to bring the price of food to p_c per unit to consumers. The lower price of food raises consumption to q_1.

Consumers would have been willing to pay area 1 + 2 + 6 + 7 + 8 + 9 for q_1, but only have to pay area 8 + 9, hence consumer surplus is area 1 + 2 + 6 + 7 — a gain of area 6 + 7 compared to the initial efficient output q_e. Subsidies plus market receipts less variable costs of area 4 + 5 + 6 + 9 for q_1 leave producer surplus of area 2 + 3 + 7 + 8 — a gain of 2 + 3 compared to the efficient output at q_e. The results for output q_1, featuring subsidies to raise producer prices to p_p and subsidies to lower consumer prices to p_c compared to efficient price p_e and output q_e, are summarized as follows:

	Area
Gain to consumers	6 + 7
Gain to producers	2 + 3
Loss to taxpayers	2 + 3 + 4 + 5 + 6 + 7
Net loss to society	4 + 5.

National income is reduced by the value of area 4 + 5 in Figure 2.7 from an attempt to serve equity by lowering prices to consumers and raising prices to producers. The effort reduced economic efficiency in this example. Economic efficiency and equity do not always conflict, however. Investments in schooling of disadvantaged youth can simultaneously serve economic efficiency and equity.

The market may be efficient in allocating resources to bring outcomes shown as p_e and q_e in Figure 2.7. If some are destitute while others are opulent at market equilibrium, the efficient market outcome is unlikely to be judged a fair outcome — hence the public sector may intervene to promote equity. Society may be more concerned about providing *equality of opportunity* to succeed than *equality of outcomes*. And *equitable* opportunities or outcomes do not necessarily mean equal opportunities or outcomes.

As noted earlier in this chapter, determining an equitable distribution of income and resources traditionally has been viewed as a value judgment to be made by the political process with little or no input from socio-economic analysts. Modern socio-economic science and modern statistical techniques are reducing subjectivity and increasing objectivity in analysis of equity issues. At issue is how the well-being of people is influenced by the distribution of income.

If an objective of an economic system is to raise the well-being of people, then resources and goods and services need to be allocated to uses where incremental (marginal) utility is greatest. Marginal utility (satisfaction derived from an additional unit of income) differs among individuals and is subjective, creating problems in measuring utility. Attitudinal scales developed and validated by psychologists and sociologists to measure well-being of individuals through personal interview surveys show promise to predict marginal utility with acceptable precision and objectivity for groups (not necessary for individuals) such as income classes because individual errors in measurement tend to average out.

Based on socio-psychological attitudinal scales utilized in personal interviews of a random sample of Americans, Tweeten and Mlay estimated the marginal utility of income to be:

$$MU = 1.3582 - .3582Y \quad (MU = 0 \text{ for } Y > 3.8) \qquad (2.6)$$

where MU is utility as a proportion of that derived by a family with median income and Y is family income as a proportion of median family income. Marginal utility differed systematically by age, education, and income, but no major differences in marginal utility were observed over time, sectors (rural-urban), or regions. Results were from a comprehensive set of socio-psychological scales relating to various domains of well-being and based on a large, random sample of one adult from each of approximately 1,500 U.S. families in 1976 and another sample of like size in 1980. Alternatively, marginal utility for various alternative income levels might be specified by the government through the political process.

Estimates of marginal utility can be used to adjust policy analysis results such as those from Figure 2.7. Suppose that weighted median income of taxpayers is 2.15 times the national median, of consumers is 1.19 times the national median, and of producers is .89 times the national median income. The marginal utility for each of these groups computed from equation 2.6 is shown in column "MU" in Table 2.1. The unadjusted policy analysis results show a net economic loss from the program or policy. Because the income redistribution was from higher income taxpayers to lower income producers, the utility-adjusted results show a net social gain rather than a loss. (Marginal utilities were simply multiplied by gains and losses under the assumption that changes in income from the project were too small to change MU.)

TABLE 2.1. Illustration of Redistribution and Net Social Loss (Gain) from Figure 2.7 Adjusted for Marginal Utility of Income

Item	Unadjusted Policy Results	Adjusted for Marginal Utility	
	($ mil.)	(MU/$)	($ mil.)
Gain to producers	160	1.041	166.6
Gain to consumers	120	.933	111.9
Loss to taxpayers	300	.588	176.4
Loss to society	20		-102.1 (gain)

Additional refinement can be introduced by assigning utility weights *by income class* to the results from Figure 2.7 to express economic gains and losses as social gains and losses. Gains and losses from subsidies to consumers and producers are distributed among income classes as shown in Table 2.2. The upper left panel shows economic gains and losses (negatives) and the upper right panel shows social gains and losses (economic outcome multiplied by marginal utility) for price interventions from Figure 2.7. Because the interventions especially benefit lower income producers and consumers with high MU at a cost especially to high income taxpayers who have a low MU, the net economic loss of $20 million is a net social gain of 102 million utility adjusted dollars.

This hypothetical sample combining the equity and efficiency dimensions to measure social gains and losses does not necessarily justify market intervention policy. The lower panel in Table 2.2 demonstrates an important principle: A pure cash transfer payment is socially more beneficial than an economically inefficient intervention such as a price ceiling or price support to redistribute income to the disadvantaged. The net social gain is $114 million with a direct transfer payment compared to $102 million by intervening in the market to promote equity.

Administrative costs are not included but are likely to be less with cash transfer payments than with price interventions. In reality, cash transfers are not fully "pure" — they also distort allocations and hence create deadweight losses. Such losses are likely to be small, however, compared to those of most market interventions designed to foster equity.

In short, the "efficient" equilibrium such as p_e and q_e in Figure 2.7 is a function of the initial distribution of resources. Allowing prices to bring about an economically efficient allocation and providing pure cash transfers to producers would have provided a larger adjusted net social gain without loss to anyone than the market interventions depicted

TABLE 2.2. Illustration of Redistribution and Net Loss (Gain) to Society from Subsidies in Figure 2.7 Adjusted by Marginal Utility of Money Among Income Classes

Family	Economic				Social		
Income ($)	Producers	Consumers	Taxpayers	MU	Producers	Consumers	Taxpayers
			(price intervention; million dollars)				
0-10,000	65	40	-40	1.2729	82.7	50.9	-50.9
10,000-20,000	50	30	-45	1.1023	55.1	33.1	-49.6
20,000-30,000	30	25	-60	.8465	25.4	21.2	-50.8
30,000-40,000	10	20	-75	.3349	3.4	6.7	-25.1
40,000-80,000+	5	5	-80	.0000	0.0	0.0	0.0
	160	120	-300		166.6	111.9	-176.4
Net Loss to Society			20				-102.1 (gain)
			(pure transfer; million dollars)				
0-10,000	65	40	-37	1.2729	82.7	50.9	-47.1
10,000-20,000	50	30	-42	1.1023	55.1	33.1	-46.3
20,000-30,000	30	25	-56	.8465	25.4	21.2	-47.4
30,000-40,000	10	20	-70	.3349	3.4	6.7	-23.4
40,000-80,000+	5	5	-75	.0000	0.0	0.0	0.0
	160	120	-280		166.6	111.9	-164.2
Net Loss to Society							-114.3 (gain)

in Figure 2.7. The overall most equitable and efficient system maximizing social gain adjusted for utility must provide for equitable and efficient distribution of resources. That is why human resource development programs and access to public services and resources are of high priority in developing countries. Efforts of developing countries to reduce farm poverty by nationwide support of farm commodity prices and to reduce consumer poverty by nationwide ceilings on food prices have sometimes placed burdens on taxpayers and the economy, ultimately slowing the economy and hurting the poor. Targeted food assistance programs such as food stamps or sale of staples consumed by the poor at subsidized prices in fair price shops have placed less burden on taxpayers and have been successful in some countries.

Estimating the level and distribution of economic gains and losses from government interventions requires considerable effort. Calculations of marginal utilities and weighting of gains and losses among income classes is often impossible. A sound approach, in many instances, is to provide data on distribution of gains and losses by income class, then leave to policy makers the decision regarding what is an appropriate redistribution.

Use of the Political Process to Correct Market Deficiencies

The political process is sometimes used by society to correct alleged or real deficiencies in the price mechanism and to form a consensus of value judgments concerning the goals to be achieved in society. In a competitive economy, the consumer is truly sovereign, voting with dollars what to produce and in the long run reaping the net social gain. In the political process, votes are one per person and not proportional to dollars, hence a quite different allocation will evolve than under the price system. Cynics may argue that power is somewhat proportional to income because those with money can spend it to advertise and sway other voters to their view.

It can be shown that even if the voting process is deemed to be appropriate, it is not necessarily consistent (Arrow, 1951). Suppose that individuals 1, 2, and 3 vote on policies X, Y, and Z as follows:

1	$X \rightarrow Y \rightarrow Z$	$X \rightarrow Z$
2	$Y \rightarrow Z \rightarrow X$	$Y \rightarrow X$
3	$Z \rightarrow X \rightarrow Y$	$Z \rightarrow Y$

The arrows are read "is preferred to." If there is a democratic vote on X versus Y, the majority (1 and 3) will favor it, so X is preferred. If there is a vote on Y versus Z, Y is favored by the majority (1 and 2). It would seem to follow that by this latter counting, since X is preferred to Y and Y to Z, then the majority would prefer X to Z. But this is not so; individuals 2 and 3 prefer Z to X. This example merely points up one of the difficulties of aggregating preferences through the voting mechanism. In a representative government, persons often are elected to the legislature based on two alternative issues such as X versus

Y and Y versus Z. Problems arise however, when in the legislature they are called upon to decide issues such as X versus Z.

OPTIMALITY IN THE POLITICAL PROCESS

Figure 2.8 analyzes the theory of political satisfaction or dissatisfaction in a democracy. Assumptions of the model are:

1. All voters can be aligned by political philosophy from left L to right R.
2. Voters are rational and cast their ballot for the political party closest to their political philosophy.
3. Elections are decided by majority vote and only one party can win the election for, say, President of the United States.
4. Dissatisfaction with a particular party comes in equal increments or decrements between any two adjacent voters along the political spectrum L to R. The area underneath the measure of dissatisfaction with party L (line LD) or party R (line RB) can be summed to form a total index of political dissatisfaction.

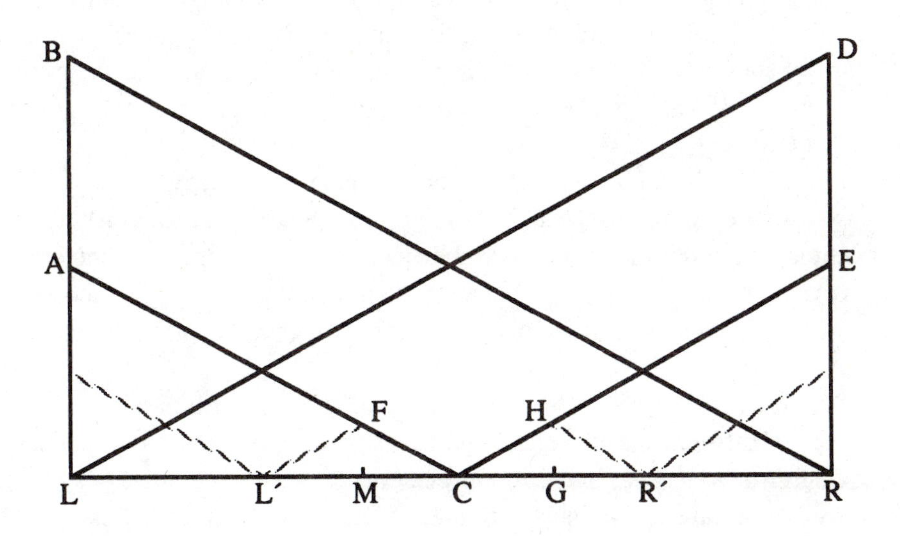

FIGURE 2.8. Model of Political Behavior in a Democracy

If extreme leftist political party L wins the election, political dissatisfaction is the area of triangle LDR. If extreme rightest political party R wins the election, political dissatisfaction is the area of triangle RBL.

Suppose that one party locates its political philosophy at the center of the political spectrum C in Figure 2.8. Suppose also that another party locates its political philosophy at a position left of center at L´. Voters to the right of point M will vote for C because their political philosophy is closer to C as apparent in the figure. Their dissatisfaction indicated by the line CA is less for C than for L' to the right of M. Voters to the left of point M will vote for L′ because the line of political dissatisfaction is less for L′ than for C to the left of M. Thus C wins the election. This case is symbolized by George McGovern's loss to Richard Nixon in 1972.

If one party locates at C and the other at R′, a similar result occurs. Voters to the left of G will vote for C because their political dissatisfaction with C, indicated by the line CE, is less than their dissatisfaction with R′, indicated by the line R′H to the right of G. Voters to the right of G will vote for R′ because their political dissatisfaction line R′H for R′ is less than their dissatisfaction CE with C to the right of G. Thus C wins the election. This situation is symbolized by Lyndon Johnson's victory over Barry Goldwater in 1964.

If the election is entirely over left-right (liberal-conservative) issues, it is clear that the middle -of-the-road party will win. Thus in a two-party system, both parties must locate at the center to have a chance to win the election. This strategy also minimizes political dissatisfaction. The area in triangles LAC + REC, the total dissatisfaction with a center-part win, is only half the area LBR or RDL of political dissatisfaction associated with a win for extremist party L or R.

The analysis is complicated by introduction of more than two parties. If the Democratic and Republican parties are respectively slightly to the left and right of C in Figure 2.8, and a liberal party enters the election at L′, the Democratic Party loses the election because the liberal vote is divided and the conservative party obtains all votes to the right of C. Such behavior of a third party is irrational however, because its position is that of a spoiler — it destroys the chances of electing a party candidate closest to the political philosophy it espouses.

The simple analysis explains several phenomena about democracy:

1. A two-party system is rational. If both parties deviate simultaneously to the right (or left) of center, a new third party can win the election by locating at the center. One of the original two parties will not survive because the political philosophy of the voter who supported it will be better served in winning elections through a second party located at the center.

2. Both parties will locate at the center of the political spectrum; there will not be a "nickle's worth" of difference between the political philosophy of the two parties. Any party that deviates from the center will lose the election.

In reality, voters are not always rational. Many Southerners, for example, have maintained allegiance to the Democratic Party although their political philosophy locates them closer to the Republican Party. Furthermore, issues are frequently blurred and a given political party may hold some positions that are more conservative and some more liberal than those of the other party. While such positions may be designed to hold the center ground of the political spectrum, such behavior creates dilemmas for voters and can reduce the effectiveness of the two-party system in representing the political philosophy of voters.

The analysis applies principally to allocation of public goods defined earlier. The "all or nothing" electoral system portrayed above is more directly applicable to presidential elections than to election of parlimentary bodies where various shades of the political spectrum can be voiced by its many members.

Coalitions may form, as when lower and middle classes combine to legislate programs to redistribute income away from the upper class. But the upper class may bribe the middle class into voting against redistribution to the lower class. While such coalitions can explain some political behavior, they serve as inexact predictive devices because coalitions are unstable.

SUMMARY AND CONCLUSIONS

The search by economists for a basis on which to make policy recommendations has led to several criteria:

1. *Pareto criterion:* A is preferred to policy B if, by changing from B to A, someone is made better off while no one is made worse off.
2. *Kaldor criterion:* policy A is preferred to policy B if, by changing to A, those who gain can compensate those who lose and still be in a better position than under policy B.
3. *Hicks criterion:* policy A is preferred to policy B if those who lose by changing to A cannot bribe those who gain into not making the change from B to A.
4. *Scitovsky criterion:* policy A is preferred to policy B if those who gain by the change from B to A can compensate those who lose, and if simultaneously those who lose cannot bribe those who gain into no making the change from B to A.

Because it is recognized that practical considerations preclude compensation, the criteria fundamentally condense to a matter of economic efficiency. That efficiency ordinarily is not measured in utility but in value of output per unit of input. Compensation in dollar terms is feasible for policies that meet the efficiency criterion, but if the disutility of a dollar to the losers far exceeds the utility of a dollar gain to those who benefit, the conventional benefit-cost measures of efficiency fail to reveal which policies increase well-being.

The efficiency criterion alone is inadequate for public policy. It is argued that the economist must be concerned about who benefits and who loses. In many basic policy issues such as rural poverty or urban slums, efficiency questions may be dwarfed by equity considerations. Economics is concerned with well-being of people which in turn depends on the level, distribution, and variability of income and on other variables.

Also relevant in this context is the statement of Colin Clark (1957, p. 1) that "economics must also take its place in the hierarchy of arts and sciences." Laws of economics coordinate and limit recommendations from the technical agricultural sciences. Economics must recognize its subordination, however, to the political process, ethics, and theology. Economists inform; the political process makes decisions. But the political process cannot make informed decisions without awareness of the impact of public policies on the level, distribution, and variability of income.

The above analysis did not directly address issues of ethics. Moralists sometimes scold economists for ignoring the alleged injustice imposed on losers in the struggle for survival in a dynamic, competitive economy. Two basic schools of ethics are the *utilitarian* school and the *moral imperative* school. Utilitarianism judges policies on the basis of whether they increase or decrease well-being of people. It is epitomized by the economic theory presented herein. The moral imperative school depends on divine revelation to know what is good or bad and what adds or detracts from well-being. It is not well suited to economics but happily most moral imperatives are judged by their advocates to increase well-being of people. Hence the utilitarian ethics used herein because it is so well suited to economic analysis is not necessarily in conflict with moral imperative ethics.

Greene (1963, p. 232) stated, "I think you economists are trying to kid us when you say you just deal with pure facts and pure theory without an orientation of your own or building toward objectives you believe to be true." The philosophic orientation of economists does indeed bias their analysis. Economists who divide economics into efficiency and equity dimensions and then throw away the equity dimension bias public policy against the disadvantaged and hence against the well-being of people. Philosophers, clergy, and social scientists who divide economics into efficiency and equity dimensions and then throw away the efficiency dimension bias public policy against economic growth and hence also against the well-being of people. Either simplistic approach is immoral by utilitarian ethical standards. Agricultural economists who show the impact of public policies on farm income but not on national income implicitly bias their analysis toward producers (especially the organized ones) and hence against the well-being of society.

Ethical issues of whether economic policies entail more gains than losses are highly complex. A policy change may have one gainer but millions of losers or vice versa. Determining whether society is better or worse off with a policy can be judged superficially by any moralist but economists know such judgments require detailed economic analysis to determine magnitudes of gains and losses to groups and society. For example, removing trade barriers may cause job losses to a vocal few but create even more job gains. Gainers "never miss what they never had," hence will not be vocal supporters of free trade. Shallow politics and moralism will say removal of barriers is an injustice whereas indepth economic analysis may show gains will far exceed losses, especially if losers are compensated. Economists employed by taxpayers to serve the public will not be adequately performing their job if they merely answer the questions being asked by politicians who serve special interests. Evaluating policy impacts on society requires extensive expertise of social scientists using procedures such as outlined in this chapter.

The world would be simple indeed if we could divide the world as Cambell (1987) did into two camps, one saying, "they believe in economics and efficiency" and the other saying, "The human element, the farmer, the family, and the rural way of life must be placed first." Before everyone sides with the latter, it is important to note that the world is not so simplistic. Economics and efficiency deal with the human element, the farmer, the family, and the rural way of life. But economics deals with much more including freedom, justice, the consumer, the nonfarm poor, the taxpayer, and the public at large. *Values cannot be separated from economics.* Before applying principles outlined in this chapter to farm problems introduced in Chapter 1, selected issues regarding beliefs and values are presented in the next chapter.

REFERENCES

Agarwala, R. 1983. Price distortions and growth in developing countries. Staff Report Working Paper No. 575. Washington, D.C.: The World Bank.

Arrow, Kenneth. 1951. *Social Choice and Individual Values.* New York: Wiley.

Campbell, Rex. 1987. Farming as a way of life may disappear in the U.S. Columbia: Department of Rural Sociology, University of Missouri.

Denison, Edward F. 1962. Education, economic growth, and gaps in information. *Journal of Political Economy* 70:124-28.

Friedman, Milton. 1953. *Essays in Positive Economics.* Chicago: University of Chicago Press.

Gardner, Bruce. 1981. *The Governing of Agriculture.* Lawrence: The University Press of Kansas.

Greene, Shirley E. 1963. Dialogue: Role of the social scientist. Pp. 232-34 in Center for Agricultural and Economic Adjustment, *Farm Goals in Conflict.* Ames: Iowa State University Press.

Harper, Wilmer and Luther Tweeten. December 1977. Socio-psychological measures of rural quality of life: A proxy for measuring the marginal utility of income. *American Journal of Agricultural Economics* 59(5):1000-05.

Henderson, J.M. and R.E. Quandt. 1958. *Microeconomic Theory: A Mathematical Approach.* New York: McGraw-Hill.

Little, I.M.D. 1950. *A Critique of Welfare Economics.* Oxford: Clarendon Press.

Marshall, Alfred, 1890. *Principles of Economics,* 8th ed. (1920). New York: Macmillan.

Mishan, E.J. 1968. What is producers' surplus? *American Economic Review* 58:1279-82.

Myrdal, Gunnar. 1944. *An American Dilemma.* New York: Harper & Row.

Reder, Melvin W. 1947. *Studies in the Theory of Welfare Economics.* New York: Columbia University Press.

Robbins, Lionel. 1935. *An Essay on the Nature and Significance of Economic Science,* 2nd ed. (1952). London: Macmillan.

Samuelson, Paul A. 1947. *Foundations of Economic Analysis.* Cambridge: Harvard University Press.

Scitovsky, Tibor. 1951. The state of welfare economics. *American Economic Review* 41:303-15.

Smith, Adam. 1776. *An Inquiry into the Nature and Causes of the Wealth of Nations,* vol. 1 (1933). London: Dent.

Tweeten, Luther. June 1985. Introduction to agricultural policy analysis: The distribution of economic costs and benefits from market intervention. Agricultural Policy Analysis Background Paper B-5. Stillwater: Department of Agricultural Economics, Oklahoma State University.

Tweeten, Luther and George Brinkman. 1976. *Micropolitan Development.* Ames: Iowa State University Press.

Tweeten, Luther and Gilead Mlay. 1986. Marginal utility of income estimated and applied to problems in agriculture. Agricultural Policy Analysis Background Paper B-21. Stillwater: Department of Agricultural Economics, Oklahoma State University.

Tweeten, Luther and Neal Walker. 1977. The social cost of poverty. Chapter 4 in Robert Coppedge and Carlton Davis, ed., *Rural Poverty and the Policy Crisis*. Ames: Iowa State University Press.

Walker, Neal and Luther Tweeten. December 1975. The impact of income maintenance programs on investment and growth. *American Journal of Agricultural Economics* 57(5):957-60.

Willig, Robert. September 1976. Consumer's surplus without apology. *American Economic Review* 66:589-97.

CHAPTER THREE

Values, Beliefs, and Politics

Two reasonable persons can look at the same facts and reach opposite policy conclusions. A major reason is because their goals, values, and beliefs differ. It has been said that economists can provide a half-dozen ways to solve the major farm problems, but progress towards a solution rests on resolution of conflicts in goals, values, and beliefs (Heady, 1961, p. vi).

Goals are ends or objectives toward which behavior is directed. *Beliefs* whether correct or incorrect are what we hold to be truth — the perceived nature of reality. *Values* — feelings of what is desirable or ought to be — are our standards of preference that guide behavior, including political behavior. Goals, values, and beliefs may be based on fact and reason, but they are often subjective and intuitive feelings deeply rooted in the psyche from accumulated experiences, from culture, from genetic disposition, and from ramblings of long-departed philosophers. Whatever the source, values and beliefs often overshadow economic fact and careful analysis in policy-making for agriculture. We cannot understand farm politics and policy unless we understand values and beliefs.

TWO CREEDS IN CONFLICT

America is of two minds about agriculture. One submits to the creed of *farm fundamentalism*, emphasizing belief in the primacy of agriculture and the family farm. The other submits to the creed of *democratic capitalism*, emphasizing belief in the primacy of free enterprise and authority of the people at large expressed through the political system.

FARM FUNDAMENTALISM

Articles in the farm fundamentalist creed include:

1. Agriculture is the most basic occupation in our society, and all other occupations depend on it.
2. Agriculture must prosper if the nation is to prosper.
3. Farmers are better citizens, have higher morals, and are more committed to traditional American values than are other people; indeed the nation's moral and social character depends on farmers.
4. The family farm must be preserved because it is a vital part of our heritage.

5. Farming is a way of life, a more satisfying occupation than others.
6. The land should be owned by the person who tills it.
7. Anyone who wants to farm should be free to do so (Paarlberg, 1964, p. 3).
8. A farmer should be his own boss — independent, rugged, and self-reliant.
9. The family farm is the ideal nuclear family unit where the family works, plays, prays, and, in general, lives together for mutual harmony and support.

ORIGINS OF FARM FUNDAMENTALISM

Farm fundamentalism has origins in the bucolic images of the Bible. God's Chosen People were pastoral and agricultural. The image was of the *Good* Shepherd. Part of the Judeo-Christian tradition was the natural law philosophy of the Catholic Church which viewed the family farm as the principal vehicle for social order, progress, and reproduction. Pope Pius XII stated in 1941, "Only that stability rooted in one's holding [farm] makes of the family farm the vital and most perfect and fecund cell of society, joining up in a brilliant manner in progressive cohesion the present and future generations" (Pius, 1943). Pius viewed urbanization and industrialization as a threat, noting with alarm in 1951 that, "Today it can be said that the destiny of all mankind is at stake. Will men be successful or not in balancing this influence [of urban-industrial society] in such a way as to preserve . . . the rural world?" (Pius, 1951).

American agricultural fundamentalism stems from a time when farming was the major source of the nation's wealth, most of the people lived on farms, and the small family farm was the most efficient economic unit. But agricultural fundamentalism became much more than that. Fundamentalism holds that farming is a divine calling and that God and man walk hand in hand to supply the physical needs of mankind (Fite, 1962, p. 1203).

Agrarian fundamentalism's most articulate and compelling progenitor in America was Thomas Jefferson. He stated:

> Those who labor in the earth are the chosen people of God, if ever he had a chosen people. . . Corruption of morals in the mass of cultivators is a phenomenon of which no age nor nation has furnished an example. It is the mark set on those who, not looking up to heaven to their own soil and industry as do husbandmen for their subsistence, depend for it on the casualties and caprice of customers. Dependence begets subservience and venality, suffocates the germ of virtue, and prepares fit tools for the designs of ambition [Jefferson, 1788, p. 175].

"Cultivators of the earth," Jefferson (1926, p. 15) wrote to John Jay, "are the most valuable citizens. They are the most vigorous, and they are tied to their country, and wedded to its liberty and interests by the most lasting bonds."

The fundamentalist theme again is apparent in William Jennings Bryan's "Cross of Gold" speech: "Burn down your cities and leave our farms, and your cities will spring up again as if by magic; but destroy our farms and the grass will grow in the streets of every city in the country."

Some academics have joined in viewing the family farm as the wellspring of virtue. The family farm according to Harvard sociologist Carle Zimmerman (1950) was the form of agriculture "in which home, community, business, land, and domestic family are institutionalized into a living unit which seeks to perpetuate itself over many generations." The family farm feeds society with virtue "as the uplands feed the streams and the streams in turn the broad rivers of life" (Zimmerman, 1950).

At an international conference on farm policy, Horace Hamilton (1946, pp. 100-113) said the family farm produced "men of strong character and moral consciousness." By way of contrast, the farm-labor camps found on large-scale commercial farms spurred "pool rooms, honky-tonks, cheap picture shows. . . , and flashy, back-slapping personalities." In a California study, Goldschmidt (1946) contended that a community surrounded by family farms was more economically and socially viable than a community surrounded by large corporate-industrial farms.

Continuity in farm fundamentalism to more recent times is not hard to find. Catholic Father Spletz (1963, p. 46) stated that, "The rural values such as reverence for the soil, love of God, love of fatherland, and willing acceptance of honest toil are fundamentally spiritual," and went on to recognize farmers as the source of such values, noting that, "even if this notion could dispense with most of its farmers, there would be the question of whether it could remain strong without the type of man agriculture produces." Maurice Dingman (1986), Catholic Bishop of Des Moines, emphasized the virtues of the family farm, and lamented that:

> The laissez faire approach has allowed the harsh forces of uncontrolled competition to drive less prosperous farmers out of agriculture. The adaptive approach goes so far as to employ the power influence of government and educational institutions, including land grant universities, to accelerate the migration of farm families from the land. This should not have been permitted. That policy has been immoral, unethical, unjust, disastrous, motivated by greed, destructive, leading inevitably to conditions similar to Central America.

PUBLIC PERCEPTIONS OF AGRICULTURAL FUNDAMENTALISM

Farm fundamentalism is very much alive. It remains a unifying symbol, a rich legacy of tradition cherished by the public. There is evidence that people think of farming not in terms of fact but in terms of deeply felt beliefs and values of what the nation ought to be.

Two recent public opinion polls probed in some depth the public perception of agricultural issues. One was the AgFocus/Gallup survey of 1,507 persons by telephone interview in 1985 (National Issues Forum, 1987). The other was a nationwide mail survey of 3,239 adult Americans conducted in the spring of 1986 by the S-198 project on socio-economic dimensions of technological change, natural resource use, and agricultural structure (see Jordan and Tweeten, 1987).

The AgFocus study concluded that:

> People's attitudes and views about agriculture are a product, not so much of their specific knowledge about agriculture, but of their overall personal and political philosophies. Few Americans appear to have brought their knowledge about agriculture to bear on their opinions of it (National Issues Forum, 1986, p. 3).

Although a surprisingly high proportion of respondents (over half) in the S-198 study had been exposed to farming through relatives or friends who are farmers, knowledge of agriculture tended to be low but varied depending on the group and issue being considered.[1] Where perceptions were incorrect, they tend to favor rather than work against the position of farmers. The implication is that better education of the public at large will not necessarily translate into greater support for farmers in the political arena.

Three-fourths of the respondents in the S-198 study agreed (incorrectly) that "today most farmers are in financial trouble." Four-fifths of the respondents agreed that "agriculture is the most basic occupation in our society, and almost all other occupations depend on it." A slightly higher percentage agreed that the family farm must be preserved because it is a vital part of our heritage. According to Table 3.1, farm fundamentalism is strongest among respondents who are women, white, aged, married, less educated, low income, conservative, religious, and close to agriculture. Thus, support for farm fundamentalism may diminish as the nation becomes more urban, affluent, educated, distant from agriculture, and as the young age (unless they change beliefs with age). The surprising conclusion from Table 3.1 is not that fundamentalism statistically differs significantly within categories for all but political party affiliation, nor that 99 percent of people on farms subscribe to farm fundamentalism. The surprise is that a large majority of people in all categories including two-thirds of those living in large cities (the lowest proportion of any category in Table 3.1) support farm fundamentalism.

That belief does not necessarily translate into special favors for farmers. Only 24 percent of respondents agreed that most consumers would be willing to have food prices raised to preserve the family farm. A 1985 poll of consumers found that 68 percent of respondents would be willing to pay an extra 1 percent of their grocery bill to preserve America's family farms (National Issues Forum, 1986, p. 39).

Respondents simultaneously subscribed to farm fundamentalism and democratic capitalism. The S-198 study found that 63 percent of the respondents agreed and only 33 percent disagreed with the statement that "the government should treat farms just like other businesses." Also more respondents agreed than disagreed with the statement that "farmers should compete in a free market without government support." However, other parts of the survey revealed strong support for Treasury transfers to farmers.

---

[1] A survey of consumers in 1985 found that 46 percent thought food raised on a corporate farm would be more expensive than food raised on a family farm (National Issues Forum, 1986, p. 30). There is no evidence that food produced on corporate farms would be any more expensive or of lower quality than food produced on family farms.

TABLE 3.1. Adherence to Farm Fundamentalism as Measured by Response to
A: Agriculture Is the Most Basic Occupation In Our Society, and Almost All Other Occupations Depend Upon It.
and
B: The Family Farm Must be Preserved Because It Is a Vital Part of Our Heritage.

Characteristic		Agree	Undecided	Disagree	(Number)
		(percent)			
U.S. Total	A	80.0	11.6	8.4	(3185)
	B	82.1	8.6	9.4	(3206)
Gender					
Male	A	79.8	8.3	11.9	(1281)
	B	79.1	10.2	10.7	(1285)
Female	A	80.5	13.6	6.0	(1866)
	B	84.1	7.7	8.3	(1875)
Race					
Black	A	76.9	17.4	5.8	(169)
	B	62.3	16.0	21.7	(169)
White	A	81.0	10.5	8.5	(2814)
	B	83.4	7.7	8.9	(2825)
Age					
Under 21	A	73.7	24.6	1.7	(57)
	B	84.3	3.0	12.7	(63)
Over 65	A	89.8	4.2	6.0	(784)
	B	87.8	8.5	3.7	(785)
Marital status					
Married	A	79.6	11.0	9.3	(1645)
	B	83.2	8.4	8.4	(1658)
Never Married	A	73.6	12.6	13.8	(404)
	B	73.0	11.1	16.0	(410)
Education					
Less than high school	A	80.1	14.8	5.1	(454)
	B	91.3	3.1	5.5	(447)
College graduate	A	69.9	16.1	14.0	(437)
	B	68.2	19.8	12.1	(440)
Income from all sources					
Under $5,000	A	83.1	15.7	1.2	(431)
	B	92.7	3.6	3.7	(432)
$60,000 or more	A	66.1	17.4	16.4	(95)
	B	68.2	10.1	21.7	(95)
Political party					
Republican	A	77.6	13.3	9.1	(739)
	B[a]	83.4	8.3	8.3	(744)
Democrat	A	83.5	10.4	6.1	(1250)
	B[a]	80.6	8.3	11.0	(1246)
Independent	A	78.7	11.0	10.3	(1084)
	B[a]	82.8	8.9	8.3	(1096)

TABLE 3.1. Continued

Characteristic		Agree	Undecided	Disagree	(Number)
		(percent)			
Political philosophy					
Conservative	A	81.6	8.6	9.9	(1081)
	B	80.9	10.0	9.1	(1082)
Middle	A	80.2	12.1	7.7	(1499)
	B	83.5	7.3	9.2	(1508)
Liberal	A	72.5	16.7	10.8	(378)
	B	77.8	8.2	14.0	(384)
Voted in 1984					
No	A	76.4	12.4	11.2	(690)
	B	81.6	11.0	7.4	(695)
Yes	A	81.2	11.2	7.6	(2441)
	B	82.1	8.0	10.0	(2449)
Frequency of religious service					
Weekly	A	82.2	12.9	5.0	(1180)
	B	86.5	6.8	6.7	(1190)
None	A	77.1	11.5	11.4	(770)
	B	79.4	8.0	12.6	(771)
Belong to labor union					
No	A	78.5	12.6	9.0	(2179)
	B	81.6	9.8	8.6	(2182)
Yes	A	79.5	12.8	7.7	(416)
	B	84.0	7.4	8.5	(421)
Have seen movie "River"					
No	A	79.8	12.7	7.6	(2497)
	B	81.4	8.5	10.1	(2512)
Yes	A	81.0	7.8	11.1	(688)
	B	84.5	9.0	6.5	(695)
Ever take course in economics					
No	A	80.5	12.8	6.6	(2056)
	B	84.2	7.5	8.3	(2073)
Yes	A	79.1	9.4	11.5	(1129)
	B	78.3	10.6	11.2	(1133)
Familiarity with agriculture					
Not at all	A	65.9	23.0	11.0	(518)
	B	74.8	15.3	9.9	(575)
Directly involved	A	89.6	4.0	6.4	(476)
	B	88.8	5.8	5.4	(477)
Close relatives own farm or ranch					
No	A	76.0	13.3	10.8	(1642)
	B	79.2	10.5	10.3	(1652)
Yes	A	84.4	9.8	5.8	(1542)
	B	85.1	6.6	8.3	(1555)

TABLE 3.1. Continued

Characteristic		Agree	Undecided	Disagree	(Number)
		---(percent)---			
Where live now					
Large city	A	63.7	25.5	10.8	(550)
	B	76.7	9.8	13.5	(551)
Town	A	83.0	9.0	8.0	(520)
	B	81.4	10.2	8.3	(526)
Country	A	85.7	6.2	8.2	(437)
	B	86.5	6.1	7.3	(428)
Farm	A	98.6	.7	.7	(174)
	B	91.6	2.7	5.7	(175)

SOURCE: Jordan and Tweeten (1987).

[a]Not significant. All other responses differ by classification at the .01 probability level or better.

The image of farming given by respondents was one of a preferred way of life but characterized by hard physical labor, high risk, and low economic returns. Other careers were more favored by respondents for themselves and for their children.

FARM FUNDAMENTALISM UNDER SCRUTINY

Most of the beliefs that constitute the creed of farm fundamentalism do not stand scrutiny. Each of the articles in the fundamentalist creed presented earlier is examined briefly below in the order in which they appeared.

1. Farming is a basic occupation, but it is one among many and is not necessarily *the* most basic. Provision of water is more basic to life than food, but society does not award the water utility industry special status. The economic value of an industry is not determined by whether it is a natural resource, whether its output is used by many people and their industries, or whether it is basic to sustaining life. Rather, its value is determined by its contribution *at the margin* to satisfactions of society. Low farm and food prices, surpluses, and production controls of recent years indicate that society places little value on additional farm output. Someone drowning in water places a negative value on having more water although some water is essential for life. A society drowning in farm commodity surpluses places negative value on additional output.

2. Agriculture is such a small part of the national economy (2 percent of gross national product) that the industry can be depressed while the nation's economy prospers as in the 1920s or 1980s. Similarly, agriculture can prosper while the nation is depressed as in 1974-75.

3. It is not possible to objectively state that farm people are superior to others. Farmers largely have lost their uniqueness. To be sure, farmers continue to have lower crime rates and divorce rates and to have higher birth rates than others. On the other hand,

there is evidence that farmers are less committed than others to virtues such as racial tolerance, sexual equality, and concern for the disadvantaged (Burchinal, 1964, pp. 160-62, 168-70). Thus it is not possible to conclude that farmers are either better or worse than others. But it is possible to say that farm goals, values, beliefs, and behavior are becoming more like those of nonfarm people. Hamilton (1946) observed that the future of family farming depended on the willingness of farmers to "value farming more than they prize competing values" such as urban living standards. However, evidence indicates that farm people are becoming more like nonfarm people, not vice versa.

A related issue, as Carlson (1986, p. 20) put it, is that "America's farms increasingly resemble old-age homes. We have exhausted their human capital. We must recognize that our nation's family farmers can no longer serve as our peasant class, reinvigorating an otherwise failing social order and providing a stream of surplus of youth for factories and offices. Put simply, the numbers are no longer there."

The belief that farm people are better than nonfarm people can lead groups to support mandatory production controls costing far more jobs in agribusiness than would be saved in farming, and to support income transfers from poor taxpayers to wealthy farmers. The destitute widowed mother of three small children receives $15,000 annually in welfare payments while the farm family with net worth of $1 million receives $60,000 of direct government payments.

4. Whether to preserve the family farm is an *option*, not a "must." The decision to preserve or not to preserve is for the political process and not for social scientists to make. Social scientists inform; the political process makes decisions whether to intervene in the market to preserve the family farm. If an *informed* society values the family farm for its intrinsic worth and is willing to make necessary sacrifices to preserve it, then social scientists have no basis to object.

5. Surveys of farm and nonfarm people reveal preferences for a farm way of life but stressful economic conditions give rise to ambivalence apparent in this song sung by farm fundamentalist activists at a farm protest rally:

Now some folks say
There ain't no hell
But they don't farm
So they can't tell.

In a 1986 feature article in *The Witness*, a religious newspaper in Iowa strongly committed to farm fundamentalism, I was sentenced to spend purgatory on a 300 acre Iowa farm — another indication of the ambivalence modern-day family farm supporters feel about the family farm. Such feelings prompted Jeffrey Pasley (1986, p. 27) to write, "Given the conditions of life on the family farm, if ITT or Chevron or Tenneco really do try to force some family farmers off their land, they might well be doing them a favor."

Sociological studies (see Coughenour and Tweeten, 1985 for review) show that farmers are more satisfied than persons in other occupations with working conditions (farm way of life), but on the whole are no more happy or satisfied than others because they feel poorly rewarded for their work. Society has intervened generously to assist farmers economically but such assistance is unusual for sectors — taxpayers do not ordinarily subsidize people because they enjoy their work. Economic instability is viewed as a hardship by farmers and they seek relief from taxpayers. Speculators in the futures and stock markets also face great uncertainty but the public would hardly consider a bailout of losers in the Chicago Board of Trade or New York Stock Exchange. President Reagan's Budget Director David Stockman raised a storm of controversy and was chastised by the President for commenting that, "I can't see why taxpayers should refinance bad debt [of farmers] that was willingly incurred by consenting adults who . . . thought they could get rich."

6. Perhaps a farm "should" be owned by the operator who tills it, but an economic farming unit of sufficient size to be efficient and provide an adequate family living now requires approximately $1.5 million of assets. Few could be owner-operators of such a unit without massive assistance from parents or other outside sources. To get started or expand in commercial farming, most operators find it advantageous to rent land.

7. Massive subsidies would be required to allow everyone to farm who wants to farm regardless of their managerial or financial capabilities. The federal government and several state governments provide concessional lending to help more persons get started in farming. Such generosity makes little sense for an industry already troubled by excess capacity to produce and by more farmers than can make a living.

8. It is nice to be one's own boss, but increasingly a farm operator must share decisions with spouse, the banker, and the government. Farmers who must depend on government for most of their net income or on artificially propped up consumer food prices (through mandatory supply control or other means) cannot call themselves rugged individualists, self-reliant, or independent. Making agriculture a public utility with prices set to provide an assured rate of return also clashes with the efficiency tradition of democratic capitalism.

9. Farm families follow life-styles pretty much like their nonfarm neighbors. Most farm operators (or their spouse) work off the farm. Kids are in extracurricular after-school activities, and car pooling is widespread. Some spouses of operators work because they have no choice financially; others work because they enjoy the discretionary spending money and the social interaction. In short, farm people are good people; so are plumbers, nurses, and their families.

Farmers may not be unique socially or economically, but they are unique in the minds of the greater public. Many more steel, auto, oil, lumber, and mining workers than farmers lost their jobs in the 1980s. Individual farm failures were detailed in feature articles, movies, and books, whereas the loss of hundreds of thousands of jobs in oil fields, steel mills, and shoe factories was little noted.

DEMOCRATIC CAPITALISM

A second creed or set of beliefs, democratic capitalism, is also prominent in farm and nonfarm America. One cannot understand farm policy without understanding it as well as farm fundamentalism. Although the two creeds sharply conflict, both are widely held, *often by the same person.* This results in a kind of individual and national schizophrenia regarding how farmers should be treated by government. Articles in the democratic capitalism creed include:

1. Farming is a business.
2. The free enterprise market is the most efficient allocator of resources and products; the market should decide the size and role of the farm sector in the nation's economy.
3. The family farm may be the most efficient mode of production, but if it is not the market should replace it by another form of economic organization better able to supply quality food at low cost in national and international markets.
4. The invisible hand of the market turns private greed into public good; government should not interfere except to ensure than the economic game is played according to rules set by society through a democratic political process.
5. Democratic capitalism is the most innovative and dynamic socio-economic system in the world, promoting growth valued for itself and also for the means to care for the disadvantaged.
6. People should be rewarded according to the value of their contribution to society. This is called *commutative justice* (Brewster, 1961). Providing equality of opportunity takes precedence over providing equality of outcomes.
7. The proper way to achieve status is to be proficient in one's chosen field.
8. One who improves his income through honest toil is worthy of respect and emulation.
9. All persons, farmers and nonfarmers, are of equal intrinsic worth and dignity.
10. The individual and his immediate family should be responsible for his economic security throughout life. To tax for purposes of redistributing wealth is institutionalized theft. An able-bodied adult who refuses to work has no inherent right to an income from taxpayers.
11. No one, however wise or good, is wise or good enough to have absolute power over others. Brewster (1961) calls this the *democratic creed.*
12. Concentration of power in government, business, or labor destroys freedom of the individual. The role of government is to avoid such concentration of power and to provide stability, security, and an atmosphere where the market will function.
13. Property rights are inviolable and essential to preserve human freedom and promote economic efficiency.
14. Public functions should be performed by the government unit closest to people, the smallest unit of government within which costs and benefits of that function are realized.
15. Profit is the rudder which guides the economic ship of progress. Markets and profits are prized because they allocate resources efficiently to serve people, increasing their well-being.

16. Everyone has a right to choose one's own occupation, to fail or succeed in that occupation without recourse to subsidies, to save and invest earnings, and to convey earnings to heirs.

It is of interest that farm fundamentalism and democratic capitalism both claim inspiration from Thomas Jefferson. Democratic-capitalists see today's farm dependent on government handouts (or food prices propped up by production controls) as proof of Jefferson's dictum "dependence begets subservience." They view reliance on the government or a controlled market as a mockery of the self-reliant, independent, and proud yeoman-farmer ideal.

ORIGINS OF DEMOCRATIC-CAPITALISM

Democratic capitalism has deep roots. First there is the Judeo-Christian culture of the Western world, which places emphasis on active mastery over nature in contrast to the more contemplative, passive, and ascetic culture of the Eastern world. There is the Reformation, which emphasized the secular worth and dignity of all persons (priesthood of all believers), democracy in governance (authority from the bottom up), and the work ethic (excellence in one's Calling). There is English democracy, with roots in Greek antiquity, the Magna Carta, and the early parliamentary system. There is capitalism with its vigor strengthened by English political stability, institutional structure (e.g., money lending for interest no longer classed as usury, the limited liability corporation, etc.), and the Industrial Revolution.

Then there is *laissez faire*. It had origins in the philosophy of John Locke which emphasized that the ideal world lies in a natural order of no collective restraints on individual actions, and in the utilitarian philosophy of Adam Smith, whereby man because of his acquisitive instinct is led to Utopia by the invisible hand of the market. There is eighteenth-century Enlightenment philosophy, which in England emphasized reason, science, empiricism, and individualism.

The American Farm Bureau Federation, by far the largest general farm organization, embodies many principles of democratic capitalism. Its 1976 policy statement read, "Freedom of the individual versus concentration of power which would destroy freedom is the central issue of all societies" (1976, pp. 1-4). The policy statement went on to assert that the Federation believes "in the American capitalistic, private, competitive enterprise system in which property is privately owned, privately managed, and operated for a profit and individual satisfaction."

Strong support comes from conservative Protestant denominations. Merrill Oster (1986), who is a farmer, agribusinessman, and member of a conservative denomination speaks for many others when he says:

> This is the kind of free society John Adams and Thomas Jefferson had in mind. The America that Adams saw is founded on certain old-fashioned concepts: that free societies produce the most good and the least evil; that theft is wrong; that work is noble and expected of everyone; that a man should be

> willing to stand on his own two feet; that he should be free to succeed or fail and keep the fruits of his labors; that it is not right to accept that which you have not earned; and that when we are successful we have responsibilities to voluntarily share our wealth with those who, because of youth or age or sickness or incapacity, cannot take care of themselves — especially our own families [p. 22].
>
> We need to defend the concept of human freedom being the basis for legal and social foundations of society. This legal and social order produces the economic framework for the free enterprise system. And we must never take for granted the fact that the fabric of civilization rests on honest money, honoring of contracts, freedom coupled with responsibility, individual property rights, limited government, individual and group integrity, and living within our means personally and nationally. The result is that free choice under the free market system has unleashed technological progress which makes the American system the most developed poverty-fighting system in the world [p. 25].

Ezra Taft Benson, who was Secretary of Agriculture under President Eisenhower, was keenly aware of the spiritual values of rural life and later became head of the Mormon Church. He noted that, "The country is a good place to teach the basic virtues that have helped build the nation," but went on to place himself on the side of democratic capitalism by contending that "Agriculture is not so much an important segment of our population as of our free enterprise system. It should be permitted to operate as such" (see Carlson, 1986, p. 18).[2]

The continuity of democratic capitalism clearly is evident from Jefferson to the modern day. Abraham Lincoln observed "my opinion of them [farmers] is that, in proportion to numbers, they are neither better nor worse than other people" (see Fite, 1962, p. 1208). Lincoln showed no special enthusiasm based on farm fundamentalism for pushing through the landmark tripartite agricultural legislation of 1862: the Morrill Act establishing land-grant colleges, the Homestead Act giving easy access to land for thousands of families, and the act establishing the U.S. Department of Agriculture. Lincoln favored these because he subscribed to democratic capitalism — he believed that passage would benefit the nation as a whole (Fite, 1962, p. 1208). Lincoln's recognition of the contribution of private property to diligence under democratic capitalism is apparent in his statement (taken from Novak, 1987):

> When those who work in an enterprise also share in its ownership, their active commitment to the purpose of an endeavor and their participation in it are enhanced. Ownership provides incentives for diligence.

---

[2]Farm activists ascribe to Benson authorship of the pejorative "Get big or get out," a phrase that lives in infamy along with Earl Butz's "Plant fencerow to fencerow." Historians have been unable to find a record of the remarks by either Secretary in speeches and writings.

Failings of Democratic-Capitalism

Democratic capitalism like farm fundamentalism is characterized by numerous myths. The most frequent is not recognizing where markets fail. As was noted in the previous chapter, markets do not function well to meet needs of society when social costs (or benefits) differ from private costs (or benefits). For example, markets alone will not ensure a safe environment in terms of soil conservation and food free from pesticide residues. The market will underinvest in public goods such as basic agricultural research and general education. The market alone will not avoid concentration of power. The market alone will not provide macroeconomic stability. Of course, market economists point out that public intervention is frequently a bigger failure than the market failure it was designed to correct. And many of the cases of so-called "market failure" turn out to be the result of government-established market distortions.

The market alone will not provide *distributive justice* to all those who are physically or mentally unable to earn a living in the market and who in an impersonal urban-industrial society cannot depend on family or private charity. As noted in the earlier articles to the creed of democratic capitalism, adherents emphasize equality of opportunity rather than equality of economic outcomes. But their commitment to *distributive justice,* defined as providing equality of opportunity and giving each able-bodied adult an equal voice in making rules of the economic gain, is weak and inherently in conflict with commutative justice which emphasizes rewards according to contribution. Economic outcomes under capitalism are skewed in favor of the wealthy minority. Some of earnings must be taxed to provide distributive justice, a violation of commutative justice. The lower-wealth majority have incentives to vote to redistribute wealth to them from the wealthy minority. Thus democracy and capitalism conflict. In world perspective, however, democracy exists only in capitalistic free enterprise countries, hence the conflict must not be decisive.

RESOLVING VALUE CONFLICTS

It is difficult to find a person or organization fitting precisely the farm fundamentalist model or the democratic capitalism model. However, many organizations are closer to one philosophy than the other. Conservative church groups, the American Farm Bureau Federation, National Cattlemen's Association, National Association of Manufacturers, U.S. Chamber of Commerce, and the Republican Party lean to democratic capitalism. Liberal church groups (including some such as the Catholic Church which are liberal on economic issues but conservative in theology), the National Farmers Union, American Agriculture Movement, National Farmers Organization, various crop growers associations, AFL-CIO, and the Democratic Party tend toward farm fundamentalism.

Farm fundamentalists do not reject the market, but believe that relieving farm economic hardship and preserving the family farm is more important than maintaining an open market. Democratic capitalists on the other hand contend that family farmers should face the impersonal rigors of the market. If family farmers are unable to compete, then society is best served in the long term by letting inefficient or unlucky farmers go the way of the blacksmith, buggymaker, and the mom-and-pop grocery store.

There is much psychological value in having a creed to live by whether that creed is myth or reality. Farm fundamentalism, for example, meets a deep-seated need in farm people for a feeling of worth and dignity and in nonfarm people for a romantic, nostalgic, unifying symbol of a simpler, less hectic bucolic life and heritage. Myths also have dangers. For farmers and groups representing them, farm fundamentalism has often meant concern for the family farm rather than for the farm family. That is, farm fundamentalists have been so preoccupied with preserving the family farm that they have not had time or inclination to be concerned with preparation for nonfarm employment by the many farm families who have found employment and residence in town or city.

The quasi-religious attitude taken by the nonfarm public toward the farm has many pitfalls. Money is thrown at the "farm problem" with distressing lack of accountability. Billions of dollars spent each year have not preserved the family farm nor prepared those who cannot survive on the farm for nonfarm occupations. General education, vocational-technical schooling, job outlook information, mobility assistance, and counseling in nonfarm job opportunities are overlooked.

Youth receive useful training in responsibility, leadership, and citizenship in the Future Farmers of America or 4-H clubs. Such training frequently distracts students from obtaining strong grounding in mathematics, English, and science so basic for achievement in the farm or nonfarm world. An atmosphere of farm fundamentalism imparts unrealistic expectations of farming. Failure to be informed of sacrifices, risks, and of the huge capital and managerial requirements for a successful commercial farm leads to too many poorly prepared entrants into farming. Once established in farming, inability to continue due to bad luck, poor management, or other reasons is viewed as great personal failure. Farm fundamentalism views life off the farm as second rate. Unable to accept the market's decision that they are unable to be a part of the most important industry in the nation, farmers too often turn to two less than adequate solutions: suicide or Washington. The suicide rate among farmers experiencing financial distress in the 1980s was high despite heroic efforts of land grant universities, churches, and other volunteer organizations to provide personal and financial counseling.

One means to handle value conflicts is to reshape economics to fit the myth. Protest movements, some mildly violent, by farmers unhappy with economic conditions have been frequent since colonial times (see Tweeten, 1979, ch. 3). Farm dissenters and activists have not been anarchists, revolutionaries, or communists. For the most part, they are populists who believe in democratic-capitalism but contend that the market is too flawed to treat farmers fairly and that the government or a collective of farmers must intervene to secure justice. A sample of the charges, mostly myth, against democratic capitalism is listed below.

1. A frequent charge is that "supply and demand don't exist anymore." Protesters really mean that perfectly competitive markets don't exist and never existed. Economists know such market competition need not be perfect but only *workable* to operate efficiently. Later chapters document that agricultural markets have worked well indeed to cull laggards

whether they lag because of misfortune or mismanagement. Activists' principal concern more properly might be that markets work too well and need to be tempered with distributive justice.

2. Activists contend that, although the best economy would be competitive throughout, the second best approach is supply management in the farm sector to counter administered and negotiated pricing in the nonfarm economy. Later chapters emphasize that no imperfections in the nonfarm sector preclude farmers' earning returns on resources comparable to returns on resources in other sectors. It is often stated that farming is the only major competitive sector in the economy, that farmers buy retail and sell wholesale, that farming is an island of competition in a sea of imperfect competition, and that farmers are the only price takers while others are price makers. In fact most Americans are price takers (e.g., laborers), most buy retail and sell their resources wholesale, most sectors have limited control over price, and nonfarm business (especially small business) has a higher failure rate than farmers.

3. Another myth is that making farmers prosperous with mandatory supply management will make the nation prosperous. Less farm output would sell at higher, controlled prices, hence farmers would be paid more to produce less and leave more resources idle — hardly a formula for national economic success in an open world economy. True, farmers have a high propensity to save and invest which ordinarily helps an economy, but it has negative value in a sector holding back resources and output to get higher prices. Activists contend farmers will spend money on consumer goods to provide a Keynesian aggregate demand stimulus. But whereas farmers spend about three-fourths of income they receive (they save and invest the rest), persons in poverty spend over 100 percent of their income. Thus if a Keynesian spending stimulus is warranted, such stimulus will be both more effective and more equitable if given to the poor than to a sector.

EXPLAINING FARMERS' POLITICAL INFLUENCE

Commercial farmers, the principal beneficiary of farm policy, with less than 1 percent of the nation's population are in no position to dictate policy by brute force of the vote. Of today's 2,443 rural counties, farm-related earnings constitute more than 20 percent of all labor and proprietor income in only 702 (Green and Carlin, 1985, p. 2). Of 435 U.S. congressional districts only 46 had one-third or more of their counties "farm dependent" as defined by the 20 percent criterion above. In 1933, over 150 members of the House of Representatives came from a district with at least 20 percent of its population on farms, a figure which dropped to 14 by 1973.

Perhaps no other group in America has wielded more political power per capita than commercial farmers. Commercial farmers' political success as measured by an unusually high ratio of government transfer payments relative to taxes paid is explained by several factors.

1. Farm Fundamentalism. A widely held complex of values and beliefs as noted earlier, farm fundamentalism constitutes a massive public reservoir of good will which farm political interests have drawn upon frequently and few politicians even from urban districts can stand against.

2. Fear of Food Shortages. Any prospect of short food supplies strikes terror among consumers. The highly inelastic domestic demand for food implies that consumers place an exceedingly high value on having enough food but place little value on having more. Consumers listen intently when farm activists contend that failure of government to provide family farm subsidies will turn farming over to a handful of large corporate industrial conglomerates which will interrupt food supplies and charge high prices. Claims that large corporations will take over farming and will interrupt food supplies and charge high prices are without substance — they are rhetorical blackmail. But consumers are willing to provide large subsidies for the insurance and peace of mind associated with adequate food supplies. Affluent consumers can afford to take even the most costly and frivolous precautions against potential food shortage.

3. Farmers Are Switch Voters. Surveys indicate that farmers are politically conservative on most issues. But they vote their pocketbook. So farmers are capitalists when times are good and socialists when times are bad. In elections frequently decided by 1 or 2 percentage points, it makes sense for a politician to promise programs to a group with a record of responding in the voting booth.

4. Farmers Are Scattered Over the Country. The fact that farmers tend to be scattered throughout the country rather than concentrated in a few areas gives them clout in the voting booth because, as noted above, elections are often decided by thin margins. Defense contractors consciously diversify plants geographically to leverage more members of Congress; for farmers geographic diversification comes naturally.

5. Farmers Are Organized. For maximum political influence, a sector must be closely organized with well recognized problems or goals giving cohesion. Commercial farmers are represented by strong general organizations and by commodity groups. Farm commodity groups are close-knit single-issue organizations with unusual power and influence. To be most effective, the sense of solidarity must be accompanied by two-way communication. Members convey their needs to government. The response of government decision makers is fed back so that grass roots members and sympathizers who can reject or support members of Congress. The demonstrated past success of farmers in obtaining help from Congress increases the expected payoff from efforts to win more government help for farmers.

6. Effective Political Representation. The tendency to reelect incumbent legislators (even after the Supreme Court reapportionment decision *Wesberry vs. Sanders* in 1964 calling for "one man-one vote" in U.S. congressional districts) and the inertia of past power structures and political alliances have maintained congressional support for farm programs even as farm constituencies diminish. In the Senate the "one man-one vote" rule does not apply. Farmers are especially abundant in the many states with small populations. That gives more senators per capita to farmers than to other major economic sectors.

7. Coalition Building. Coalitions, though requiring compromise of individual group goals, provide political power. The most enduring farm coalition has been with the urban-based hunger lobby favoring Food Stamps and other food and nutrition programs. The Farm Bloc, a bipartisan coalition in Congress and led originally by the American Farm Bureau but with cohesion from other groups, was instrumental in initiating the great commodity programs of the 1930s. For the most part, however, farmers and the groups representing them have had difficulty forming a united farm front; it has been even more difficult to form coalitions with agribusiness groups. All major farm groups have not pulled together politically since the darkest days of the Great Depression. Part of the reason is that farm groups have not often needed to compromise.

8. Farmers Have Recognized, Clearly Defined Problems. Social scientists in land grant universities and government agencies, part of the "agricultural establishment," articulate farm problems intellectually and the farm press publicizes them. It is irrelevant that the "farm problem" is not as well defined as people imagine it to be or as curable as people suppose with commodity programs. What matters is that people think they are dealing with a well defined and curable farm problem.

Conflicting interests of groups associated with food and agriculture dilute the political power of farmers. In general, higher farm income comes at the expense of taxpayers, agribusiness middlemen, or consumers. Consumers want low food prices, taxpayers want low taxes, and farmers want high incomes. It is not possible to satisfy these wants simultaneously. That means that farm interests inherently conflict with those of the other groups. Within farming, interests of feed producers conflict with those of livestock producers. Agribusiness firms want high volume; farmers want lower volume and higher prices. Consumers and agribusiness firms see no reason to pay farmers more than enough to bring forth adequate supplies of food and fiber.

In short, the ability of farmers to convince the larger public that farm problems are real, that the farmer is victimized by conditions beyond his control, and that the federal government can and should take measures necessary to assure farm income and survival gives farmers political influence. Widely held belief in farm fundamentalism is the unique element that allows farmers to turn their economic appeals into government largess while other sectors with at least as great needs cannot. A nonfarm public holding such views leaves little choice for urban legislators — vote for farm programs or face a constituency unhappy that their congressman or senator has failed to respond.

TURNING INTENTIONS INTO REALITY THROUGH THE POLITICAL PROCESS

The process of turning political support into specific farm legislation is illustrated by formulation of the comprehensive Food Security Act in December 1985. Principal participants in the process included the intellectual community, agribusiness firms, general farm organizations, farm commodity organizations, consumer and environmental organizations, Congress (especially the House and Senate Agriculture Committees), and the Executive branch (especially the U.S. Department of Agriculture). Farm groups, agribusiness firms, the agriculture committees, and the U.S. Department of Agriculture were most active.

Academics began an educational process soon after the 1981 farm bill was passed. A large number of nonpartisan academic papers were prepared, mostly citing shortcomings of existing legislation and possible revisions. This effort intensified after 1983 with a number of symposiums, conferences, and seminars.

It is traditional for the Administration to take initiative in forming farm and food policy. The debate became official in early 1985 when the Administration submitted a proposal to phase out deficiency payments, cut price supports sharply, and eliminate production controls. Ordinarily, an extreme position in the direction desired for legislation constitutes a useful initial bargaining position from which the Administration can compromise with Congress to obtain a farm bill close to its liking. Such standard political strategy did not work in 1985 because the farm financial crisis was unfolding in such serious proportions and with so much media attention and public sympathy that markedly lower farm price supports suddenly seemed out of the question. The Administration's bill, described as "dead on arrival," had so little support in Congress that it did not provide a bargaining base.

The Republican controlled Senate Agriculture Committee was allowed by the Democratic controlled House Agriculture Committee to take the lead in formulating a bill because anything acceptable to Senate Republicans and later approved by the House Committee would either have to be accepted by the Administration or Congress would override a veto with support from Democrats and Republicans. Although the Administration's benchmark bill was largely irrelevant to deliberations, the Administration still had influence because of veto power and technical expertise to provide data and to analyze implications of proposed programs. The Administration was a frequent and influential witness before agricultural committees, with those committees constantly modifying legislation to get as much as possible for farmers while avoiding a veto. The real initiate for farm programs rests with Congress because its members must answer to powerful farm interest groups. Congress wants generous farm supports more than does the Administration and must go on the offensive to get it. The Administration plays the defensive role of advise and consent, mostly trying to curb the excesses of Congress and preserve as much discretionary power as possible for the Secretary of Agriculture in administering the bill.

Early on, the broad features of the bill were established to maintain target prices so as to sustain farm income while reducing loan rates and providing export subsidies so as to allow U.S. farm exports to compete in international markets. The subcommittees began to work on the 17 titles of the comprehensive bill. Separate but somewhat similar bills were passed by the House and Senate; a joint House-Senate Conference Committee worked out differences and the bill was signed into law by President Reagan in December 1985.

The Administration got part of what it wanted (competitive loan and market prices), Congress and farm groups got most of what they wanted (high target prices and deficiency payments to maintain farm income), environmentalists got what they wanted (a 45 million acre general Conservation Reserve program and a provision to eventually exclude a farmer from receiving any program benefits if an approved conservation plan was not followed on highly erodible land), and agribusiness got what it wanted (competitive pricing and no more than an option for mandatory controls they were confident the Secretary of Agriculture would not exercise). Taxpayers and the public at large did not get what they wanted — reduced federal spending and economic efficiency. Achieving the objectives of those who dominated the hearings was estimated initially to cost $35 billion for the first three years but cost $20-25 billion *each* of the first three years of the program.

The poor representation by taxpayers as witnesses and the dominance of farm and agribusiness interests is apparent in Table 3.2 summarizing by affiliation the numbers and proportions of witnesses before agricultural committees regarding the 1985 farm bill. The fact that agribusiness and producers had nearly equal numbers of witnesses veils the far greater influence of producers. Agribusiness lacks power because it lacks grass roots support. That is, a politician can count on few votes from agribusiness which accounts for 90 percent of the agricultural and food industry. Furthermore, the agribusiness middleman receives no sympathy support from the broader based electorate which has a decided tinge of populism — a distrust of "big business." Meanwhile, farmers carry a massive sympathy vote from an electorate committed to farm fundamentalism. Of course, agribusiness firms provide campaign contributions — but so do farm groups.

TABLE 3.2. Witnesses, Congressional Agriculture Committees, by Type of Organization, 1985

Affiliation	House	Senate
	number (percent)	number (percent)
<u>Producers</u>	95 (28)	73 (28)
General farm organizations	38 (11)	30 (11)
Commodity organizations	48 (14)	36 (14)
Individual farmers	7 (2)	2 (1)
Miscellaneous farm groups	2 (1)	5 (2)

TABLE 3.2. Continued

Affiliation	House number (percent)	Senate number (percent)
Agribusiness	89 (27)	77 (29)
Output processing and marketing	44 (13)	47 (18)
Farm/output marketing and processing	19 (6)	4 (1)
Input supply	16 (5)	11 (4)
Other agribusiness	10 (3)	15 (6)
Environment	21 (7)	15 (6)
Soil, water, forestry, fish, and wildlife	9 (3)	11 (4)
General environmental	12 (4)	4 (2)
Government and public institutions	87 (25)	40 (15)
Members of Congress	38 (11)	20 (8)
Federal, state, and local government	34 (10)	19 (7)
Land grant universities	15 (4)	1 (small)
Technical	21 (6)	29 (11)
Academic/professional	12 (3)	21 (8)
Federal, state, and local government officials	2 (1)	1 (small)
Agricultural policy consultants	7 (2)	7 (3)
Other (nonfarm)	25 (7)	28 (11)
Consumer	4 (1)	3 (1)
Hunger/relief	8 (2)	10 (4)
Nonfarm advocacy	10 (3)	11 (4)
Foreign countries	3 (1)	4 (2)
Total	338 (100)	262 (100)

SOURCE: Guither (1986, pp. 32-36).

FINAL COMMENTS ON THE POLITICAL PROCESS

Farm commodity programs have retained their same basic structure for over a half century. Despite the political power of producers to influence legislation, few persons or groups defend them openly — the programs have been excessively costly to taxpayers, have not preserved family farms, and have allowed stocks to accumulate to unreasonable levels where they hang like a dark cloud depressing the farm economy. The farming economy continues to be troubled by problems listed in Chapter 1.

Farm legislation has been fundamentally flawed and a disappointment relative to objectives in part because of value conflicts among farm groups as noted above, and in part because of characteristics of the political process listed below.

1. Farm commodity legislation carries inertia. Major farm legislation reforms occur only in crisis. The best predictor of new farm bill content is old farm bill content.

2. Events more than ideology determine farm policy. The Administration's initial proposal to substantially phase out commodity programs from 1985 to 1990 was ignored in formulating the 1985 farm bill because of the deepening farm financial stress. On the other hand, when times are good the tendency is for lawmakers to gain cheap political support from farmers by passing legislation to raise supports to levels that presumably will not be tested. Such expectations are often wrong. Farm program costs are perennially underestimated.

3. As noted earlier, support is stronger in Congress than in the Executive branch for commodity legislation. The Administration can take a passive "advice and consent" role in commodity legislation because Congress can be depended on to take initiative. That means the legislation is dominated by special interests.

4. Constituencies behave as economic man, expending resources to secure programs to the extent they gauge personal benefits will exceed costs. A group receiving large benefits can afford to invest heavily in efforts to retain programs. Such spending diverts resources from higher value uses to lobbying.

5. Agricultural legislation and policies cannot be separated from other legislation and policies — monetary, fiscal, trade, environment, etc. Monetary-fiscal policies heavily influenced the farm economy in the 1970s and 1980s. Efforts to improve commodity programs cannot be separated from success in improving other policies.

6. The economic fortunes of farmers are closely tied to international markets. When exports are favorable, farm receipts are favorable; when exports fall, farm receipts fall. This slants farm commodity programs toward export crops where instability is especially large. And because export crop constituencies are so politically powerful, their claim on income amounts to an entitlement. Consequently, government is indirectly obsessed with export expansion at unreasonable cost to reduce Treasury outlays for commodity programs.

7. The power base for any commodity program is commodity organizations, the congressional subcommittees for that commodity, and the executive branch agencies which administer the program. This power base shifts from time to time. Activity to influence farm legislation depends on pressures of the moment. Consumer groups motivated by the

high food prices of late 1972 and 1973 took great interest in 1973 farm legislation. Agribusiness firms, stung by the massive payment-in-kind diversion programs of 1983, took great interest in 1985 legislation. But interest groups with the general welfare in mind tend to have little interest in or influence on farm policy.

8. The number of agencies outside the Department of Agriculture taking an active interest and wielding some authority in agricultural policy has grown in recent years as issues of budget, international trade, and food prices have become more prominent.[3] This makes it more difficult to coordinate legislation (Abel, Daft, and Earley, 1983, p. 45).

In addition to the U.S. Department of Agriculture, executive branch agencies with an interest in agriculture include:

Cabinet Departments
State
Treasury
Commerce
Executive Office of the President
Office of Management and Budget
Council of Economic Advisors
National Security Council
Special Trade Representative

Five of the seven agencies cited are involved because of the internationalization of U.S. agriculture. The Council of Economic Advisors is especially interested in food prices; the Office of Management and Budget is primarily concerned with the federal cost of programs.

Interagency coordination of agricultural policy has been stifled by lack of either authority or competence (Abel, Daft, and Earley, 1983). That is, a meeting of top-level people with authority to make decisions will not have participants (except USDA) who know agriculture. On the other hand, those who know agriculture in agencies outside USDA will be lower level functionaries not empowered to make decisions.

That is one reason why many decisions regarding commodity program administration, export embargoes, food price ceilings, and budget are poorly conceived, and are viewed by farmers as inappropriate. The executive branch is blamed for pursuing what farmers derisively call a "cheap food policy." Examples are export embargoes damaging the nation's reputation as a reliable food and feed supplier and causing food importers to turn to self-sufficiency policies or to other exporters for long-term diversification. Another example is price ceilings which when imposed on beef by the Nixon Administration caused feeders to hold back suppliers only to have large backed-up supplies released on the market to drive prices unreasonably low after controls were lifted.

---

[3]Many other agencies provide information but lack authority. In the legislative branch, the Congressional Budget Office, General Accounting Office, Office of Technology Assessment, and Legislative Reference Service (Library of Congress) provide objective information. But the most comprehensive objective information from government comes from the Economic Research Service, USDA, in the executive branch. Land-grant universities, of course, also contribute in a major way.

The Office of Management and Budget has sometimes used poor judgment in refusing sufficient acreage diversion to reduce excessive stocks, saving budget outlays in the short run only to be more than offset by higher budget costs for storage in the longer run. Life is made more difficult for farmers when non-USDA agencies delay acreage diversion decisions for wheat until the prospective supply-demand outlook becomes more clear. Meanwhile farmers in the South already have planted winter wheat.

Because agricultural policy affects more than farmers, involvement of non-USDA agencies in decisions is inevitable. Mistakes will be made and farmers will be made worse off from time to time. Farmers can seek and sometimes will obtain justice — unlike producers in a large number of developing countries of the world where food price ceilings and export taxes are perennial. The U.S. does not follow a "cheap food policy" found in many developing countries chronically depressing farm commodity prices with export taxes and food price ceilings. However a "low-cost food policy" emphasizing applications of science and technology to reduce food production cost and food prices makes sense for any society.

9. Because agriculture is dynamic, uncertain, and complex, it is impossible for Congress to legislate sufficient detail in commodity programs. The result is large discretion left to the Secretary of Agriculture who can impose virtually any kind of farm program he chooses — subject to lower-bound legislative constraints on loan and target prices. Many farmers desire longer-term legislation for better farm planning decisions rather than the typical four to five year duration.[4] That hope seems futile even in the unlikely case that Congress is willing to give even greater discretion to the Secretary of Agriculture. Congress amended the basic farm policy every year from 1977 to 1985, often to force the Administration to take action it is empowered to take but chose not to (Abel, Daft, and Earley, 1983, p. 22).

10. Finally, the most important and unfortunate change in government in recent times is the decline of *encompassing institutions*. The major encompassing institutions are the Presidency, political parties, and congressional leadership including committee chairpersons. Encompassing institutions tend to view legislation from the perspective of the nation at large and hence the public interest. Other institutions view legislation from the standpoint of special interests — the local district of a Congressman, the political action committee (PAC) which supplies funds for a specific purpose, or any other special interest which provides benefits to the decision maker. Encompassing institutions are accountable to the public at large and are especially sensitive to nationwide popular vote; nonencompassing institutions are more often accountable to special interests, including the local congressional district or state.

Any well-functioning democracy must maintain a balance between encompassing institutions concerned about the nation's welfare and nonencompassing institutions ensuring that legislation is tailored to legitimate local needs and circumstances. When

---

[4]A Farm Board patterned after the Federal Reserve Board to give continuity and coherence to farm policy has been proposed. Although farm policy is now more a political liability than an asset, Congress is reluctant to delegate authority to a Board it cannot control (Flinchbaugh and Eddleman, 1984).

encompassing interests dominate, local needs are overlooked, injustices to small groups are commonplace, and many persons are alienated. When nonencompassing institutions predominate, the *fallacy of composition* or *micro-macro inconsistency* dominates. Legislation that is favorable to each special interest group separately is damaging to the public interest and also causes alienation.

The presidency remains the most powerful of encompassing institutions but has lost some of its force to serve the public interest even as it has gained overall power by virtue of its technocratic expertise. The reasons for the decline are many (see Bonnen, 1984). In former times, the President was nominated by and his campaign was financed by the party. Local party organizers assured an audience on the campaign circuit. The party had long-term perspective and had a major stake in a candidate whose good governance would build a favorable nationwide party image so that the party could prosper and place more persons in office in the long run. But PAC funding has replaced party funding, television has replaced the local party organizer, and conventional delegates bound to primary election results replace party professionals. Who will screen out a candidate willing to sacrifice the distant interests of children and grandchildren to finance generous programs for today's parents?

The President today can take office obligated to no one with the long-term public interest at heart. Thus, the President's policies tend to reflect the special interests which got him elected and his own short-run perspective — at most eight years rather than the indefinitely long horizon of the party. He can pursue a profligate fiscal policy of the nation living beyond its means until credit runs out (after he leaves office), leaving a burdensome legacy of debt for future generations.

The situation in Congress is similar except that Congress is inherently nonencompassing. That is, it makes sense for a congressman to aggrandize his constituency knowing that it will pay only 1/435 = .2 percent of the cost (assuming his district is of average size and wealth).

In former times, party discipline helped to ensure that private greed did not bring national fiscal ruin for which the party and its congressional leadership would be accountable. But Congress today is all but unable to serve the public interest or even legitimate private interests in the face of the rise of single issue interests and television, with decline in the power of congressional leadership (committees no longer appointed by leaders but by caucus), with a sharp rise in congressional staff numbers (good contacts for special interests), and with proliferation of subcommittees easily commandeered by special interests. The answers to the problem are not obvious except to return more influence to encompassing institutions. Some suggestions include:

1. Greater funding of elections by the public rather than by special interest groups, and a sizable portion of funds provided to parties which in turn will allocate funds to candidates. Free media, especially television converge in-depth on substantive issues, could help. PAC money could go only to parties and not to individuals.

2. Strengthen professional expertise of staffs of the Congressional Budget Office, General Accounting Office, Office of Technology Assessment, Library of Congress (Legislative Reference Service), and congressional committees.

3. Tighten party discipline and leadership influence in Congress. Reduce the number of subcommittees.

4. Divide states into four groups at random and hold primary elections in each group one month apart. The intent would be to have each grouping represent a cross section of interests so that a candidate would have to appeal to broad interests rather than to special interests to obtain the nomination.

5. Give the President the line-item veto.

A balanced budget amendment to the constitution has been proposed but is a drastic response to fiscal irresponsibility in government. It is pathetic if cluttering the constitution with an essentially procedural matter is the only way to restore fiscal discipline to government. The Budget and Impoundment Act of 1974 established "strict" datelines for budget-making and procedures to ensure that specific tax and spending measures conform to established macroeconomic fiscal needs of the economy. The Act has not brought fiscal responsibility but the situation might have been worse without it.

The above suggestions are intended to be indicative rather than definitive. The need for reform is more apparent than the specific measures which should be taken to restore a proper balance between encompassing and special interest institutions.

SUMMARY AND CONCLUSIONS

Farm policy is at least as much the product of personal and group values as it is of economic analysis. Two value systems have dominated: agricultural fundamentalism and democratic capitalism. The two were compatible in Jefferson's day; they now sharply clash. If left to market forces, the family farm will fade away albeit very slowly. On the other hand, interventions to save the family farm contradict not only free enterprise principles but also some parts of the farm fundamentalism creed.

Diminishing numbers of farmers have not dampened willingness of society to support farm commodity programs. In worldwide perspective, farmers are favored inversely to their proportion of the population. One reason is that the affluence gained from shifting resources out of agriculture provides a dividend which is used by society to buy food security and preserve family farms. Economists cannot label such policy as wrong if the public is informed of the tradeoffs when making decisions and if the political process is representative. The evidence suggests that the farm policy decision process is not well informed and is not politically representative of the public interest. It is highly doubtful that the public is getting its money's worth out of current farm policies in reaching objectives such as adequate, stable, quality food supplies at reasonable Treasury and market cost, preserving the family farm, earning foreign exchange, or providing equitable distribution of tax dollars. Providing information to help improve farm policy decisions is the task of the remainder of this book.

REFERENCES

Abel, Martin. 1986. *U.S. Agricultural Policy Process and the Role of Policy Analysis.* Staff Paper No. 3. Washington, D.C.: Abt Associates.

Abel, Martin, Lynn Daft, and Dan Earley. 1983. *Scope of Food and Agricultural Policy.* Report for agricultural policy analysis project. Washington, D.C.: Abel, Daft, and Earley.

Bonnen, James. 1984. U.S. agriculture and the instability of national political institutions: The shift from representative to participatory democracy. Tucson: Department of Agricultural Economics, University of Arizona.

Brewster, John. 1961. Society values and goals in agriculture. Ch. 6 in Center for Agricultural and Economic Development, *Goals and Values in Agricultural Policy.* Ames: Iowa State University Press.

Burchinal, Lee G. 1964. The rural family of the future. Pp. 160-62, 168-79, 184-89 in James Copp, ed., *Our Changing Rural Society.* Ames: Iowa State University Press.

Carlson, Allan. Winter 1986. Should America save its peasant class? *Small Farmer's Journal*, vol. 10 no. 1.

Coughenour, Milton and Luther Tweeten. 1985. Life perceptions and farm structure. Proceedings of the symposium on *Agricultural Change: Consequences for Southern Farms and Rural Communities* held at Atlanta, Georgia. Boulder, Colorado: Westview Press.

Dingman, Maurice. 1986. What does Christian theology have to do with the farm crisis? Proceedings of the third annual conference, *Is There a Moral Obligation to Save the Family Farm.* Ames: Religious Studies Program, Iowa State University.

Fite, Gilbert. 1962. The historical development of agricultural fundamentalism in the nineteenth century. *Journal of Farm Economics* 44:1203-11.

Flinchbaugh, B.L. and Mark Edelman. 1984. The changing politics of the farm and food system. FS26 in *The Farm and Food System in Transition.* East Lansing: Cooperative Extension Services, Michigan State University.

Goldschmidt, Walter. 1978. *As You Sow.* Mountclaire, New Jersey: Allenheld, Osmun, and Company.

Green, Bernal and Thomas Carlin. 1985. Agricultural policy, rural counties, and political geography. ERS Staff Report No. HHES850429. Washington, D.C.: Economic Development Division, ERS, USDA.

Guither, Harold. 1986. Tough choices: Writing the food security act of 1985. Washington, D.C.: American Enterprise Institute.

Hamilton, Horace. 1946. Social implications of the family farmer. Pp. 110-13 in J. Ackerman and M. Harris, eds., *Family Farm Policy: Proceedings of a Conference on Family Farm Policy.* Chicago: University of Chicago Press.

Heady, Earl. 1961. Preface. Pp. v-vi in *Goals and Values in Agricultural Policy.* Ames: Iowa State University Press.

Jefferson, Thomas. 1788. *Notes on the State of Virginia.* Philadelphia: Prichard and Hall.

Jefferson, Thomas. 1926. Letters to Jay, August 23, 1785. In J.G. Hamilton, ed., *The Best Letters of Thomas Jefferson.* Boston and New York: Houghton Mifflin Co.

Jordan, Brenda and Luther Tweeten. 1987. Public perceptions of farm problems. Research Report No. P-894. Stillwater: Agricultural Experiment Station, Oklahoma State University.

National Issues Forum. 1987. The farm crisis: Who's in trouble, how to respond. Briefing Book No. 2. Helena, Montana: AgFocus, Office of the Governor.

National Issues Forum. 1986. America looks at agriculture: An analysis of contemporary attitudes on some basic issues. Briefing Book No. 1. Helena, Montana: AgFocus, Office of the Governor.

Novak, Michael. February 1987. Cash income and the family farm. In Gary Comstock, ed., *Are Farmers Exploited by Agribusiness Corporations?* Ames: Religious Studies Program, Iowa State University.

Oster, Merrill. 1987. Do religious values suggest "family farms" are more socially desirable than "corporate farms"? In Gary Comstock, ed., *Is There a Moral Obligation to Save the Family Farm?* Ames: Iowa State University Press.

Paarlberg, Don. 1964. *American Farm Policy.* New York: Wiley.

Pasley, Jeffrey. December 8, 1986. The idiocy of rural life. *The New Republic*, pp. 24-27.

Pius XII. 1951. Problems of rural life. In *Christianity and the Land.* Des Moines: National Catholic Rural Life Conference.

Pius XII. 1943. La solennita della pentecoste, June 1, 1941. In H.C. Hoenig, ed., *Principles for Peace.* Washington, D.C.: National Catholic Welfare Conference.

Rohwer, Robert A. December 1951. Family farming as a value. *Rural Sociology* 16:330-39.

Speltz, George. 1963. Theology of rural life: A Catholic perspective. Ch. 4 in Center for Agricultural and Economic Development, *Farm Goals in Conflict.* Ames: Iowa State University Press.

Tweeten, Luther. 1979. *Foundations of Farm Policy.* Lincoln: University of Nebraska Press.

Zimmerman, Carle C. September 1950. The family farm. *Rural Sociology* 15:211-21.

CHAPTER FOUR

Explaining Alleged Chronic Low Returns

It is conventional wisdom that farmers earn low returns on resources. Yet the five principal farm problems listed in Chapter 1 did not include chronically low rates of return on resources. Whether commercial farmers are predestined under the market chronically to earn low rates of return on resources is the single most basic and controversial issue in all of agricultural policy.

Do markets work for farmers? Are markets efficient, using and acting on available information to allocate resources so that returns on farm resources are as great as returns on resources elsewhere? Do markets reach equilibrium in a reasonable period of time and if so does that equilibrium provide returns equal to opportunity costs? Or must government price and income support interventions be a continuing presence in farming to assure farmers economic justice and consumers adequate food supplies? If farm markets work efficiently to adjust to equilibrium and to dissipate disequilibrium, not only is perennial government intervention unnecessary to earn fair returns, it is ineffective because program benefits will be bid into fixed assets and lost to the new generation of farmers.

As noted in Chapter 1, the price and income problem in agriculture is often described as a cost-price squeeze. But a cost-price squeeze does no damage if increased productivity allows farmers to cover all costs at ever lower real prices and if rates of return are adequate. Poverty and low income do not define economic problems for commercial farmers because most have substantial owned resources to earn a large income. But commercial farmers contend they are *underpaid*, hence the issue is low *rates of return* on owned resources rather than low wealth, low income, or poverty. In analyzing the issue whether there are characteristics of agriculture and its economic environment that cause chronically low rates of return on farm resources, the hypotheses to be tested are that market returns are perennially low in farming because (1) forces create disequilibrium at too rapid a rate for farmers to adjust, or (2) farmers do adjust and catch up to equilibrium but that equilibrium features lower returns than elsewhere.

In testing these hypotheses, it is important to recognize that equilibrium is established at the margin by adequate-size, reasonably well-managed farms. The issue relates to commercial farms, not small farms. It is also essential to recognize that equilibrium is apparent in long-term *average* returns on reasonably efficient farms equal to returns elsewhere. There may be substantial instability and cyclical variation around the average.

Two formal theories — Willard Cochrane's technological treadmill theory and Glenn Johnson's fixed asset theory — play the central role in testing the hypothesis. Because of the central role of technology in these theories, it is necessary to devote considerable attention to that topic. These theories focus on forces creating disequilibrium and forces inhibiting adjustments to that disequilibrium.

The next section showing that these theories do not explain nonexistent chronic low returns is followed by a section advancing two theories. The decreasing-cost theory and cash-flow theory of the farm problem explain why returns appear to be low even in equilibrium when they are not in fact low.

TECHNOLOGY AND OTHER SOURCES OF DISEQUILIBRIUM

As noted in Chapter 1, many forces create disequilibrium including macroeconomic policies (to be discussed further in Chapter 6), economic growth, political decisions, and weather. The most widely recognized source of chronic disequilibrium in agriculture is technology. The *treadmill theory* of the farm problem has been built around it.

TREADMILL THEORY

Paraphrasing the treadmill theory attributed to Willard Cochrane (see 1965, p. 116 for succinct summary), (1) improved technologies are turned out year after year; (2) these are profitable and productive and hence rapidly and unavoidably adopted by farmers; (3) supply expands against an inelastic and slowly expanding aggregate demand for food, thereby reducing prices and receipts because demand is inelastic in the short run. (4) Farm conventional resources need to be reduced in response to lower returns but are not responsive to prices, resulting in (5) price depressing chronic surplus production. (6) Before the surplus is eliminated and profitability returned, new technology is released to renew the process.

Thus the treadmill theory not only formalizes the role of technology in creating disequilibrium but also recognizes the slow adjustment to disequilibrium explained later by fixed asset theory. It is well to digress at some length to examine in some depth the role of technology in farm problems and opportunities.

TECHNOLOGY, PRODUCTIVITY, AND DYNAMIC EFFICIENCY

Technology refers to methods or techniques relating inputs and outputs. Technology may be classified as mechanical (e.g., machinery), biological (e.g., hybrid corn), or chemical (e.g., pesticides). Technology is not free, it is a form of capital. It requires scarce resources to produce and is durable, giving off services over several periods. Hence the decision to produce or use technology is economic. Technology, productivity, and efficiency are not the same thing, however.

Although the Great Depression and bad weather slowed progress and delayed the full impact, the scientific revolution began to hit American agriculture with some force in the 1930s. At that time the scientific efforts of the land-grant universities established under the Morrill Act of 1862, the agricultural experiment stations established under the Hatch Act of 1887, and the cooperative extension service established under the Smith-Lever Act of 1914 made it possible to increase farm output without much increase in real aggregate production input as measured by the constant dollar inputs of farm labor, land, and purchased inputs. The vast industrial sector developed since the industrial revolution began about 1850 also contributed mightily. Improved seed varieties, hybrid corn, low-cost commercial fertilizers, pesticides, tractor power, and electricity were the highlights of the farm technological revolution that took place after 1920 (Tweeten, 1979, ch. 4).

From 1920 to 1940, the principal source of increased farm output was the replacement of horses and mules by tractor power. This not only improved timeliness of operations and tillage practices but also released for other uses, between 1920 and 1965, an estimated 86 million acres formerly providing feed for work animals.

The 48 conterminous states used 21.7 million tons of principal plant nutrients in 1985. If each ton added production equal to 10 unfertilized crop acres, then fertilizer added the equivalent of 217 million acres to cropland. Thus the release of cropland from production of power and the use of commercial fertilizer added production capacity equivalent to 303 million cropland acres. The above factors partly explain our ability to feed twice as many people on 334 million cropland acres in 1985 as were fed on nearly the same acreage in 1910. A preposterous 7 billion acres of cropland would be required in 1987 to provide today's food and fiber on land of 1910 productivity!

Several measures of farming productivity are shown in Table 4.1. The spectacular increases are in output per man-hour and in persons supplied per farm worker. Considering that the number of persons supplied per farm worker increased only 72 percent in the hundred years from 1810 to 1910, the gains since then are indeed impressive. The measure of efficiency is misleading, however, because over half the farm inputs in 1985 were supplied by the nonfarm sector — the average farm worker received much outside help to feed 75 persons in 1985!

Output per man-hour increased 10-fold since 1910. This measure too is inadequate, because it imputes to labor the output that arises from capital investment. The way to make labor productive is to use so much capital that the latter's return is zero. That would be economically efficient only if the cost of capital is zero, and it is not. The rising output per man-hour is in one sense a measure of lack of labor efficiency. Low productivity and attendant low returns for labor in agriculture caused capital to replace labor. Because output was maintained or even increased by the process, the gains in output per unit of labor were spectacular. Measured by gross output per draft horse, think how productive the few remaining farm draft horses are today! Measuring efficiency by output per unit of all production inputs partially circumvents this problem of inefficient use of one resource and is the best measure of farming efficiency in Table 4.1.

All of the foregoing measures show gains in *relative* efficiency over time rather than absolute efficiency in agriculture. Substantial inefficiency could have characterized the use

TABLE 4.1. Selected Measure of Farming Performance, 1910-85

Year	Farm Output	Production Input	Output per Production Input	Crop Production per Acre	Livestock Production per Man-hour	Output per Man-hour	Persons Supplied per Farm Worker
			-------------------- (1977 = 100)			-------------------	(no.)
1910	36	87	42	18	12	13	7.1[a]
1920	42	99	43	19	12	14	8.3
1930	43	102	42	17	13	15	9.7
1940	49	102	49	20	14	18	10.7
1950	61	106	58	22	19	31	15.5
1960	76	99	76	41	32	42	25.8
1970	84	96	87	70	64	74	47.9
1980	104	103	101	105	129	109	75.8
1985	119	94	127	136	175	140	74.7
Annual Increase							
1910-85 (percent)	1.6	.1	1.5	2.7	3.6	3.2	3.2

SOURCES: U.S. Department of Agriculture (April 1987 and earlier).
[a]Persons supplied at home and abroad per U.S. farm worker were 4.12 in 1820, 3.95 in 1840, 4.53 in 1860, 5.57 in 1880, and 6.95 in 1900.

of resources in all periods. The foregoing data show only dynamic efficiency in that progress has been made. The multifactor productivity index shows that technology has shifted the aggregate farm supply curve to the right at a rate of 1.5 percent per year since 1910 and at about 2 percent per year since 1960.

TECHNOLOGY AND STATIC EFFICIENCY

Static economic efficiency can be measured in various ways but all methods use variants of the concept shown in Figure 4.1. Nonconventional inputs such as research and extension C shifting the supply curve for farm crops and livestock from S to S' change equilibrium price from p to p' and output from q to q'. Net social benefits are summarized as follows:

Gain to consumers	1 + 2 + 3
Loss to producers	1 - (4 + 5)
Loss to taxpayers	C
Net gain to society	2 + 3 + 4 + 5 - C.

Because demand for farm output tends to be inelastic in most lengths of run considered, productivity gains bring losses to producers; that is, area 1 loss is greater than area 4 + 5 gain. As exports become a larger share of demand and make it less inelastic, that conclusion of loss to producers from productivity gains becomes less certain.

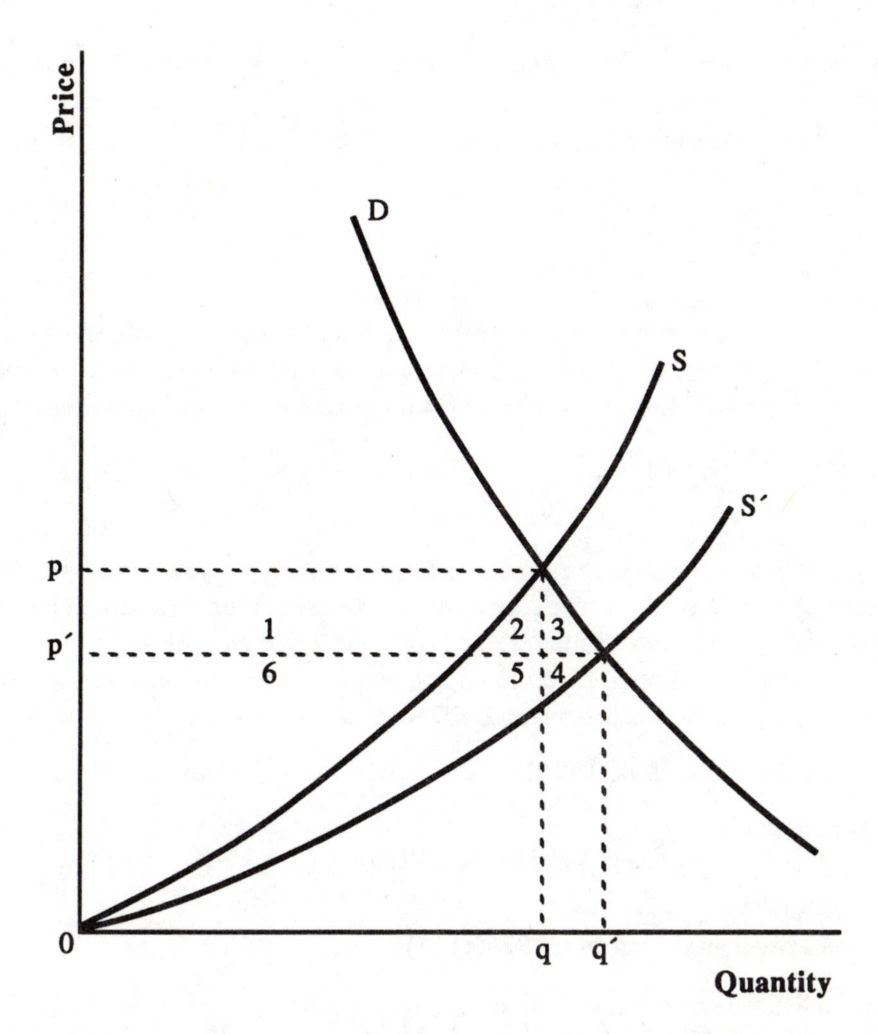

FIGURE 4.1. Hypothetical Economic Impact of Productivity Advancing the Farming Industry Supply Curve from S to S´

In reality, the gains to society from investment in C are spread over many years. One dollar invested in C today tends to have its peak payoff in 8 years and is obsolete in 16 years (Braha and Tweeten, 1986).

Three means of expressing static economic efficiency from investment in nonconventional inputs are widely used, recognizing that area 2 + 3 + 4 + 5 constitutes benefits in individual years designated b_i. Future benefits must be discounted to the present to correct for the lesser values of a benefit received in the future. The net present discounted value NPV is

$$NPV = \sum_{i=1}^{16} \frac{b_i}{(1+r)^i} - C_0. \qquad (4.1)$$

If C_0 is \$1, a typical value of $\Sigma b_i/(1+r)^i$ is \$4 when the discount rate r = .10 or 10 percent, hence NPV = \$3.

An alternative measure is the benefit-cost ratio B/C

$$\frac{B}{C} = \frac{\left(\frac{\Sigma b_i}{(1+r)^i}\right)}{C_0}. \quad (4.2)$$

A typical benefit-cost ratio for agricultural research and extension outlays is \$4 using a discount rate r = .10 or 10 percent. The most widely used procedure is to compute the internal rate of return, IRR, the value of r which makes the net present value zero, i.e.,

$$IRR = r: \left(\frac{\Sigma b_i}{(1+r)^i}\right) - C_0 = 0. \quad (4.3)$$

Internal rates of return for C range from 25 to 50 percent, implying that interest rates of 25 to 50 percent could be paid on funds used to improve agricultural productivity and just break even on that investment (Ruttan, 1982; Braha and Tweeten, 1986).

Typical internal real rates of return on conventional and nonconventional resources used in farming summarize static economic efficiency are as follows:

Nonconventional Inputs	Percent
Research (Ruttan, 1982; Braha and Tweeten, 1986)	25-50
General education (Tweeten and Brinkman, 1976, ch. 5)	10
Vocational-technical education (Tweeten and Brinkman, 1976, ch. 4)	15
Conventional Production Inputs[1]	
Large farms (Over \$500,000 sales)	15 to 20
Medium to large family farms (\$100,000 - \$500,000 sales)	5 to 15
Small farms (\$40,000 - \$100,000 sales)	0 to 5
Rural residences (under \$40,000 sales)	Negative

The foregoing data appear to indicate static economic inefficiency in agriculture: too little input devoted to developing profitable, productive, and environmentally safe new

---

[1]Calculated using basic data from U.S. Department of Agriculture (November 1986 and earlier issues).

inputs and practices as indicated by the very high internal rates of return on research; and too much conventional investment in farms too small to realize economies of size. Considering both static and dynamic measures of efficiency including social as well as private rates of return, however, the overall economic performance record of American agriculture is favorable based on the above numbers applying to the 1960s to 1980s:

1. The rate of return on public and private agricultural research and extension, typically 25 to 50 percent, is indeed high and indicates economic inefficiency in that too little was invested. But the resulting induced increase in agricultural productivity, near 2 percent per year, was rapid. Analysis (Braha and Tweeten, 1986) indicates the public research and extension outlays optimally could expand at 4 percent rather than their historic 3 percent annually to lower the rate of return on research to a more conventional 10 percent. The more rapid rate of increase would shift the aggregate supply curve rapidly to the right relative to demand. The social cost of farmers adjusting to a higher rate would have been large and perhaps politically unacceptable.

Real farm prices and income would fall at rates farmers could not adjust to without considerable trauma. Except in the 1950s and 1980s, productivity has not expanded at rates significantly in excess of demand growth. As noted in Chapter 1, productivity of farm resources approximately tripled since 1910-14 while real farm prices of output fell by half. Thus the price-benefits per unit of production inputs increased by about 50 percent since 1910-14 with productivity advances! The aggregate farm production input volume has changed very little for decades although the mix of inputs has shifted. Technological change has been steady, predictable, and at a pace farmers could absorb. Productivity growth exceeded demand growth through much of the 1980s because of failure of demand due to a depressed world economy, high dollar, and other factors rather than because of an acceleration in productivity. Later we observe that scale-biased, labor-saving technology has been the principal source of farm structural problems but the *major* capital-labor substitution process has run its course.

2. The negative rate of return to resources on small farms is based on all resources including operator and family labor and management earning opportunity cost returns. Most families on small part-time farms are there for reasons other than profit maximization. The small farm offers tax savings and a way of life to consume like a sports car and to pay for out of nonfarm income. As indicated in Chapter 1, the Census of Agriculture indicates that numbers of farms less than 50 acres in size increased 17 percent between 1978 and 1982. If persons on these farms viewed their full social returns as being too low, such farms would not have increased in numbers.

IMPACT OF TECHNOLOGICAL CHANGE

The impact of technological change on agriculture depends in part on whether the technology is factor-neutral or factor-biased, scale-neutral or scale-biased, and whether it is mainly cost-saving or output-increasing. To be economically attractive to the firm, a technology must reduce costs, increase output, or both, hence reduces unit cost. Thus resource- and scale-neutral or biased technology will be cost-saving or output-increasing. Firms adopt technology because it increases profit or reduces loss. Early adopters are attracted by profits which compensate for risk. After enough firms have innovated to reduce product prices or raise fixed input cost, latecomers must adopt to avoid losses. Firm profits are increased in the short run but not in the long run by technological change in a competitive industry.

Pure forms of technology are difficult to find. Reduced tillage or biological nitrogen fixation in grasses are mainly cost-saving technologies, hybrid seed is an output-increasing technology, the four-wheel drive tractor is a labor-saving and scale-biased technology, and the one-man round baler is a somewhat factor-neutral technology. The microcomputer tends to be a commodity-neutral technology and the broadleaf herbicide a commodity-biased technology. Public research has emphasized output-increasing biological technology; private industry has emphasized labor-saving mechanical technology.

In Figure 4.2 panels A and B, output q is considered to be constant between time period 0 and 1 ($q_0 = q_1$) when technology reduces firm cost from C_0 to C_1. The slope of the isocost line C_0 is determined by the prices of capital and labor. With factor-neutral technology in Figure 4.2A, labor m and capital k inputs are reduced by similar proportions from m_0 to m_1 and k_0 to k_1, respectively. With cost-saving technology biased toward labor-saving in Figure 4.2B, the technology which leaves output unchanged and reduces costs from C_0 to C_1 results in a major reduction in labor from m_0 to m_1 and expansion in capital from k_0 to k_1 for the individual farm firm.

Figure 4.2 panels C and D illustrate the impact of output-increasing technology. Technology increases real revenue (output) from R_0 to R_1 with constant input volume $x_0 = x_1$. With commodity-neutral technology, wheat and soybean outputs increase in equal proportions from w_0 to w_1 and s_0 to s_1, respectively (Figure 4.2C). With commodity-biased technology, the shift in the product transformation curve with a given volume of input results in a relatively greater increase in wheat than in soybeans as shown in Figure 4.2D.

Figure 4.3 illustrates impacts on the farm firm and industry from technological change. Industry demand and supply are assumed to be inelastic but the representative farm firm faces a perfectly elastic demand curve. The firm's supply curve is short-run marginal cost SMC lying above the average variable cost curve (not shown). Economies of firm size to output q_0 are apparent in the declining long-run average cost curve LAC in Figure 4.3A. Increasing returns to size prevail as LAC decreases and constant returns to size characterize the flat portion of the long-run average cost curve. In the U.S., most farm firms are in the declining average cost portion of the curve but most farm output is produced by larger firms in the nearly flat portion of the average cost curve. Empirical studies have rarely found a rising LAC but it undoubtedly exists for extremely large farms.

First consider the case of output-increasing technology which is assumed to be scale-neutral (no factor-bias favoring larger firms) in Figure 4.3A. The short-run marginal cost is initially SMC_0, short-run average cost is SAC_0, long-run average cost is LAC_0, aggregate demand is D, and aggregate supply is S_0. (SAC_0 includes fixed and variable costs. The long-run marginal cost curve is omitted.) For the farm firm, average revenue

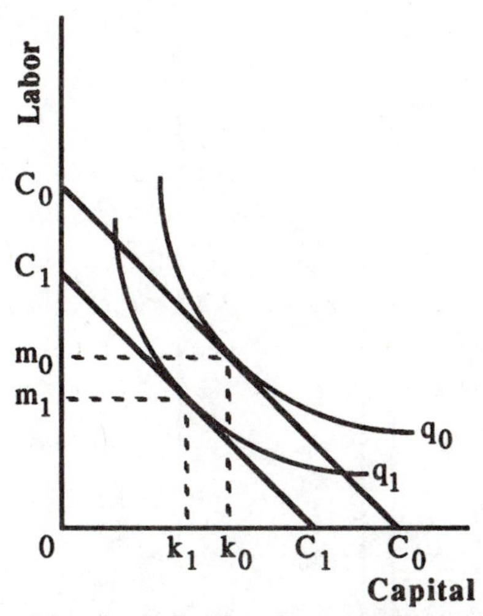

A. Cost-Saving Factor-Neutral Technological Change

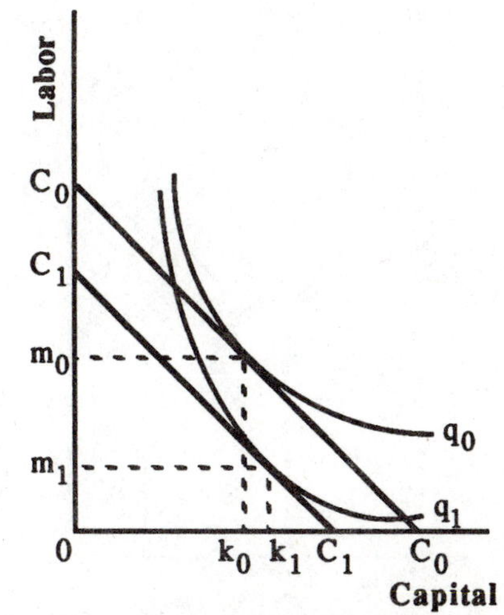

B. Cost-Saving Factor-Biased (Labor-Saving) Technological Change

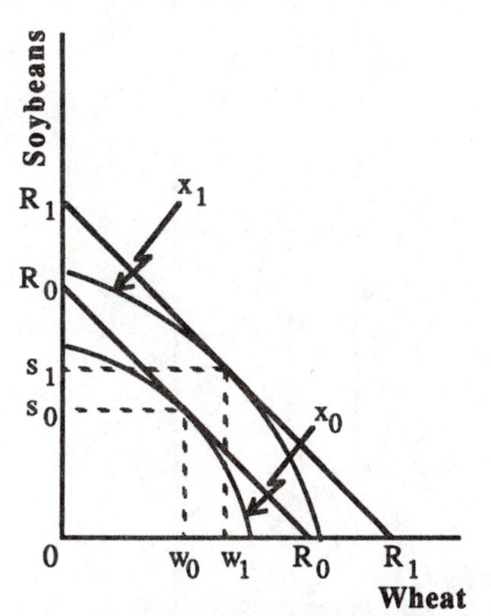

C. Output-Increasing Commodity-Neutral Technological Change

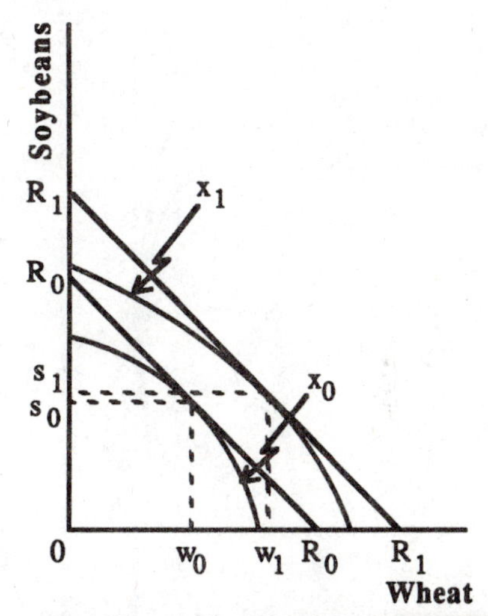

D. Output-Increasing Commodity-Biased Technological Change

FIGURE 4.2. Firm Factor and Product Impact of Changing Technology

(demand) is equal to marginal revenue at initial price p_0. Equilibrium firm output is q_0 and industry output is Q_0.

The early adopter of the successful technology is assumed to maintain plant size but faces cost curves SAC_1 and SMC_1. The firm will increase output from q_0 to q_1 so that MR $= p_0 = SMC_1$. The firm's pure profit is $(0q_1)(p_0 - m)$.

As other farmers adopt the new technology, the supply curve shifts from S_0 to S_1. Price falls to p_1 and industry output increases to Q_1. With inelastic demand, total revenue

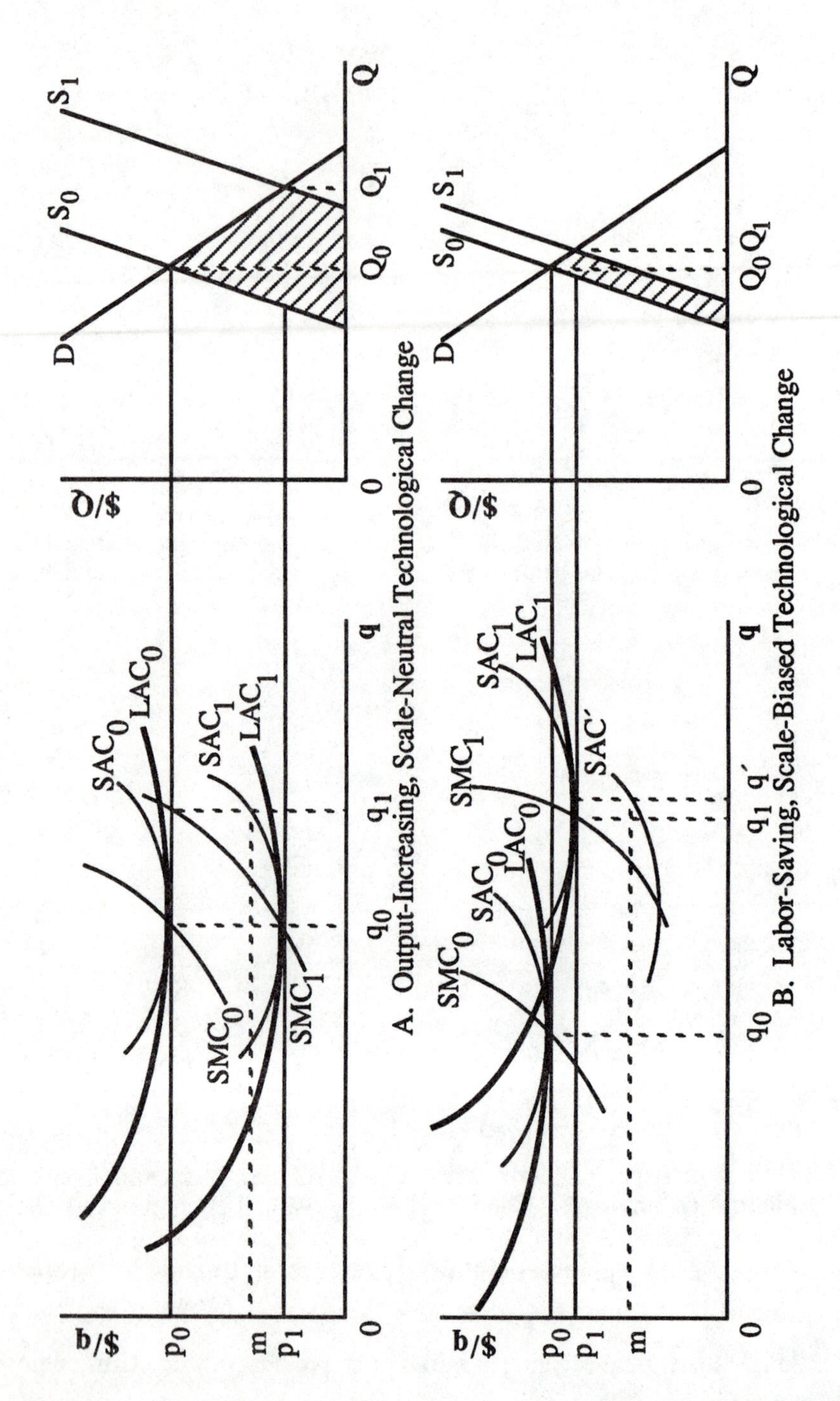

FIGURE 4.3. Farm Firm and Industry Impact of Changing Technology

falls and redundant resources begin to shift out of agriculture. Firms not adopting the new technology will receive low returns and may be forced out of business. After what may be long and painful adjustments, equilibrium again returns at firm output q_0. Normal but not pure profits are made and all costs are covered. The net dollar value in terms of additional goods and services produced and consumed from the new technology is the shaded area between S_0 and S_1 in Figure 4.3A. Because the technology was scale-neutral, no notable changes occur in farm size and numbers. However, firms enter to increase industry output while some not adopting the new technology drop out. Figure 4.3A presumes that increasing yields are offset by reduced acreage so farm size (output per farm) remains unchanged. In reality, increasing yields will increase farm size modestly. It should be noted that another new technology may restart the adjustment process before long-run equilibrium is reached on the long-run average cost curve LAC_1.

Now turn to the case of labor-saving technology assuming little aggregate output-increasing or cost-saving effect. In Figure 4.3B, the firm is originally in equilibrium at SMC_0, SAC_0, LAC_0, price p_0, firm output q_0, and industry output Q_0. The representative firm adopting labor-saving technology shifts its marginal cost to SMC_1 and average cost to SAC'. Firm output expands to q' and pure profit is $0q'(p_0 - m)$.

Two impacts occur in the long run as many firms adopt the new technology. One is that supply expands from S_0 to S_1. This changes equilibrium industry output to Q_1 and firm output to q_1. The second major impact is that pure profits become capitalized into the most fixed resource, farmland, as real estate passes from one owner to another. The implication is that laggards in adopting technology will earn low rates of return on their resources valued at opportunity cost even with no change in industry output Q_0 or price p_0. The total short-run average cost curve increases to SAC_1 and the long-run average cost curve becomes LAC_1 with full adjustments. Net gain to society in terms of additional real value of goods and services is the shaded area between S_0 and S_1 in Figure 4.3B.

Structural impacts are also apparent from labor-saving technology. Firms entering farming which do no adopt the technology after land prices are bid up will experience losses and may go out of business. As the curve is drawn in Figure 4.3B, the net result of labor-saving technology is to double the size of the average farm firm and to halve the number of farms — if total land in farms is somewhat fixed as in American agriculture. It is notable that the principal beneficiary of scale-neutral output-increasing technology is the consumer. It is also notable that with an inelastic demand output-increasing technology tends to be hostile to non-adopters while labor-saving technology tends to be more passive. That is, non-adopting owner-operators and landlords are not forced out of business and actually gain from land prices bid up by adopters of labor-saving technology. (Of course, non-adopting *new* owner-operators and renters fail to cover costs.) Non-adopters of output-increasing technology are subject to lower product prices and are made worse off by the new technology.

Demand sometimes expands at a sufficient rate to compensate for output increasing and labor-saving technology so that no resources need adjust out of farming. At other times, sometimes for a series of years, technological change outpaces growth in demand and depresses resource returns.

It may be noted in Figure 4.3 that producers benefit from output-increasing technology when industry demand is perfectly elastic whereas consumers benefit when demand is perfectly inelastic. To the extent that greater reliance on export markets makes demand more elastic, producers benefit more from technological change.

A highly elastic supply and demand would imply that the farming industry adjusts quickly to technology shocks. Unfortunately, supply and demand elasticities are low in the short run. Supply and demand (including exports) are much more responsive to price in the long run, helping the industry to recover from low returns caused by introduction of major new technology.

In U.S agriculture from 1950 to 1985, output per unit of labor rose about 4 percent annually and output per unit of production inputs increased about 2 percent annually. Because demand and production input productivity increased at about the same rate, the total volume of production inputs has remained at nearly the same level for three decades. That means that if technological change were factor- and scale-neutral, farm size and labor input would have remained essentially unchanged for the last three decades. Thus compared to labor-saving technology, output-increasing technology had modest impact on farm structure.

In fact, capital substituted for labor on a massive scale as noted in Chapter 1. The full-time small-farm operator has nearly vanished. Farm population fell from 23 million in 1950 to less than 5 million in 1987. In short, technological change has much more followed the scale-biased pattern depicted in Figure 4.3B than the scale-neutral pattern depicted in Figure 4.3A. Whether adequate-size, well-managed farms adjusted to technological change in the past to maintain income and rates of return is an empirical question to be examined later.

FIXED-RESOURCE THEORY

Classical economic theory stipulates that an efficient allocation occurs when the value of marginal product of a given resource is equal to its market price in all uses. The concept must be revised considerably when applied to the dynamic environment of the actual agricultural economy. Not one price but four prices are relevant for the farm labor resource in Figure 4.4. P_A is the acquisition price for labor (Hathaway, 1963, ch. 4; Johnson and Quance, 1972). It may be regarded as the return needed to attract farm operators into agriculture from the nonfarm sector, or the wage rate necessary to attract hired labor to farms from employment elsewhere. Optimum employment of labor in agriculture is X, given P_A and MVP_1.

The second price P_O in Figure 4.4 is the opportunity cost of farm labor. It is what labor currently employed in agriculture would earn in the nonfarm sector. Farmers tend to have less formal education and tend to be older than nonfarm workers. If farmers were employed in the nonfarm sector, their average earnings would be below that of nonfarm workers.

General education imbues skills that are somewhat readily transferable between occupations. Farmers have unusually high management skills not easily learned by outsiders. But when considering the salvage price of farm labor, the value of these skills *in*

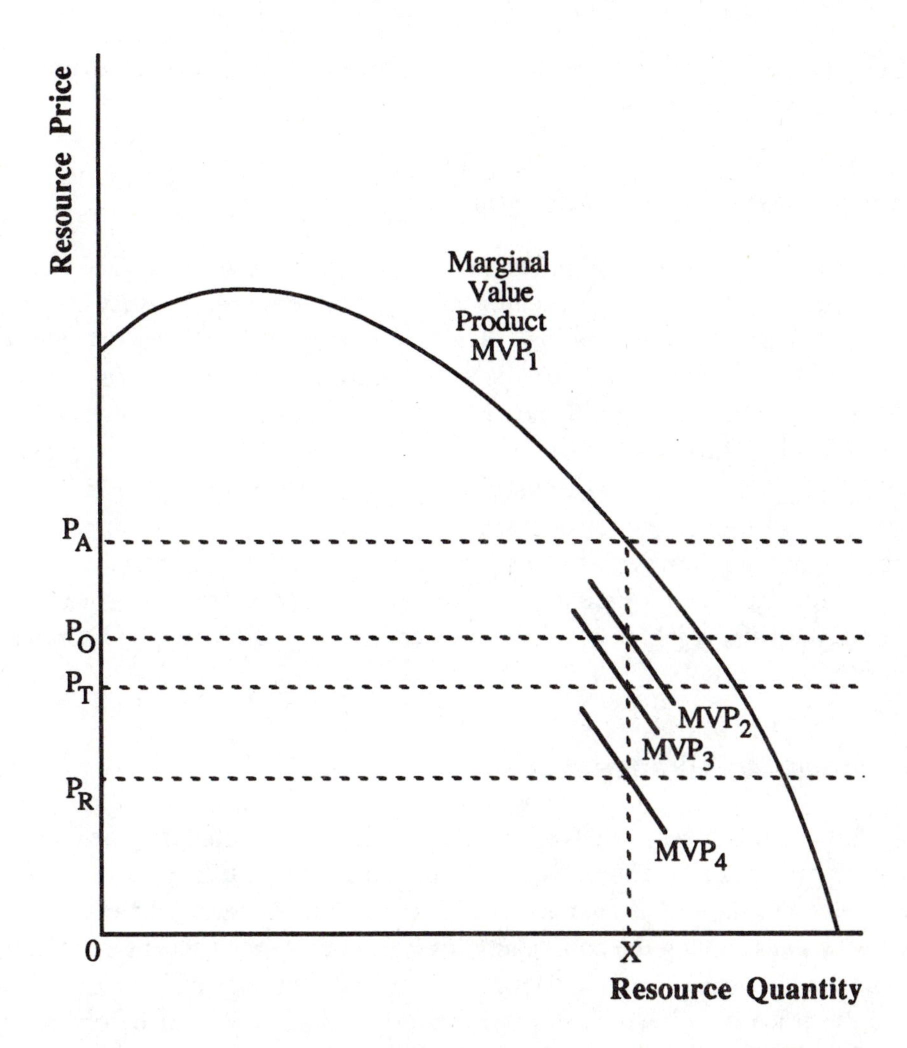

FIGURE 4.4. The Marginal Value Product of Labor in Agriculture

the nonfarm sector is important. Unfortunately, these skills often have limited value outside of agriculture. For farm workers who have few transferable skills, the alternative to farm employment is often urban unemployment. For them the opportunity price P_O either is zero or the value of unemployment compensation.

The age distribution of farmers limits mobility. The fact that farm employment is characterized by relatively high proportions of very young (mostly hired) workers and very old workers who have relatively low earnings and few opportunities for outside employment contributes to a low opportunity price at which to salvage farm labor. The median age of farmer operators is 51 years.

Data presented in Chapter 1 showed that most farms received a low return on resources when all farm resources were valued at their opportunity price. Gross farm income did not cover all costs of farming when farm resources were valued at their opportunity price — what they would likely earn if employed elsewhere. We must look beyond the opportunity price to explain the difference between actual returns and the acquisition price of farm resources.

THE TRANSFER-COST HYPOTHESIS

Wage rates in economic equilibrium differ because of transfer costs. The wage rate in agriculture must drop below the opportunity value less transfer cost to justify economically the transfer of labor from agriculture. The opportunity price P_O less transfer cost is designated P_T in Figure 4.4. Transfer costs are unlikely to be a large part of potential earnings in the nonfarm sector. If moving costs are $2,000 and the migrant will earn $20,000 for each of 20 years in new employment, moving costs are only .5 percent of future earnings. Thus monetary transfer costs appear to be too small to explain the discrepancy between the earnings of comparable farm and nonfarm resources. Transfer costs may loom large, however, for individuals who because of old age will have few years of productive employment ahead or who must search long and perhaps unrewardingly for a new job or who have little capital or credit base on which to support the transition.

THE ENDODERMAL HYPOTHESIS

The price that induces outmovement of labor from agriculture may be considerably lower than P_T for workers who value highly the farm way of life. Brewster (1961, pp. 129 ff.) labeled this the *endodermal hypothesis* explaining the farm problem. If alternate employment means leaving the community, the farm family may accept a low return for labor rather than leave relatives and friends. The reservation price of farm labor is labeled P_R. It is the price below which the farm labor earnings must fall before alternative employment will induce release of labor from agriculture, given potential earnings in nonfarm employment, transfer cost, psychic satisfaction from farming, and expectations of future farm earnings and of living in the city.

Given that favorable economic conditions such as from 1973 to 1975 (indicated by MVP_1) have generated farm employment X at labor acquisition earnings P_A, suppose that prices received by farmers fall because supply is increasing faster than demand. This can occur because government supports are curtailed, because of favorable weather for crops, or because demand contracts due to a depressed national economy or declining farm exports. If farm labor responded strictly to opportunity costs in the absence of transfer costs and special values on farm employment, then outmovement of farm labor would begin with the drop of the marginal value product just below MVP_2 (Figure 4.4). Reduction of farm employment would begin with a drop in the marginal value product to just below MVP_3 if transfer costs are included. And the marginal value product would need to fall below MVP_4 to induce lower farm employment when the psychic satisfactions of the farm way of life, imperfect knowledge, and other factors reflected in P_R are considered. The marginal value product could range from MVP_1 to MVP_4 without changes in farm employment! The farm economy would be in "equilibrium" in a period requiring fewer resources only if all farm labor were valued at its reservation price, P_R. Use of any other price, P_A, P_O, or P_T, would show that the gross income in agriculture would not cover all farm costs.

EFFECTS OF RESOURCE FIXITY ON FARM OUTPUT AND EARNINGS

Before appraising the actual role of fixed-resource theory in explaining persistence of low returns, it is necessary to examine the fixity of farm production resources in addition to labor. How fixed are farm resources? To answer this question it is useful to divide farm resources into five categories (1) financial resources, (2) operating capital, (3) durable capital, (4) real estate, and (5) labor. It is also necessary to consider two basic dimensions of fixity — durability of resources and the relationship of the marginal value product to the four prices listed in Figure 4.4.

Financial resources such as currency, bank deposits, and bonds are quite readily transferable to nonfarm investments and should cause no extended problems of fixity and attendant low returns.

Operating capital such as fertilizer, fuels, seed, repairs, pesticides, and hardware supplies have generally low opportunity prices for salvage once they are committed to farming. These supplies, purchased on the basis of an expected product price, are likely to be used even if the product price falls sharply during the production period. But the inputs are consumed quickly, most in one production period. Hence the farmer can correct his error by purchasing fewer operating capital inputs in the next production period. Furthermore, the farm MVPs for these inputs have tended to lie above acquisition prices. That is why agricultural chemical (fertilizer, pesticide) use remained high from 1979 to 1985 even as aggregate input use fell 11 percent and farm real prices fell 26 percent (Table 4.2). It follows that operating capital poses no input fixity problem.

TABLE 4.2. Farm Input Volume, 1970 and 1985 as Percent of 1979

Input or Output	1970	1979	1985
		(percent of 1979)	
All inputs[a]	91	100	89
Nonpurchased	106	100	87
Purchased	78	100	90
Farm labor	113	100	86
Farm real estate[a]	102	100	94
Mechanical power and machine	82	100	80
Agricultural chemicals	61	100	100
Feed, seed, and livestock	84	100	100
Taxes and interest	99	100	93
Miscellaneous	79	100	103
Output	76	100	107

SOURCE: U.S. Department of Agriculture (April 1987).
[a]Diverted acres included.

Durable capital items such as farm machinery give off services over an extended period and have low salvage value outside of the farming industry. Estimates (U.S. Department of Agriculture, November 1986) show an annual depreciation rate for farm machinery of about 16 percent. Even if services decline at one-fifth this rate, without adding to the stock of machinery the machinery input would fall one-tenth in three years — hardly a serious degree of fixity. If other less fixed resources reacted similarly, excess production and resource capacity higher than the 9 percent found in the peak year 1986 could be eliminated in only three years. Farm-produced durables such as breeding stock would seem not to create serious problems of fixity because the acquisition price varies proportionately with farm economic conditions and because they can be salvaged at slaughter prices.

The fourth input category, farm real estate, is durable and has a low opportunity price for salvage in nonfarm uses (except around cities). Input including diverted acres of real estate fell 6 percent as total farm inputs fell 11 percent from 1979 (the peak input year) to 1985 (Table 4.2). Land fixity is so great, however, that its price is largely determined by economic conditions within agriculture. Because the land price does not determine farm output, excess capacity and associated low returns in farming cannot be blamed on the land price. Analysis shows that the land market is efficient, using available information to establish land values consistent with expectations of future discounted earnings from land (Tweeten, 1986).

Because the input life is short or the MVP exceeds the acquisition price — or because of the pricing structure — the above nonlabor resources accounting for over three-fourths of all farm inputs do not support the asset fixity theory as an explanation of low farm returns over an extended period of time.

We must return to labor as the "problem" resource if fixed-resource theory is to provide an explanation of low farm resource returns. Labor is a durable resource, and, thanks to high birth rates, unlike other resources it increases in supply even though farmers do not purchase more from the nonfarm sector. Because a change in the labor input itself has little effect on farm output, it would appear that resource fixity would not explain excess capacity in agriculture. Labor is more important than its contribution to output might suggest, nevertheless, because capital inputs are often used in fixed proportions to labor on family farms, and because labor is very important as a *denominator* in expressing returns per *unit* of labor.

Some farmers could increase their earnings by employing their labor resource elsewhere. Questions remain of whether earnings below P_T represent inefficiency and a cost to individuals and society, and whether in fact the disequilibrium ascribed to fixity of labor even exists.

Whether earnings below P_T represent economic disequilibrium depends on the source of the gap. Given no disassociation of private and social costs (or returns), earnings of labor below P_T would indicate economic inefficiency, and the product of society could be raised by closing the gap. If the gap arises because of reasonable inertia in making adjustments to dynamic forces that constantly generate disequilibrium, and the inertia can be overcome only with great trauma to workers, the gap may not represent a social cost. If the gap arises only because farmers, though fully informed of alternatives, place a high value on farming as a way of life, then the gap measured by appropriate economic yardsticks does not represent a social cost. If the gap arises only because farmers are not informed about employment opportunities elsewhere or about the city way of life, then the gap is a social cost as well as a private cost unless the cost of providing such knowledge exceeds benefits. And if the gap arises because of racial discrimination and labor unions that inhibit mobility, then it may represent a private and social cost but no disequilibrium. That is, farm workers may not close the earnings gap because they cannot obtain the higher-paying nonfarm jobs which they are able to perform.

Large numbers of persons who appear to be receiving earnings below opportunity cost in farming are off-farm workers who reside on a farm for its consumption value. They do not expect to receive full pay for such "recreational" activity. The diminishing excess labor in farming and rising proportion of variable inputs relative to fixed inputs diminishes the ability of fixed-resource theory to explain chronically low resource returns in agriculture. The pressures are now great to force excess labor from farms: these pressures come from high land costs and other capital costs, large know-how requirements, and other barriers to entry. Also, the farm population is becoming such a small portion of the national population that a sizable percentage of the farm population can take nonfarm jobs without turning the economy upside down. For example, an exodus of 2 percent of the farm labor force constituted only .1 percent of the U.S. labor force in 1986 but would have been .3 percent of the U.S. labor force in 1950.

Tyner and Tweeten (1966, p. 613) estimated that farm labor was in excess supply by two-fifths in the 1952-61 period. Farm labor input fell 55 percent from 1961 to 1986 in response to market forces. Relatively little excess labor remains in farming. Growing numbers of large farms and part-time small farms give no indication of excess labor there. Approximately 50,000 farm operators retire or die each year, reducing farm population approximately 2.4 percent per year, other things equal. Farm population fell 2.4 percent per year in the 1980s, hence the number of farm entrants almost exactly offset the number of operators forced out because of the severe financial stress in the decade. The rate of decline in number of farms and farm population was much higher in the 1950s and 1960s than in the financial stress years of the 1980s, providing compelling evidence that excess labor is no longer massive in agriculture. To be sure, redundant labor is still found on many mid-size farms, but numbers are very few relative to the size of the national labor force,

In short, fixed resource theory and supporting empirical evidence explains why farm resources do not adjust to shocks in the short-and intermediate-run — it doesn't pay farmers to adjust. Fixed resource theory helps explain cyclical low returns in the farm economy associated with the persistence of farm output even as receipts fell in the 1980s. Data in Table 4.2 indicated that farm resources are responsive to changing economic conditions. Fixed resource theory and empirical evidence do not explain chronic low returns in farming.

Technological change mostly has shifted the supply curve in a steady and predictable manner. That makes it easier to make adjustments. By focusing on *only* demand or *only* supply, it is easy to underestimate ability of the farming industry to adjust to release of excess supply capacity or to other shocks. The ability to dissipate excess capacity is determined by the *sum* of the supply and demand elasticities. Suppose that excess capacity defined as production in excess of what the market will absorb at current support prices is 9 percent as in 1986. The following lengths of run; supply, demand, and excess capacity elasticities; and fall in price required to alleviate excess capacity of 9 percent are as follows:

Time Period (Years)	Assume Price Elasticity of: Supply E_s		Demand E_d		Excess Supply E_c	Price Decline Required to Eliminate Excess
	E_s	+	$\lvert E_d \rvert$	=	E_c	$\frac{9 \text{ percent}}{E_c}$
2	.1	+	.2	=	.3	30 percent
4	.3	+	.3	=	.6	15 percent
10	1.0	+	.5	=	1.5	6 percent.

Given the elasticity of demand, release of 9 percent of output removed annually from the market by the government would reduce prices at least 30 percent which would alleviate the excess capacity in two years based on the assumed demand and supply elasticities. That would be traumatic, and a wise approach would be to draw down government stocks by (say) paying farmers in kind to cut production so that prices would fall only 15 percent for 4 years — the second case above. Or payment-in-kind could be used to phase-out excessive stocks, production capacity, and deficiency payments in 10 years — by dropping prices 6 percent and decoupling payments from production incentives.

The point is that by allowing lower prices to at once increase demand quality and reduce supply quantity, rather large shocks can be absorbed without drastic price reductions. The phase-out of payments and attendant decoupling is important because past target prices and deficiency payments have encouraged farmers to produce for the programs rather than for the market.

Empirical Evidence of Favorable Returns to Reasonably Efficient Commercial Farms

The foregoing analysis indicates that supply, demand, and land price adjustments combine to dissipate disequilibrium in a relatively short time in the farming industry. If that analysis is valid, we would not expect to find low rates of total return for extended periods on adequate-size, well-managed farms. Wealth or net worth reported for farmers in Chapter 1 is several times that of nonfarmers. But farmers may be poorly rewarded for their wealth. Rates of return averaged over all farms appear to be perennially below rates of return elsewhere as apparent in the following data showing residual returns to farm assets (subtracting all production costs including operator and family labor and management from gross income) as a percent of assets:

Current Rate of Return to Farm Assets
(percent)

1970-71	3.1
1972-74	5.9
1975-79	2.7
1980-84	2.0
1985	3.3
1986	3.0

Source: Johnson et al. (1987)

Laypersons (and many economists) compare these data with real rates of return elsewhere (typically 10 percent) and conclude farm resources are underpaid. But such comparisons are flawed for several reasons:

1. Corporate returns are based on "book values" more closely related to a low purchase price rather than current market price.

2. Farm *current* returns do not include capital gains which were large from the 1930s to 1981. However capital gains can be negative as in the 1980s.

3. Most importantly, returns are averaged for large and small farms, for commercial farms and hobby farms, for poorly-managed and well-managed farms. From 1965 to 1980, current returns on farm equity averaged 4.6 percent and real capital gains averaged 6.1 percent for a total real return to farm equity of 10.7 percent (U.S. Department of Agriculture, 1981, p. 51). The latter compares very favorably with a .1 percent real return on common stock and a -3.6 percent total real return on long-term bonds. None of these rates is typical — averaged over a longer period total real rates of return on stocks and bonds have averaged 5-12 percent.

Another measure of farm returns is the real estate returns which can be fairly easily calculated. The annual return to farm real estate owners consists of the annual operating return (e.g., cash rent) and the change in land value. Expressing the return as a proportion of the current land value provides a measure of farmland's performance, the annual percent return. Other assets, such as common stocks and bonds, have similar annual return components: a current return, such as dividends or interest, and a change in asset value. Future returns are highly uncertain; however, past returns may help in forming expectations of future returns. Obviously, farm real estate returns were poor in the 1980s; however, long-term performance offers a different perspective:

Annual Returns to Assets, 1926-86

Asset	Mean	Standard Deviation
	percent	
Farm Assets	7.78	8.59
Common Stocks	12.12	21.04
Corporate Bonds	5.33	8.44
U.S. Treasury Bonds	4.71	8.49
U.S. Treasury Bills	3.51	3.34
Consumer Price Index	3.14	4.84

SOURCE: Irwin and Rask (1987, p. 12)

Farm assets, of which farm real estate is the primary component, had returns nearly 8 percent per year from 1926 to 1986. Although common stocks had higher returns than farm assets, year-to-year variability in returns were much larger with common stocks. While farm asset owners experienced financial risk, common stock owners experienced over twice as much risk during this period. From this long-run perspective, farm assets performed *better* than interest bearing assets (corporate bonds and U.S. Treasury bonds). That is, returns to farm assets were higher and variability was about the same as those of interest bearing assets. The real rate of return is the mean return less the inflation rate.

Farm capital gains were large and positive in the 1970s and large and negative in the 1980s, adding to the instability of farm economic conditions. Current rates of return (residual income to equity or assets divided by total equity or assets) are a more reliable indicator of longer-term real returns on farm resources. (The residual method includes a charge for farm operator labor and management; hence, a low return to assets could also be interpreted to imply a low return to labor and management.)

Impact of Size. An example of current rates of return calculated from actual farm income and expense data is shown below:

Source	Concept	Year	Sales Class ($1,000)	Percent Rate
Hottel and Reinsel (1976)	Current rate of return to farm equity	1970	$100+	6.9
			40 - 100	5.9
			20 - 40	4.4
			10 - 20	2.9
			Smaller	Negative
Tweeten and Mayer (forthcoming)	Current rate of return to farm assets, excluding dwelling	1984	$500+	16.0
			250 - 500	5.2
			100 - 250	2.7
			Smaller	Negative
Melichar (1985)	Current rate of income return to farm assets	1984	$500+	18.2
			200 - 499	6.3
			100 - 199	4.2
			40 - 99	1.7
			Smaller	Negative
Tweeten (for this study)	Current return to farm assets, including dwelling, and with improved labor estimates	1986	$500+	20.1
			250 - 500	8.7
			100 - 250	4.4
			Smaller	Negative

Hottel and Reinsel (1976) estimated a current rate of return to equity averaging 2.1 percent over all farms but 5.9 percent or approximately triple the average on farms with sales of $40,000 to $100,000 in 1970. Similarly, Tweeten and Mayer (forthcoming) estimated a current rate of return to equity on "adequate-size" farms in 1984 of 5.2 percent or five times the average return on all farms of 1 percent. Melichar's (1985) estimate was slightly above that; it was 6.3 percent on farms with sales of $200,000 to $500,000. Computations for other years revealed similar results.

The conclusion from the above data is that if current returns average 3 percent on all farms, they average 4-6 times that level or 12 to 18 percent on adequate size farms. Results for large farms show higher returns but are excluded for comparison purposes because they are not comparable in structure to other farms. Using procedures similar to those above, Knutson and Richardson (1986) found similar patterns of declining costs per unit of output for wheat farms, sorghum farms, corn farms, and cotton farms based on 1982 data. Magnitudes differ, but "engineering" studies (see Office of Technology Assessment, 1986) of economies of size by enterprise display the same pattern as shown above of efficiencies for larger farms.

Impact of Management. The influence of management on returns can be gauged using data from two states, Kansas (Parker and Langemeier, 1987) and Illinois (Sonka et al., 1987) from comprehensive farm financial records. In Kansas in 1986, 84 percent of farms keeping records had gross farm income in excess of $50,000, hence most were commercial farms. Rates of return on resources were not given, but net income of the top 25 percent of the farms averaged 3.8 times the net income of $17,965 averaged over all farms. Sizes of the two groups (top 25 percent and overall average) of commercial farms were similar.

The Illinois study was based on a sample of 179 cash grain producers from 1976 to 1983. Because the sample was dominated by commercial-size farms, performance measures were little influenced by farm size. The top one-fourth of farms in management return per acre over the entire period averaged four times the management return for all farms of $5 per acre. The top one-fourth of farmers ranked by rate of return on nonland assets averaged 2.9 times the rate of return, 14 percent, realized on the average by all farmers over the 1976-83 period. Of particular interest is that the farmers who were in the top management categories on the average for the entire period were rarely in the top ranking in any one year. The implication is that performance of the good managers is steady over the years and they tend to avoid risks that give them spectacular returns in any one year.

The conclusion is that the top quarter of farm managers realize rates of return four times that of farms with average management. Hence if average farms were earning 3 percent in 1986 as noted above, we would expect well-managed farms to earn returns of 12 percent.

In theory, the adequate-size, well-managed categories combine in a multiplicative manner, hence we might expect rates of return up to 16 times as large on adequate-size, well-managed farms as on the average farm. Recognizing that farm size was not fully controlled when estimating the impact of good management and that management was not fully controlled when gauging the impact of farm size (good mangers tend to be on larger farms), it is probably safe to conclude that reasonably well-managed farms of adequate size realize current rates of return six times that of the average farm. Evidence is compelling that returns on assets of reasonably efficient commercial farms averaged at least 12-18 percent in the 1980s and earlier decades — numbers comparable to what resources earn in other sectors.

SUMMARY COMMENTS ON ADJUSTMENT CAPABILITY

Growing evidence such as that above points to considerable capacity of the farming industry to adjust to technology and other changes. The three major sources of adjustment are:

1. Supply Response. The elasticity of supply is low in the short run, typically .1 to .3, creating large fluctuations in price when demand shifts — if buffer stocks are low. But long-run supply is responsive to price. That elasticity is near 1.0 for aggregate output and is higher for several individual commodities (Tweeten, 1979, pp. 272-276; Henneberry, 1986).

2. Demand Response. Like supply, demand is highly inelastic in the short run, causing large price movements when supply shifts in the absence of buffer stocks. But with international demand included, demand quantity is more responsive to price in the long run and approaches -1.0 for several commodities (see Tweeten, 1983).

3. Land Price Response. Land earnings are a residual claimant on net farm income because land tends to be the most fixed resource. When net farm returns rise or fall investors in adequate-size, well-managed farms which account for a small portion of all farms but for over half of farm output bid land prices to levels that provide a rate of return comparable to rates on alternative investments. They cannot pay less or land will be bid away; they are unwise to pay more if investments are expected to earn more elsewhere (Tweeten, 1986).

We have not and will not encounter a theory or empirical evidence that explains *chronic* low rates of return to resources on reasonably well-managed, adequate-size farms. That is because rates of return are not low on such farms which account for over half of farm output but for relatively few farms. If farm resources do not perennially earn low returns, it follows that long-term returns on farm resources will average as high as returns elsewhere, that farm markets including resource markets are efficient, that government commodity price and income support programs are not needed to maintain farm returns, and persistent maintenance of such programs will merely result in bidding of benefits into land values by an efficient market so that new owner-operators will lose the intended benefits.

An alternative interpretation of the above analysis is that adequate-size,well-managed farms earn equilibrium returns on the average because of responsive government programs rather than because of well-functioning markets. This issue cannot be answered definitively, but the massive adjustments in farm resources noted in Table 4.2, in land prices, and in other adaptations suggest it is responsiveness of farmers and land investors to market-incentives rather than responsiveness of government to farm needs that has been paramount, although both adjustments have played a role. There is reason to believe that given time for adjustment (about five years), these progressive commercial farms would earn an equilibrium return with or without government programs.

Annual and cyclical instability would remain a problem, however, especially for commercial farmers. Even small changes in incomes and interest rates can lead to large changes in real wealth as experienced in the 1980s.

THEORIES EXPLAINING WHY FARMING RESOURCES APPEAR TO BE EARNING LOW RETURNS EVEN WHEN THEY ARE NOT

The foregoing analysis indicating that rates of return have been favorable to efficient commercial farmers in past decades, even the 1980s, raises the question: Why is the conventional wisdom so widespread and deeply held that commercial farmers chronically earn low returns? Why do rates of return in commercial farming appear to be low when they are not? The following *decreasing-cost* and *cash-flow* "equilibrium" *theories* explain the phenomenon. The theories indicate that attempts to create disequilibrium by raising average rates of return for all farms to nonfarm standards will be self-defeating.

DECREASING COSTS AND INCREASING RETURNS TO FARM SIZE

Numerous studies have documented the existence of decreasing average costs and increasing returns to size of farm firms (Office of Technology Assessment, 1986; Knutson and Richardson, 1987). The concept is rarely related to farm problems. The term "size" rather than "scale" is used. Returns to scale refer to the impact on farm output of a given change in the level of all resources in fixed proportion. Because farming units do not expand all resources in fixed proportions, the pure concept of returns to scale is academic and without relevance to farm policy. The expansion in the farm firm is generally characterized by increasing the proportion of capital to labor, and of variable capital to fixed capital. These changes result in a sizable reduction in cost per unit of production.

Evidence of decreasing costs per unit (increasing returns to size) is readily apparent in Figure 4.5. The curve is an average within an economic class of farm but is a marginal curve among classes. The output-input ratio is defined as gross income from farming divided by the resource cost of all farm inputs including the opportunity cost of equity capital and all labor. The cost of all inputs (including the opportunity cost of equity capital and of operator and family labor) per unit of output (including receipts from farm commodities, nonmoney income, and government payments) averaged $2.4 on farms with sales of $10,000 to $20,000 and 83¢ on farms with sales of over $500,000. Most economies of size are achieved by family farms, and unit costs decline very slowly beyond an annual output of $100,000 per farm. Still, 10 percent or more of economies of size extend beyond family farms with $250,000 of sales. The very largest and smallest class of

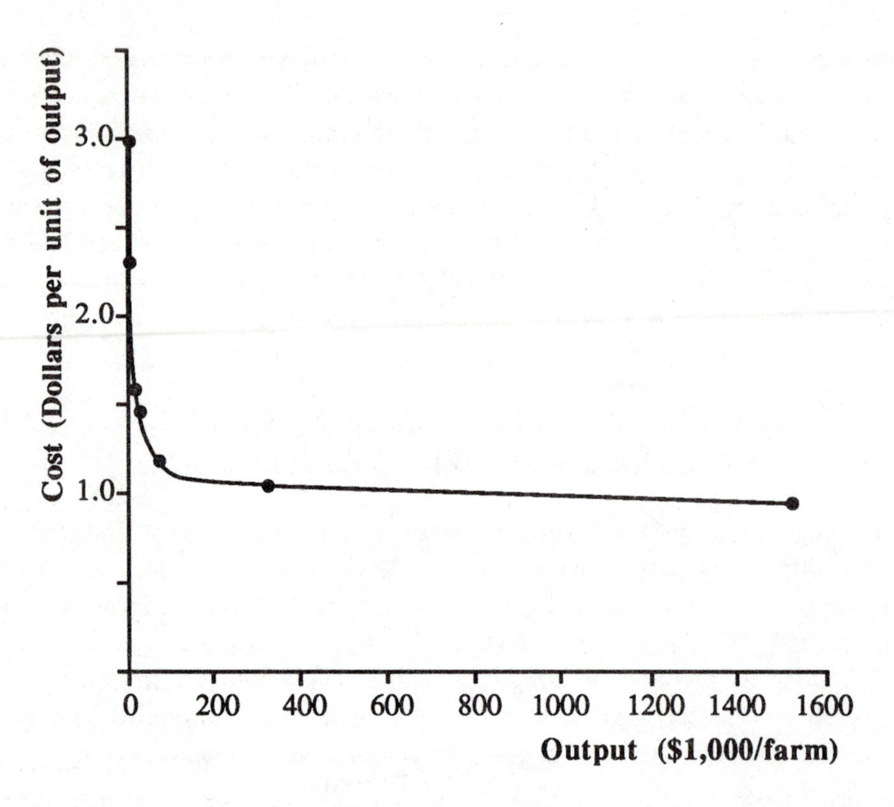

FIGURE 4.5. Resource Cost per Unit of Output by Size of Farm, U.S., 1984

farms could be omitted in Figure 4.5 because they have disproportionately large shares of cattle, calves, fruits, and vegetables. Omitting those, economies of size remain apparent and agree with results from "engineering studies" for individual enterprise-types of farms (see Office of technology Assessment, 1986). However, Figure 4.5 accounts for more market economies and for high commuting transportation costs on small farms than do engineering studies. It is notable that Figure 4.5 cost curves constructed for 1960, 1965, 1970, 1975, 1980, 1982, and 1984 bear striking similarity. This strongly supports the argument of equilibrium tendencies in farming because adequate size farms were essentially breaking even with all resource costs *each* year valued at opportunity rates.

The marginal and average revenue curves are indicated by an imaginary horizontal dotted line at $1 in Figure 4.5. The classes of farms with the revenue curve lying below the cost curve on the average lost money and did not cover all production costs. Small farms survived by accepting a low money return on their labor and equity capital.

It may be said that small farms earned low returns because they paid too much for their land. Land tends to be a complementary input with farm size. There is pressure to expand farm acreage to achieve the economies of size so apparent in Figure 4.5. The savings through greater efficiency are bid into the price of land. The actual price of land tends to be that price which will make all costs, including real estate interest, equal to the value of all farm receipts *on an economic-size unit.*

Competition in the land market tends to bid the land price to the point where the return on land will be equal to the return on capital in other uses. The residual return to land tends to be greatest on large farms. The law of one price and the large potential number of investors will ensure that the "high" price for land on large, efficient farms will be the market price of land applicable to all farms.

The small farmer must pay this price or land will be bid away from him by an investor who has or can achieve an economic-sized unit. Thus the small farmer tends to actually incur losses if he paid the current land price. And the small farmer who has full equity in land is losing money if a charge is made for the opportunity cost of his owned land valued at the current price of farm real estate. The tendency in some localities to bid up the price of farmland for potential nonfarm uses also contributes to apparent persistent low returns on resources assumed to be used only for farming.

Figure 4.5 suggests that land is *not overpriced* at the margin because large farmers were earning returns needed to hold capital in farming. But land is clearly overpriced for the vast majority of farmers — those with gross sales under $100,000.

The heterogeneous size structure of farms helps to explain why farmland is overpriced for most farmers and why returns are low on farm resources of most farmers. It also is a partial explanation for technological change increasing output per unit of input. The continued trend toward larger farms increases farm output, other things equal, because given inputs produce more output. However, the decreasing-cost theory depends more on labor-saving technology and differing efficiency among farms than it does on output-increasing technology to explain apparent low resource returns in farming. The term "apparent" is used because small farms are compensated by taxes and farm way of life for low earnings, pay for these benefits with off-farm income, and were growing in numbers from 1978 to 1982. Full-time small farms have largely exited farming.

The decreasing-cost theory is especially instructive in pointing out the permanency of the problem of low returns. An increase in the product price results in a larger residual return of land. The higher land return causes land prices to rise to the point where farm and nonfarm investors can realize an equilibrium return on their investment. Thus land price is determined at the margin by adequate size units. It follows that small farms, which constitute a large majority of all farms, will receive a low return on all resources when valued at their imputed opportunity costs even with higher farm product prices. When low rates of return on small farms are averaged with near break-even rates of return on adequate-size farms, the result is low rates of return when averaged over all farms whatever the price given time for market adjustments. The majority of farmers will appear to be underpaid for their resources whatever the parity ratio!

Cash-Flow Theory

The cash-flow farm problem was introduced in Chapter 1 but is presented here in greater detail to further explain why farm returns appear to be low even when they are not. The theory centers on real estate which accounts for 70 percent of farm assets.

It is assumed that land market participants expect land earnings or rents R to increase or decrease at a constant rate i + i' in the future,

$$R_t = R_0 e^{(i + i')t} \tag{4.4}$$

where the current year is t and the initial year is 0, e is the base of natural logarithms, i is the expected inflation rate, and i' is the expected real rate of increase in land earnings or rent. Assuming rents continue to infinity and are discounted at a rate α (real desired or market equilibrium rate of return on land investment) plus i to compensate for inflation, then the present value of land is:

$$P_0 = \frac{R_0}{(\alpha - i')}. \tag{4.5}$$

Market participants will not pay more because the return would be higher investing elsewhere; they cannot pay less or land will be bid away by those satisfied with return α. If transaction costs are nominal, the present value (4.5) applies even if land is sold because the current owner expects to receive all remaining discounted benefits in the land value at time of sale. Thus equation 4.5 applies not just to year 0 but to any year t. It follows that because P_t and R_t are related by the constant α-i', the price of land is:

$$P_t = P_0 e^{(i+i')t} \tag{4.6}$$

and R and P both increase at the rate i+i'. The capital gain is defined as the increase in P. Thus the real capital gain rate is i' and the pure nominal capital gain rate is i. The current rate of return on farmland is $R_0/P_0 = \alpha - i'$ from equation 4.5. The total nominal return is current return α-i' plus capital gain i+i' or α+i and the real rate of return α is nominal return less the inflation rate as defined earlier. The cash-flow deficit rate i+i' from interest alone on a fully indebted acre is the bond interest rate assumed to be α+i (if the real interest rate equals the real return on land) less the current return on land α-i'. Each of the concepts is illustrated in Table 4.3 under alternative scenarios.

Although the key variable for rational investors, the real rate of return on farmland α, arbitrarily is assumed to be 4 percent in the three scenarios, the cash-flow situation differs markedly. When land earnings behave like a bond, the current return is equal to the interest rate so the cash-flow deficit is zero and the land price is $P_0/R_0 = 1/(\alpha - i') = 10$ times earnings.

In the three decades preceding 1980, land rents at least kept up with inflation so scenario 2 is labeled "land." With land earnings expected to be constant in real terms, the assumed parameters yield a current return of 4 percent, a nominal capital gain and cash flow deficit of 6 percent, and the land price is 25 times rent.

Under growth stock scenario 3, real land earnings are expected to increase 3 percent annually. The result is to reduce the current rate of return to 1 percent and to raise the cash-flow shortfall and capital gain to 9 percent — of which 6 percentage points are nominal and

TABLE 4.3. Illustration of Cash-Flow Theory Applied to Land, Assuming a Real Rate of Return α = 4 Percent, and an Inflation Rate i = 6 percent

Current Return $\alpha - i'$	Capital Gain and Cash-Flow Shortfall $i + i'$	Total Nominal Return $\alpha + i$	Total Real Return α
Bond Scenario 1: Land earnings constant in nominal value, i.e., $i' = -i = -6$			
10	0	10	4
Land Scenario 2: Land earnings constant in real value, i.e., $i' = 0$			
4	6	10	4
Growth Stock Scenario 3: Land earnings increase in real terms, i.e., $i' = 3$			
1	9	10	4

3 percentage points are real capital gains. Land price is 100 times rent! The earnings from 100 acres are required to pay the mortgage interest on one acre. Although land appears to be overpriced, it is yielding a real return rate α.

The notable conclusion is that given scenarios 2 or 3, comparison of current return to land with interest rates or returns on many other investments will give the impression that farm owner-operators are not adequately compensated for their resources and motivates political pressure to raise commodity supports or lower interest rates. Although the impression of low real returns is an illusion, *the cash-flow problem is real.* Its intensity is proportional to the inflation rate.

The cash-flow theory of the farm problem helps to explain *financial stress* problems in the 1980s. Suppose the current earnings or rent R_0 is \$100 per acre and land is viewed as a growth stock with $i' = .03$, that is, real earnings are expected to increase 3 percent per year as in scenario 3 representing the 1970s. The present value of land is then \$100 ÷ (.04 - .03) = \$10,000 per acre if the desired real return is 4 percent. On the other hand if land earnings are expected to remain constant in real terms, then the appropriate land price is \$100 ÷ .04 = \$2,500 per acre. And if inflation is expected to be 6 percent and land is expected to behave like a bond so earnings remain constant in nominal terms, the equilibrium market land price is \$100 ÷ (.06 + .04) = \$1,000 per acre. Thus the shift in expectations that land earnings will remain constant in nominal terms rather than grow 3 percent annually in real terms decreases the land price from \$10,000 to \$1,000 per acre or by 90 percent!

Alternatively, if land earnings are expected to remain constant in real terms but real mortgage interest rates and desired real rate of return triple from the past expected levels, then the land price falls from \$100 ÷ .04 = \$2,500 to \$100 ÷ .12 = \$833 per acre, a drop of two-thirds. These two factors, declining expectations for land earnings and rising desired real returns to cover high real interest rates, explain financial stress caused by farmland depreciation in the 1980s. More will be said of this in Chapter 6.

DISCREDITED THEORIES

A discredited theory is that farmers supply a greater quantity of products in response to lower prices. This backward-bending supply curve theory is grounded on three assumptions: (1) the correct assumption that the demand curve faced by individual farmers is perfectly elastic, (2) the correct assumption that farmers have certain fixed obligations in the form of living expenses and payments on mortgages and other loans, and (3) the incorrect assumption that farmers can raise income to meet these fixed obligations by increasing output in response to falling prices (demand). If farmers had no variable costs and chose only to attain a gross income equal to their fixed obligations, then they would in fact expand output as prices fall and decrease output as prices rise. This behavior would give a backward-bending aggregate supply curve, a rectangular hyperbola with a price elasticity of -1.0. This specious reasoning ignores variable (marginal) costs, diminishing returns, and the profit motive. The concept is supported neither by economic theory nor empirical results, which show that the farm aggregate supply curve has a positive slope.

Suppose the farmer faces the perfectly elastic (horizontal) demand curve at p_0 and marginal cost is MC (Figure 4.6). MC slopes upward to the right (displays the principle of diminishing returns) because producing more output requires use of less productive soils, stretched managerial capabilities, or reduced timeliness of operations. The equilibrium quantity is q_0 at price p_0 and producers surplus (return to fixed resources or "profit") is area 1 + 2 + 3. Now suppose that price falls to p_1. If the farmer cuts back output to equilibrium quantity q_1, net return will be area 1, the maximum possible. If the producer increases output to q_2 to cover fixed expense obligations, his net return will be only 1 - 4. He reduced net income by area 4 by increasing output in response to a lower price. Such action is irrational.

If the producer was originally producing q_1 when the price is p_2 he is foregoing net income of 3 + 5 attainable by producing at q_2. If the price falls to p_0 and he responds by increasing output from q_1 to equilibrium q_0, he will increase net income by area 3. So increasing output when price falls is rational only if the starting position was irrational.

A similar line of reasoning has been applied to the farm labor-supply curve which in theory could bend backward if the income effect overshadows the substitution effect. Laborers, it is said, adjust employment in inverse proportion to the wage, thus providing a fixed labor income to meet living expenses. The supply curve for labor would be backward sloping with an elasticity of -1.0 under these assumptions. Again this conclusion is not supported by empirical evidence, which shows that the labor supply quantity rises in response to higher wage incentives (cf. Heady and Tweeten, 1963, ch. 8). These two theories provide no explanation of farm problems.

A final explanation of the farm problem is the *imperfect competition theory* which states that a competitive farm sector sandwiched between two oligopolistic or monopolistic agribusiness sectors will earn low returns. This theory also cannot stand scrutiny and is dealt with in Chapter 8.

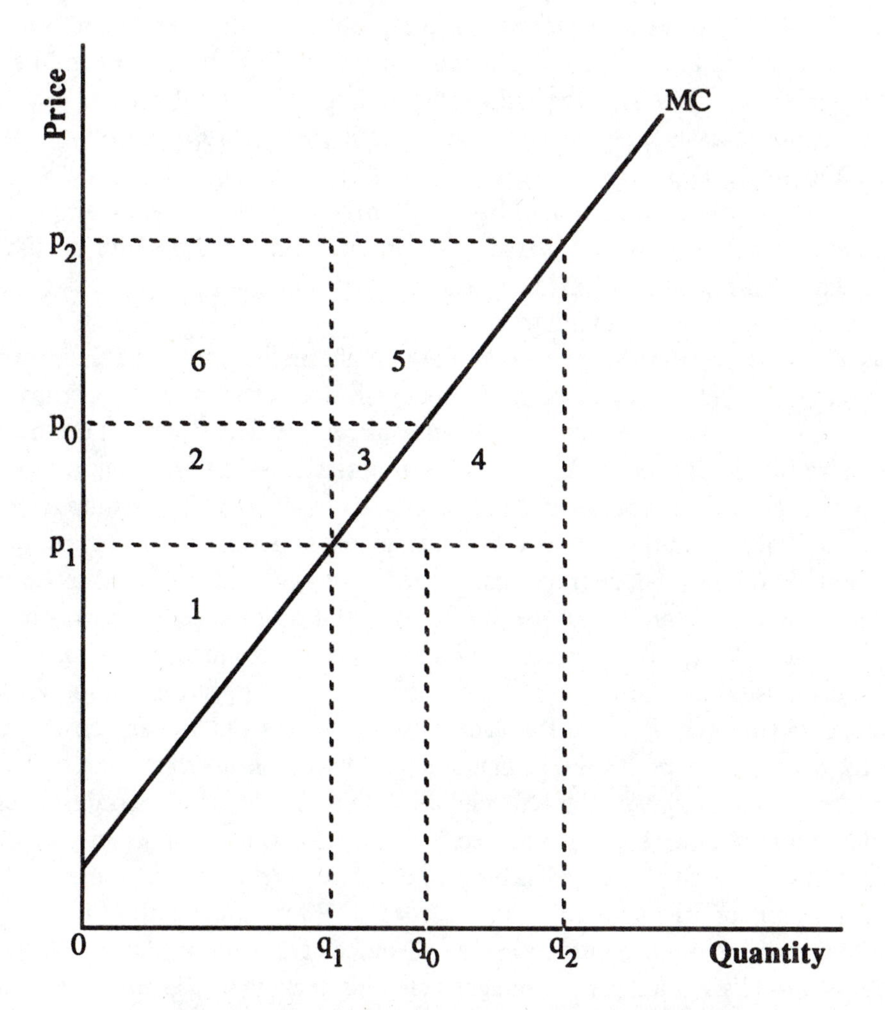

FIGURE 4.6. Illustration of Response to Price

SUMMARY AND CONCLUSIONS

Neither theory nor empirical evidence supports the hypothesis that commercial farms are chronically predestined to earn low returns in farming in the absense of government interventions. The farming industry demonstrates substantial resiliency in adapting to the uncertain environment in which it operates. This is not to deny that even well-managed commercial farms do not receive low incomes, low rates of return, and capital losses some years, and sometimes for several consecutive years. The treadmill theory and fixed resources theories of farm problems help to explain cyclical instability characterized by intermittent periods of low farm income and rates of return on resources. But neither those theories nor does the evidence support the widespread myth of persistent low returns in farming. Evidence is compelling that returns on resources are at least comparable to those elsewhere.

The conclusion of this chapter is that the problem of low rates of return on farm resources is not a problem of chronic *disequilibrium* as defined by the treadmill and fixed asset theories. Rather it is an *apparent problem* of chronic *equilibrium* as explained by decreasing-cost and cash-flow theories. It is quite normal for resource returns to be low for farms too small or poorly managed. Cyclical low returns and capital losses are a serious problem but one of instability that might justify occasional government interventions to avert unnecessary transitory adjustments. Farms with persistent low returns need farm resource adjustment assistance rather than a continual government presence to prop up income for all farmers.

The decreasing-cost theory advances the hypothesis that farm land, labor, and other resources are in equilibrium at the margin. Adequate-size farms are earning returns as high as the opportunity cost on resources. Because prices are determined at the margin (by adequate-size farms), it follows that resources are priced too high for majority of farmers who operate small, inefficient farming units. The marginal cost of farm output tends to be equal to marginal revenue. But with a decreasing cost structure, marginal cost lies below average cost. It follows that average cost exceeds average (and marginal) revenue. Thus when the average costs are multiplied by output, total cost exceeds total revenue in the farming industry. Average rates of return will appear to be low in equilibrium.

The decreasing-cost theory takes into account the size of farms and the valuation of real estate at the margin. Because the slack between returns and nonland costs tends to be absorbed by the land price, it follows that land prices will adjust to make returns equal to resource costs at the margin on the efficient-sized units. As land values adjust to changing conditions, economic-sized farms will receive an equilibrium (opportunity-cost) rate of return on resources within a considerable range of prices received by farmers. It follows that whether product prices are high or low, roughly 90 percent of all farms will receive low returns when the opportunity cost of all resources are taken into account. However, amenities of rural living and tax advantages compensate so that total *social* returns are not below equilibrium for most small farms. Farm bargaining power and government income-support programs offer only temporary reprieve. These programs raise both total cost and total revenue so that the unit cost and revenue curves remain somewhat stable as the benefits become capitalized into the control instruments used to raise prices.

The decreasing-cost theory of farm problems is related to technology, imperfect competition, and fixed resources. The ability to expand farm size is linked to the ability of farmers to purchase more land. And the ability to buy land and consolidate farms is linked to the rate at which a neighbor can find employment outside of agriculture. The process of adjustment to economic farming units is slowed if operators of inefficient units are unaware of better alternative uses for their resources, if education is inadequate to equip farm people for the exodus, if low-income farmers prefer farming as a way of life, and if work rules and high national unemployment inhibit mobility. Technology, reflected especially in farm machinery and farm management, has given the shape to the unit cost curve and is continually and slowly shifting it to the right.

Successful treatment of farm economic ills requires proper diagnosis. Annual and cyclical instability, the most serious problem of commercial agriculture, should not be confused with a nonexistent chronic low return problem. The treadmill and fixed resource theories of the farm problem help to explain cyclical low returns, instability, capital losses, financial stress, and the cash-flow squeeze. But reasonably well-managed, adequate-size farms which account for most farm output but for relatively few farms adjust to equilibrium rather quickly. Perennially supporting farm income with commodity programs to avoid low rates of return on farm resources not only is unneeded but also fails to maintain rates of return. Benefits are merely bid into land values and lost to the new generation of farmers. If commodity programs are deemed necessary to address effectively the very real instability, cash-flow, and other problems of commercial agriculture, the problems will need to be drastically redesigned.

One can argue that the findings of this chapter only indicate that government responds quickly to provide commercial farmers with adequate returns. Such a conclusion overlooks the equilibrium tendency of just over half of the farming industry not covered by commodity programs and also overlooks the massive resource adjustments that farmers and land market participants have made to changing real interest rates and exports.

Instability is not a prima facie case for government transfer payments to farmers. Restaurant owners, speculators in futures markets, Wall Street investors, and many other sectors of society face greater risks than farmers but receive no government subsidy. Neither can the subsidies and government distortion in farming be justified by fear of inadequate food supplies — there will be sufficient farmers under any foreseeable future conditions to avert food and fiber shortages and to provide workable competition. Nonetheless, the problem of instability is of such importance to commercial farmers and society that the next chapter is devoted to it.

REFERENCES

Braha, Habtu and Luther Tweeten. 1986. Evaluating past and prospective future payoffs from public investments to increase agricultural productivity. Technical Bulletin T-163. Stillwater: Agricultural Experiment Station, Oklahoma State University.

Brewster, John M. 1961. Society values and goals in respect to agriculture. Pp. 114-37 in Center for Agricultural and Economic Development, *Goals and Values in Agricultural Policy*. Ames: Iowa State University Press.

Cochrane, Willard. 1965. *The City Man's Guide to the Farm Problem*. Minneapolis: University of Minnesota Press.

Hathaway, Dale E. 1963. *Government and Agriculture*. New York: Macmillan.

Heady, Earl. 1949. Basic economic and welfare aspects of farm technological advance. *Journal of Farm Economics* 31:293-316.

Henneberry, Shida. 1986. A review of agricultural supply responses for international policy models. APAP Background Paper B-17. Stillwater: Department of Agricultural Economics, Oklahoma State University.

Hottel, Bruce and Robert Reinsel. 1976. Returns to equity capital by economic class of farm. Agricultural Economics Report No. 347. Washington, D.C.: Economic Research Service, U.S. Department of Agriculture.

Irwin, Scott and Norman Rask, eds. 1987. *Outlook Guide*. Volume 25. Columbus: Cooperative Extension Service, Ohio State University.

Johnson, Glenn and C. Leroy Quance. 1972. *The Overproduction Trap in U.S. Agriculture*. Baltimore: Johns Hopkins University Press.

Johnson, James, Mitchell Morehart, and Kenneth Erickson. 1987. Financial conditions of the farm sector and farm operators. *Agricultural Finance Review* 47:1-18.

Knutson, Ronald and James Richardson. 1987. Technology as a force of change. AFPC 87-25. College Station: Agriculture and Food Policy Center, Texas A&M University.

Melichar, Emanuel. 1985. The incidence of financial stress. (Paper presented at Agricultural Seminar, Congressional Budget Office.) Washington, D.C.: Board of Governors of the Federal Reserve System.

Office of Technology Assessment (OTA). 1986. Technology, public policy, and the changing structure of American agriculture. Washington, D.C.: Congress of the United States.

Parker, Leonard and Larry Langemeier. 1987. Preview, 1986 farm management associations. Manhattan: Extension Agricultural Economics, Kansas State University.

Quance, Leroy, and Luther Tweeten. 1972. Excess capacity and adjustment potential in U.S. agriculture. *Agricultural Economics Research* 24(3):57-66.

Ruttan, Vernon W. 1982. *Agricultural Research Policy*. Minneapolis: University of Minnesota Press.

Sonka, Steve, Robert Hornbaker, and Machael Hudson. 1987. Managerial performance and income variability for a sample of Illinois grain producers. (Mimeo). Urbana-Champaign: Department of Agricultural Economics, University of Illinois.

Tweeten, Luther. Fall 1986. A note on explaining farmland price changes in the seventies and eighties. *Agricultural Economics Research* 38:25-30.

Tweeten, Luther. 1983. Economic instability in agriculture. *American Journal of Agricultural Economics* 25:922-31.

Tweeten, Luther. 1981. Farmland pricing and cash flow in an inflationary economy. Research Report P-811. Stillwater: Oklahoma Agricultural Experiment Station, Oklahoma State University.

Tweeten, Luther. 1979. *Foundations of Farm Policy*. Lincoln: University of Nebraska Press.

Tweeten, Luther and George Brinkman. 1976. *Micropolitan Development*. Ames: Iowa State University Press.

Tweeten, Luther and Leo Mayer. Forthcoming. Long-term agricultural development and price and income policy. (To be published in essays honoring Earl O. Heady.) Ames: Iowa State University Press.

Tweeten, Luther and Leroy Quance. 1969. Positivistic measures of aggregate supply elasticities. *American Journal of Agricultural Economics* 51:342-52.

Tyner, Fred and Luther Tweeten. 1966. Optimum resource allocation in U.S. agriculture. *Journal of Farm Economics* 48:613-31.

U.S. Department of Agriculture. April 1987. Economic indicators of the farm sector: Production and efficiency statistics, 1985. ECIFS 5-5. Washington, D.C.: Economic Research Service, USDA.

U.S. Department of Agriculture. November 1986. Economic indicators of the farm sector: National financial summary, 1985. ECIFS 5-2. Washington, D.C.: Economic Research Service, USDA.

U.S. Department of Agriculture. November 1981. A time to choose. Washington, D.C.: Office of the Secretary of Agriculture.

CHAPTER FIVE

Farm Problems: Instability

As noted in Chapter 1, commercial farmers on the average are far from poverty and they have substantial net worth. Their major problem is annual and cyclical instability. Small farmers too experience instability from farming but for the majority that instability is buffered by off-farm income. Price and income instability reemerged as a major problem of agriculture in the 1970s and 1980s. The problem dates to the very origins of commercial agriculture but was obscured in modern times by low income problems of the 1930s, high farm prices in the 1940s, and government commodity programs of the 1950s and 1960s. Ability of commercial agriculture to adjust and earn adequate returns as noted in the previous chapter creates optimism for the long-term average level of farm income. But the food balance is expected to oscillate between excess supply and excess demand, creating instability in an economic environment of inadequate buffer stocks and other shock absorbers.

This chapter begins with definitions and measures of uncertainty and instability, then examines the relationship between instability and economic efficiency. The chapter concludes with analysis of buffer stock and other policies for dealing with instability.

DEFINITIONS AND SOURCES OF INSTABILITY

Following Frank Knight, we define *risk* as a situation in which outcomes are variable (random) but the distribution of outcomes is known. *Uncertainty* is also defined as a situation in which outcomes are variable, but the distribution is unknown or meaningless in the mind of the decision-maker. Because the degree of knowledge of the parameters of the distribution of future events ranges in a continuum from nearly perfect to highly imperfect, and because the decision-maker is likely to have *some* subjective concept of the distribution of outcomes, the distinction between risk and uncertainty is too blurred to be useful. The terms are used interchangeably herein.

The theory of games applied to decisions was prominent in the 1950s (see Walker et al., 1960). The analytical framework views decision in a two-way classification with outcomes established for each state of nature ("nature" including markets and weather) and possible action by the decision-maker as below:

Farmers' Alternatives	States of Nature	
	S_1	S_2
	(dollars of net income)	
A_1	10	50
A_2	30	20

The minimum gain if the farmer pursues risky alternative A_1 is $10 and if he pursues alternative A_2 is $20. A very cautious farmer expecting the worst from nature or who must have an income of at least $20 to survive financially pursues strategy A_2 because it maximizes his minimum gain, assuring him of at least a return of $20. If he assumes each state of nature is equally likely, then the expected gain from A_1 is .5($10 + $50) = $30 and from A_2 is .5($30 + $20) = $25, making A_1 the preferred action. Suppose that in fact the probability of S_1 is .4 and S_2 is .6. Then the expected gain from A_1 is .4($10) + .6($50) = $34 and from A_2 is .4($30) + .6($20) = $24. The less cautious farmer follows strategy A_1 to increase net income. The loss in net output is $34 - $24 = $10 over the long run by pursuing the cautious minimax strategy. Alternatively, up to $10 could be invested in outlook information and other means to perfect decisions of the cautious farmer.

The assumption thus far is that either A_1 or A_2 must be selected and followed consistently. If greater flexibility is built into the farming operation allowing shifts between A_1 and A_2 and coupled with accurate outlook information, then gains compared to those from the cautious strategies can be larger. If S_1 occurs .4 proportion of the time and S_2 .6 proportion of the time, then with an exact forecast gains could be as high as .4($30) + .6($50) = $42. This simple example illustrates costs of inaccurate forecasts and risk-averting behavior as well as potential benefits from measures to reduce risk in an uncertain environment.

Not all variation in the economic system is uncertainty or a source of economic inefficiency. A trend or cycle of outcomes which decision-makers anticipate and adjust to with a high degree of precision is not uncertainty. Changes in prices that induce resources and commodities to move to higher value uses enhance economic efficiency. Random events (and uncertainties associated therewith) creating loss of value of goods and services produced and consumed are of concern. Whether the direct loss of goods and services entails overall economic inefficiency requires further analysis. Measures to stabilize outcomes are pursued to the point where the additional gains from stabilization are just offset by the additional cost of measures to reduce variation in outcomes. Because stabilization devices are not costless, it does not pay to remove all variability. Some variation and losses from uncertainty remain even in an efficient economy.

Modern decision theory emphasizes Bayesian analysis with its subjective probabilities and prior information that enable the decision-maker to choose the course of action maximizing expected outcomes or utility (see Walker and Nelson, 1977, for review). In farm management research, modern decision theory stresses the utility function of the decision-maker and the technical production-marketing opportunities confronting him. The technical opportunities are portrayed as a frontier showing tradeoffs between expected value (E) and the inverse variance (V). In the typical case for farm management, the E-V frontier slopes downward to the right. This implies a tradeoff — average net income and output must be sacrificed to obtain greater certainty (less variation) in net income from one production period to the next. Because for the most part farm managers' utility functions exhibit positive satisfactions from high net income and certainty, their indifference curve also slopes downward to the right on the E-V graph. The utility maximizing strategy is at the point where their E-V indifference curve is tangent to the technical E-V frontier.

This point is generally neither maximum net income nor minimum risk. Because farmers sacrifice some output or net income for stability, aggregate farm output (and national output to the extent similar analysis applies to other sectors) is reduced by uncertainty. This conclusion does not necessarily mean that real national output and efficiency would be enhanced by programs to reduce instability. Gains from more output forthcoming by measures to reduce uncertainty may be more than offset by costs of the measures to reduce uncertainty. Nonetheless, a strong case can be made that the rate of return on investment in measures to reduce uncertainty is high, and reduction of uncertainty would increase efficiency of the nation's economy.

SOURCES OF INSTABILITY

Instability arises from nature and man. Man-made sources of instability to farmers include (1) business and inflation cycles, examined in the next chapter; (2) commodity cycles; and (3) domestic and foreign government actions such as export embargoes, price controls, environmental restrictions, and commodity price support or production controls.

Although government actions are sometimes designed to buffer instability arising from nature, the actions often add to uncertainty for agriculture. The rise of the consumer influence in U.S. farm and food policy creates uncertainty to farmers manifest in price controls and export embargoes. While such actions may stabilize prices, farmers who planned production based on a price rise will find their resource commitments and output inefficient in the light of realized prices.

The shift from fixed exchange rates to flexible exchange rates was a man-made institutional change. Flexible exchange rates permit short-term changes in terms of trade that influence the real price of U.S. farm commodities to foreign countries and cause fluctuations in export demand. While causing more short-term price fluctuations, flexible exchange rates dampen longer-term adjustments in export demand caused by periodic major changes in fixed exchange rates in response to chronic balance-of-payments problems. Flexible exchange rates, by dampening need for domestic monetary-fiscal policies to correct foreign exchange imbalances, may reduce the uncertainty from changes in domestic policies that effect markets for farm inputs and outputs. Thus the net effect on economic uncertainty for farmers from the shift to flexible exchange rates is difficult to judge.

Instability from nature arises from weather, disease, and from insects and other pests. Sometimes these natural sources have man-made origins. Man-made technology made corn plants over a wide area susceptible to the blight of 1970. Man-made increases in debt-equity ratios made farmers more vulnerable to financial crisis from fluctuations in economic outcomes from weather and macroeconomic policy.

Much has been written about the existence of weather cycles. While tree rings and other sources reveal long-term weather or climate cycles, meteorologists are unable to predict cycles with useful reliability. Evidence for short-term weather cycles is less apparent. Luttrell and Gilbert (1976) examined 1866-1970 annual yield data of several crops for the United States, individual producing states, and major producing areas. Statistical tests revealed almost no evidence of nonrandomness, bunchiness, or cycles except that accounted for by uneven application of high-yielding inputs such as fertilizer and hybrid seed.

Reduced world output of wheat and coarse grains in marketing year 1965-66 reduced world grain stocks by 35 million tons mainly because of a short crop in the Soviet Union and India. Stocks fell 44 million tons in 1972-73 mainly because of the short grain crop in the Soviet Union (Trezise, 1976, p. 1). The impact on world trade and prices was nominal in the first case but dramatic in the second case. The world's margin of cereal reserves at the *beginning* of the 1972-73 shock was not much larger than it was at the *end* of the 1965-66 shock. It was for this reason that prices hardly rippled in the former year and rose sharply in the latter years. Low stock reserves increase the vulnerability of world food prices to shocks. Holding of large reserves chronically depresses farm prices, however.

Firch (1977) found that the business cycle was a major source of the variance in farm receipts in the 1930-39 period but had almost no influence in the 1946-65 period. Farm output was the only notable source of variation in receipts in the 1956-65 period but export instability caused by changes in exchange rates and other factors was the single largest source of instability in the 1966-75 period. In the 1960s supply shifts were the principal source of variation in the aggregative supply-demand structure. In the 1970s and 1980s demand shifts principally due to exports became the dominant source of instability. Based on commodity-by-commodity analysis, Firch (p. 13) concluded "that stock changes generally have been very effective in stabilizing income relative to changes in production." Large stocks make supply highly elastic.

Highlights of a study of sources of instability include (Tweeten, 1983, pp. 930, 931):

1. A major proportion of annual variation in farm product utilization was accounted for by exports in recent years in contrast to the dominance by domestic utilization in earlier decades.

2. Relative and absolute variation in real marketings, real own-price, and real net income (from either farm, off-farm, or all sources) showed no consistent increase over time periods considered in this study despite instability introduced by exports. The destabilizing influence of exports on total farm income was buffered by a number of factors including (a) a rising price elasticity of demand, (b) decreased domestic utilization in response to greater exports and higher prices, (c) off-farm income, and (d) increased exports associated with increased production.

3. Government payments helped to stabilize net farm income but such programs probably slowed private initiates to reduce instability. That is one reason a rigorous test of

the hypothesis that government programs stabilize the farming industry was not possible from the study.

4. Commodity buffer stock requirements to stabilize excess demand and prices have increased in absolute value but only modestly in relation to utilization for the farming industry. Relative to utilization, buffer stock requirements to cope with a shortfall with specified probability have declined for wheat but increased for feed grains.

5. Net income from farm sources was relatively much more unstable on small farms than on larger farms; but total net income variance of small farms was low because of rather stable, sizable income from off-farm sources since 1970. The coefficients of variation of net farm income and of total income were not greater for large farms than for medium-size farms.

6. Relative variation in real net farm income from all sources appeared to be declining for the farming industry. A major reason is the stabilizing effect of off-farm income. Off-farm income is influenced by rural development and monetary-fiscal policy.

7. Major emerging sources (exports, input prices, and the general price level) of variation in net farm income are strongly influenced by monetary-fiscal and trade policies. A conclusion is that the traditional singular focus of agricultural interest groups on commodity programs needs to be broadened to encompass rural development, monetary-fiscal, and trade policies influencing the emerging sources of farm income instability.

UNCERTAINTY AND ECONOMIC INEFFICIENCY

A case is made in the following pages that uncertainty shifts the "average" supply and demand curves for agricultural products to the left. Given the average demand and supply curves, uncertainty also causes prices and output to deviate from equilibrium. Each of these elements is discussed below.

IMPACT OF UNCERTAINTY ON AVERAGE DEMAND

The case that instability reduces the demand curve for agricultural commodities rests on the concept of diminishing marginal utility (DMU) of income. DMU is often taken as axiomatic in economics, with passing reference to declining demand curves and diversified consumption by consumers as observed manifestation of unobserved DMU. More rigorous evidence of DMU was noted in Chapter 2.

Findings (Tweeten and Plaxico, 1974) that consumers are willing to pay slightly higher prices on the average to obtain more stable prices and that stock and bond buyers (Heifner and Mann, 1976) are willing to surrender between one-half and two units of average interest or dividend returns (rates) for each unit reduction in the standard deviation of return also lend support for DMU. Some agricultural economists (Officer and Halter, 1968; Lin, Dean, and Moore, 1974) found evidence from Bernoullian utility curves for

DMU among farmers, although conclusions are highly tentative because of small sample size and estimation problems.

Figure 5.1 illustrates how DMU relates to diminished demand for a commodity. Although a range 0a of increasing marginal satisfaction from greater income may exist, the weight of evidence is on the side of DMU for most consumers as shown by the utility curve of income (above a) that increases at a decreasing rate.

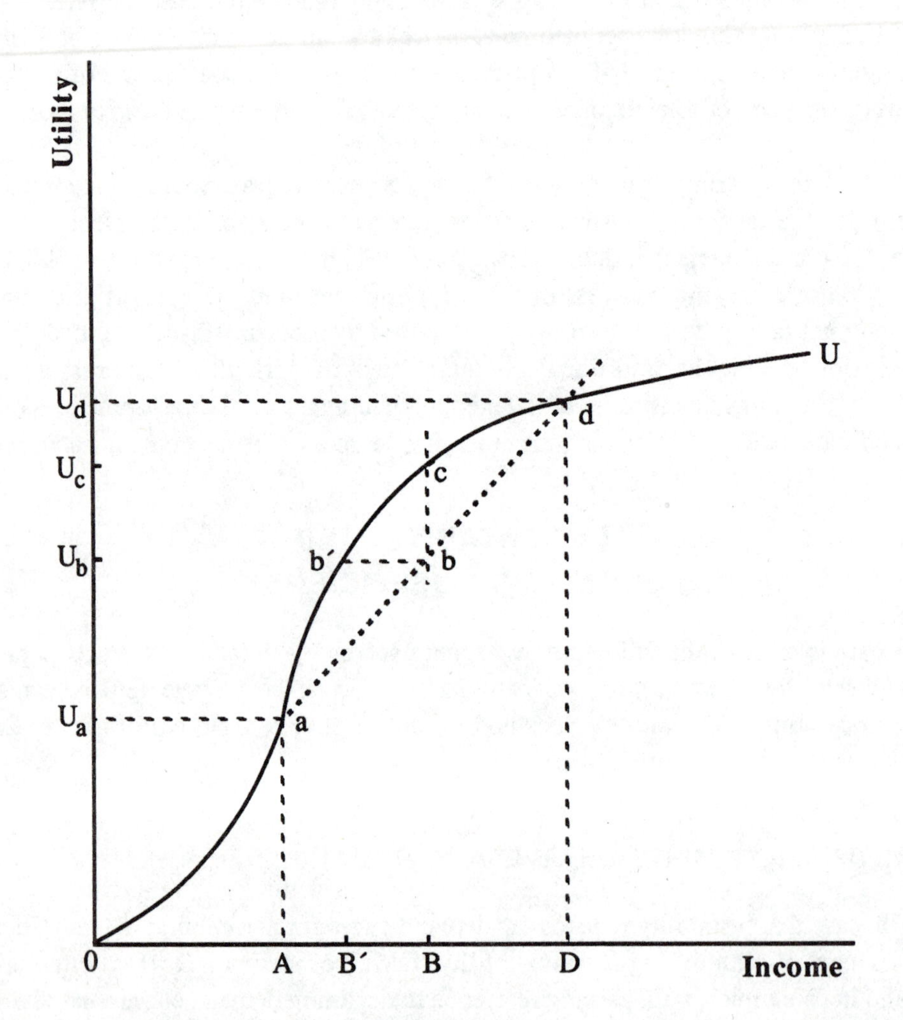

FIGURE 5.1. Hypothetical Curve Relating Income of a Representative Individual to Utility

Fluctuating food prices influence real income. Higher food prices reduce real income because dollars buy fewer goods and services. Assume that food prices vary randomly from a high price reducing real income to A and a low price increasing real income to D. Events A and D bring respective utilities U_a and U_d. Expected (average) total utility U with A and D occurring randomly with probability p and 1-p respectively is $U = pU_a + (1 -$

p)U_d. The locus of utilities associated with probabilities of A ranging from 0 to 1 is shown by the dotted straight line ad. If the probabilities of A and D are each .5, then average utility is U_b and average income is B. If B were received with certainty all the time, utility would be U_c. Hence U_c - U_b of utility is sacrificed because of fluctuating food prices. The representative individual could pay a risk premium up to B - B' = b - b' on a program to stabilize food prices and be as well off with certainty-equivalent income B' as with risk income B.

Consumers maximize utility if the marginal utility per dollar's worth of food consumed equals the marginal utility per dollar's worth of nonfood items consumed. Loss in utility from instability in food prices causes consumers to shift consumption to items with less price variability, reducing the demand for food.

IMPACT OF UNCERTAINTY ON AVERAGE SUPPLY

We now turn to uncertainties that cause the average supply curve for agricultural output to be shifted upward and to the left. These uncertainties arise mainly from unpredictable variation in prices and yields at the firm level. Compared to a situation of certainty, these uncertainties cause loss in output or excessive use of inputs through choice of levels and combinations of (1) products, (2) inputs, (3) production processes, and (4) scale of firm. Farmers pursue strategies stressing diversification, flexibility, and liquidity to reduce risks, but such strategies reduce output from a given amount of inputs. Heady (1952, p. 740) asserts that, "Although the extent of this force has never been measured quantitatively, it is likely that inefficiency [in farm production] growing out of economic uncertainty is more important than any other single source of inefficiency."

Given a fixed quantity of resources that can be devoted to production of commodity A or B, the total variance σ_T^2 in income from the combined enterprise is defined as

$$\sigma_T^2 = q^2\sigma_A^2 + (1 - q)^2\sigma_B^2 + 2\rho q(1 - q)\sigma_A\sigma_B,$$

where q is the proportion of income from A, 1 - q is the proportion from B, σ_A^2 and σ_B^2 are variances in income from enterprises A and B, and ρ is the correlation coefficient between income from A and B. If incomes from A and B are independent so that $\rho = 0$, and half of income comes from A and B, then the above equation reduces to

$$\sigma_T^2 = .25\sigma_A^2 + .25\sigma_B^2.$$

Total variance in income will drop by adding B until equal proportions of income come from A and B (compared to only income from A) if the variance of B is less than three times the variance of A. That is, the variability of income is substantially reduced by diversification despite adding an enterprise B that by itself has more variability than the original enterprise. The variance of total income is reduced even more if income of the two enterprises is negatively correlated. Thus impetus exists for farmers to diversify, but that diversification tends to reduce net farm income. A Midwest farm yielding highest income over time with cash grain and hog production might include a dairy operation to reduce variation in income. Other farmers diversify by off-farm employment, but Chapter 4 indicated that full-time farmers produce more efficiently than part-time farmers.

In addition to diversification, farmers also seek flexibility and liquidity to cope with uncertainty. To obtain flexibility in changing enterprises from year to year, a farmer may forego large investments in specialized, efficient beef or pork production facilities. He may rent equipment and hire custom harvesters. A strategy to reduce risk through flexibility tends to induce inefficient input ratios such as excessive labor relative to capital. Internal and external capital rationing caused by uncertainty regarding ability to repay loans accentuates this tendency. To assure ability to repay loans and other fixed expenses, financial liquidity may also be increased by uncertainty. Excessive cash reserves reduce efficiency because these reserves could be invested in productive and profitable inputs in the absence of uncertainty.

Many of the above inefficiencies would exist even if prices could be accurately predicted. But these inefficiencies are accentuated by inability to predict prices with accuracy.

Figure 5.2 shows foregone output or wasted resources at the firm level arising from inability to predict prices even though average price expectations are correct. Inefficiencies illustrated in Figure 5.2a to 5.2d with normal shapes for the technical curves emerge respectively from:

5.2a — improved combinations of product A and B from a given amount of resources, production processes, or techniques;

5.2b — imperfect combination of resources X and Y used to produce A;

5.2c — imperfect levels of input X used to produce A; and

5.2d — imperfect levels of product A.

To save space, all components of Figure 5.2 are discussed as one example, although each component may in fact depict very different levels of price instability and inefficiency. Furthermore, the firm supply curve associated with the marginal cost curve in 5.2d is moved to the left by several influences — by inefficient combinations of inputs X and Y and by production such as at t and r rather than at s.

Suppose the actual realized price ratio at harvest is tangent at s to the respective curves in 5.2a, 5.2b, and 5.2c and at the price level s in 5.2d. Farmers, however, do not know this price when planning production. Assume farmers accounting for half of all output (or input as in 5.2 b) expect the price ratio or level to be at t and accordingly produce $a_t b_t$, use input combination $x_t y_t$, input-output combination $a_t x_t$, and output a_t. Farmers accounting for the other half of all output expect the price ratio or level at r, giving rise to combinations and levels of inputs and outputs indicated at point r in Figure 5.2. Average output or input is at point u. If all farmers would have correctly anticipated prices and produced to maximize profit, input and output would have been at or near point s. Thus s - u is the approximate cost of price uncertainty in forgone output or wasted resources at the firm level. When firm supply curves are aggregated to form the national industry supply curve, the result of uncertainty is an average supply curve shifted upward and to the left in comparison to the industry supply curve in the absence of price risk.

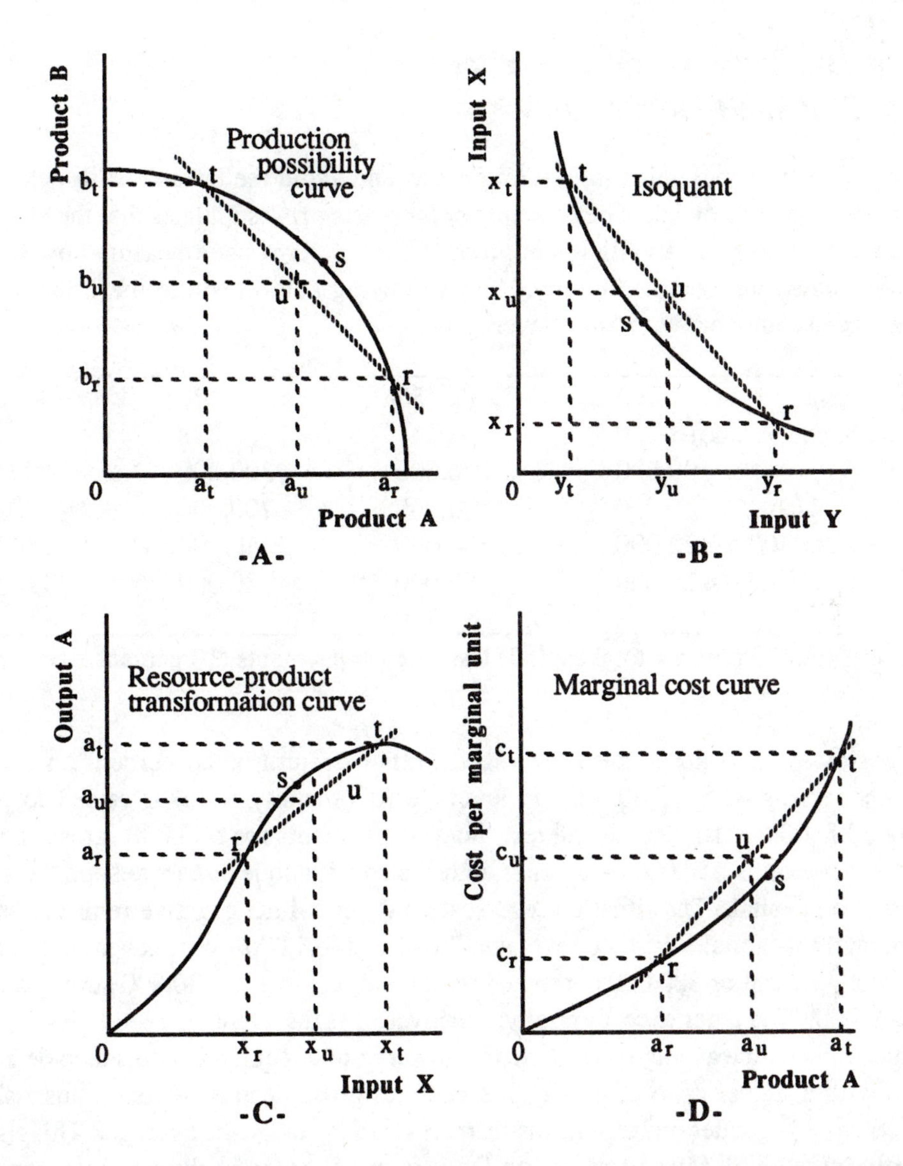

FIGURE 5.2. Illustrations of Costs of Uncertainty in Farm Production at the Firm Level

Just (1974) investigated the importance of risk in decisions concerning field-crop supply response in California. Results indicated that stabilization associated with government programs and other sources might have seriously offset the acreage-reducing effects of voluntary acreage restrictions. While such results suggest that reduced risk increases farm output, it reveals little or nothing about whether that additional output comes from application of more resources or from applying existing resources more efficiently.

Other elements of uncertainty discussed below also reduce supply.

Impact of Uncertainty on Scale of Farms and on the Farming Industry

If farmers are to maximize long-term returns and avoid the trauma of foreclosure, survival of the firm is essential. The *principle of increasing risk* stipulates that the chances of insolvency increase from a variation in product price as farm size (measured by owned assets) is expanded with a fixed equity. The following data illustrate the principle of increasing risk as equity of $100,000 is leveraged:

Equity percentage	100	50	33	25
Leverage (debt-equity ratio)	0	1	2	3
Total capital ($)	100,000	200,000	300,000	400,000
Borrowed capital ($)	0	100,000	200,000	300,000
Net returns @ 20% ($)[a]	20,000	30,000	40,000	50,000
Net returns @ -20% ($)[a]	-20,000	-50,000	-80,000	-110,000

[a]Gross returns (rate of return x total capital) less interest payments (10 percent x borrowed capital).

With no borrowing and hence zero leverage, a rate of return of 20 percent on capital provides net returns of $20,000. If the leverage (debt/equity ratio) is raised to 3 by borrowing $300,000 at 10 percent interest, interest payments are $30,000, gross returns $80,000 and net returns $50,000 — sharply higher than with no leverage, despite the same rate of return on capital. The situation reverses, however, with a negative return (loss) of 20 percent on total capital. With no leverage, the loss is $20,000. With leverage equal to 3, the interest payment of $30,000 combined with the direct loss of $80,000 sum to a total loss of $110,000. The latter exceeds equity; the investor is insolvent.

Large farms on the average operate with a lower ratio of equity to assets than do small farms and with a higher ratio of production costs to gross farm receipts. Thus risk of financial collapse is greater on large farms than on small farms on the average. This risk is offset partly by the high rates of return on large farms as noted in the previous chapter. Tweeten and Schreiner (1970, p. 54) showed that the capitalized value of reduced income that would have followed termination of government programs in 1965 (farm receipts were estimated to drop by 10 percent in 1965) would have been greater than average equity on each economic class of farms with sales in excess of $10,000. These data not only highlight the vulnerability of commercial farms to price and income fluctuations but also point to public policy as a source of uncertainty.

Vulnerability to economic setbacks is particularly acute for young farmers, many very efficient but operating with high debt-equity ratios. Ordinarily, unfavorable economic conditions in an unstable economy are expected to weed out the inefficient producers, but to the extent that price variability causes a preponderance of farm liquidation among young, efficient farmers rather than less efficient established farmers, it is a source of inefficiency.

Another source of inefficiency arises to the extent that farms are restrained in obtaining optimal size because of such uncertainties.

After successive years of favorable farm prices, farmers expect continued favorable prices and make investments in land, machinery, and capital improvements accordingly. Because these investments are specialized to farming, they are not released to the nonfarm sector when low prices occur (see Johnson and Quance, 1972, for a useful theoretical and historical perspective). These resources become fixed to agriculture. Production continues to be high even in years when the market signals through low prices that less output is desired. The result is considerable financial hardship to farmers who are making high interest and principal payments relative to earnings. Compared to a situation of costless flexibility in resource use, commitment of durable assets causes inefficiency apparent in too little input use and output in years of favorable prices and too much input use and output in years of unfavorable prices.

External capital rationing by lenders and internal capital rationing by borrowers often limit investments in risky enterprises and activities giving favorable rates of return. Because the risk and required rate of return may be much less when viewed from the standpoint of society than when viewed from the standpoint of a financially vulnerable firm, the result can be private investment below a socially optimal level. External capital rationing is likely to be less for durables providing collateral that can be confiscated by the lenders in case of default than for investments in human resources and in operating inputs that are used up in the production process. Underinvestment may occur in very long-term capital investments such as land-forming because long-run prices are discounted heavily for risk by borrowers and lenders.

The argument that instability reduces investment in agriculture below optimal levels is not entirely convincing, however. If farmers consume out of their expected permanent income and save and invest out of remaining, transitory income, variable income may result in a smaller permanent component used for consumption and a larger residual component that is saved and invested in farming. Furthermore, the fear of financial failure at a later time may increase internal saving and investment to enhance chances of firm survival.

Farm land is often viewed by society as a hedge against inflation. Economic instability from business cycles and inflation felt by agriculture along with other sectors of the economy may cause nonfarm people to shift capital to agriculture, particularly land, as a secure investment in "indestructible" assets. In short, instability can cause over- or underinvestment in the agricultural industry. It can cause excessive bunching of investment among time periods and too many small, inefficient farms. A compelling case can be made that uncertainty reduces output per unit of input for the farming industry. The impact of uncertainty on aggregate average input volume cannot be determined, however.

IMPACT OF UNCERTAINTY ON PRICING EFFICIENCY

Farmers' price expectations do not average the actual price as assumed in Figure 5.2 but instead the mean expectation falls above or below the actual prices paid and received.

When farmers collectively form "naive" expectations based on past prices, the result is production cycles as apparent in livestock markets. The commodity cycle is an example of man-made instability. Because such cycles are a result of ignorance, one would expect cycles to disappear with improved methods of forecasting outcomes and presenting results to producers.

Figure 5.3 illustrates a two-period cycle arising from farmers' expectations that prices of the past production period will prevail for the commodity for which they are making production plans. The cycle may begin with the equilibrium price $\bar{p}$ and quantity $\bar{q}$ disturbed by bad weather, causing output to be q_L and price p_U. Expecting the price next

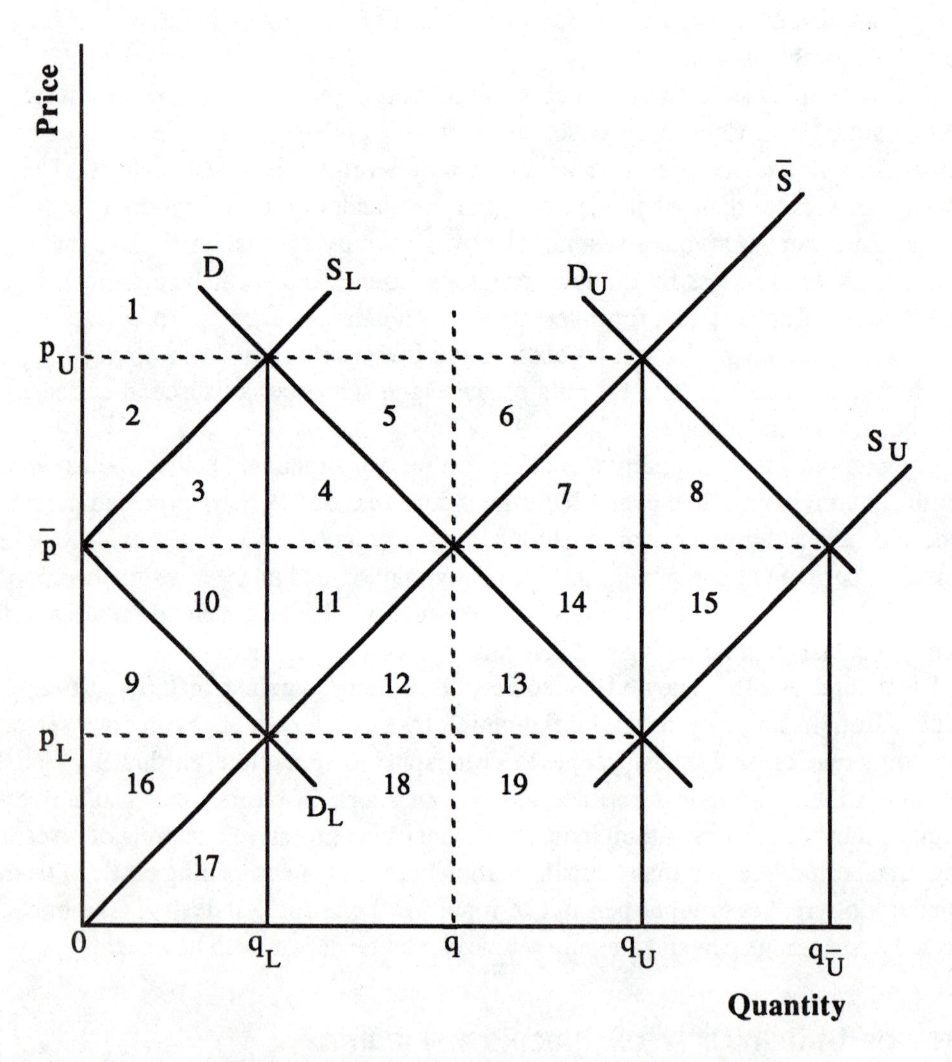

FIGURE 5.3. Hypothetical Aggregate Demand and Supply Curves for a Farm Commodity to Illustrate Impacts of Uncertainty

year to be the current price p_U, farmers produce q_U, given the normal industry supply curve $\bar{S}$. This results in a low price p_L along industry demand curve $\bar{D}$. The low price is expected next period by producers, so they produce q_L in the next period and this brings a price p_U, given normal demand $\bar{D}$ and supply $\bar{S}$. Because the supply and demand curves in Figure 5.3 have the same slope, the cycle is continuous — it does not dampen or explode. In theory, if the slope of the supply curve is steeper than the slope (absolute value) of the demand curve, the cycle will eventually converge to equilibrium $\bar{p}\bar{q}$. The length of the cycle depends on the time required to adjust production. For beef production, the period from conception to consumption is long and the cycle is approximately ten years. The cycle for hogs is approximately four years and for crops is approximately two years. The social cost (value of goods and services foregone) is area 4 + 11 in Figure 5.3 when output is low and is 7 + 14 when output is high, hence on the average is .5(4 + 11 + 7 + 14).

Countercyclical buffer stock changes and improved crop and livestock price forecasts can reduce this social cost. Because provision of information and buffer stocks is costly, the optimal solution is not to totally eliminate cycles but rather to expend additional outlays for stabilization to the point where such outlays are just offset by social benefits from stabilization at the margin. A social rate of return is obtained from more accurate statistical reporting of crop and livestock output only if the adjustment cycle is converging (Hayami and Peterson, 1972, p. 124). Because of no real world evidence of diverging cycles, the assumption that "ignorance is bliss" is rejected.

Hayami and Peterson (1972) estimated the marginal social benefit-cost ratio for reduction in sampling error in crop and livestock estimates made by the Statistical Reporting Service (SRS) of the U.S. Department of Agriculture. Based on a multiple frame sampling technique, the SRS goal is to reduce sampling error to 2 percent. Reducing the sampling error from 2.0 to 1.5 percent would bring an estimated $106 of benefits for each dollar of sampling cost. The economically optimal sampling error, defined as the point where an additional dollar spent on sampling accuracy would be offset by an additional dollar gain in benefit, occurred for an average sampling error of less than .5 percent. The implication is that more funds can profitably be spent to improve crop and livestock reporting unless funds are limited and buffer stock operations, research, or other activities bring even higher returns per dollar of outlay. The foregoing benefit-cost ratios are for improved estimates of *existing* crop acreages and livestock numbers at a point in time. Benefit-cost ratios for improved *forecasts* have not been computed with useful reliability.

In summary, some variability in economic outcomes is essential to efficiently allocate products and resources in response to changes in the economic system. But the weight of theory and empirical evidence is on the side of net economic losses from instability and uncertainty. More variability exists in the economy than necessary for efficient production of goods and services and the rate of return is favorable on some measures to reduce this variability. One such measure, a buffer stock policy, is discussed at length below.

SOCIAL BENEFITS AND COSTS OF STABILIZATION WITH BUFFER STOCKS

A case was made in the foregoing pages that uncertainty reduces average demand and supply. We now turn to costs of fluctuating supply and demand curves and benefits of buffer stock policies to stabilize markets. Buffer stocks increase economic efficiency (reduce social cost) in a direct way by acquiring reserves in years when a commodity has low marginal value for release in years when a commodity has high marginal value in consumption — benefits illustrated in this section. But by stabilizing price, buffer stocks also reduce uncertainty at the farm and consumer level, thereby reducing social costs by shifting average supply and demand curves forward toward the position they would hold in the absence of uncertainties discussed in previous sections. Buffer stocks are held to bring stability *among* production periods in contrast to *seasonal* stocks held to bring stability *within* a production period, evening out consumption from one harvest to the next. The classical concept of producers surplus (net returns to producers' fixed resources) and consumers surplus (value consumers are willing to pay for a commodity over what they are required to pay) is used to measure benefits and costs of stabilization in Figure 5.3.

PRODUCERS SURPLUS WITH FLUCTUATING DEMAND

With an upward sloping and stable industry supply curve $\bar{S}$ and demand shifting randomly with probability .5 each for D_L (with price p_L) and D_U (with price p_U), producers gain from an unstable market (Just, 1975). With p_L, producers surplus on the average is area 16. With P_U, the producers surplus is 2 + 3 + 4 + 5 + 6 + 9 + 10 +11 + 16. With variable price, producers surplus on the average is .5(2 + 3 + 4 + 5 + 6 + 9 + 10 + 11) + 16. With stabilization of supply quantity $\bar{q}$ through storage, the price is $\bar{p}$ and producers surplus is area 9 + 10 + 11 + 16. Because producers' loss 2 + 3 + 4 + 5 + 6 from stabilization when demand is D_U is obviously greater than the gain 9 + 10 + 11 from stabilization when demand is D_L, it follows that producers surplus is greater with variable prices. They will prefer instability to raise net incomes. This conclusion assumes that farmers are not averse to risk per se and that they can predict price and adjust immediately to it. Tisdell (1963) has shown that if prices cannot be predicted, then producers surplus will be no greater with variable than with stabilized prices. It is also notable that $\bar{q}$ is that quantity giving the greatest producers surplus if producers hold output constant in the face of inability to predict price.

NET BENEFIT TO SOCIETY FROM BUFFERING FLUCTUATING DEMAND

Because gains to consumers can be losses to producers, it is well to examine the *net* impact of stabilization on society. First consider a fixed supply curve $\bar{S}$, demand varying randomly with probability .5 each for D_L and D_U, and a costless storage program that holds price at $\bar{p}$ in Figure 5.3. If price is increased from p_L to $\bar{p}$ through storage when D_L

occurs, producers gain 9 + 10 + 11 and consumers lose 9. Net gain to society is 10 + 11. If price is decreased from p_U to $\bar{p}$ through storage when D_U occurs, producers lose 2 + 3 + 4 + 5 +6 and consumers gain 2 + 3 + 4 + 5 + 6 + 7 + 8. Net gain to society is 7 + 8. The average net gain to society is .5(7 + 8 + 10 + 11). It would be zero if supply were perfectly elastic. Although as noted earlier, producers lose and consumers gain from stabilization with a fluctuating demand curve, the existence of net gains to society means that consumers can compensate producers for losses and both groups can be better off with stability. Stability is achieved by storing quantity $\bar{q}$ when demand is D_L and releasing $\bar{q}$ from storage when demand is D_U. Thus a storage policy minimizing social cost with variable demand actually increases variability in the demand quantity from q_Lq_U to $0q\bar{U}$! However, supply quantity is stabilized at $\bar{q}$. If the assumption of costless storage is relaxed, the allowed variation in demand quantity is decreased and in supply quantity is increased.

Net Benefit to Society from Buffering Fluctuating Supply

Now consider a fixed demand curve $\bar{D}$, supply varying randomly with probability .5 each for S_L and S_U, and a costless storage policy that holds price at $\bar{p}$ in Figure 5.3. It can be shown (see Just, 1975) that if the demand curve is fixed at $\bar{D}$ and supply shifts randomly with a probability of .5 from S_L to S_U, then consumers surplus will be greater with instability rather than a stabilized quantity $\bar{q}$ through a storage program. Again, gains to consumers are eroded if prices cannot be predicted and adjustments cannot be instantaneous.

If price is increased from p_L to $\bar{p}$ through acquisition of stocks when S_U occurs, producers gain 9 + 10 + 11 + 12 + 13 + 14 + 15 and consumers lose 9 + 10 + 11 + 12 + 13. Therefore, the net gain to society is 14 + 15. If price is decreased from p_U to $\bar{p}$ through release of stocks when S_L occurs, producers lose 2 and consumers gain 2 + 3 + 4. Hence, the net gain to society is 3 + 4. The average net gain to society over the long run is .5(3 + 4 + 14 + 15). Although consumers lose and producers gain from stabilization when supply varies, the presence of net gains to society means that producers can compensate consumers to make both groups better off. It is interesting to note that this storage policy which minimizes net social cost (maximizes real value of goods and services produced and consumed) increases variability in supply quantity from q_Lq_U to $0q\bar{U}$ while stabilizing demand quantity at $\bar{q}$. Relaxing the assumption of costless storage, higher storage costs reduce the optimal quantity stored and variation in supply quantity compared to the situation depicted above. If demand were perfectly elastic, the net gains from storage in the face of fluctuating supply would be zero. If supply is more elastic than demand for agricultural commodities, it follows that, other things equal, buffer stocks that dampen fluctuating supply are more vital than those that dampen fluctuating demand.

NET BENEFIT TO SOCIETY FROM BUFFERING FLUCTUATING SUPPLY QUANTITY

In reality, costs of producing a given supply quantity are unlikely to be known in advance, hence the entire supply curve is unlikely to shift randomly as above. Taking the other extreme, the planning supply curve (marginal production cost) is fixed at $\bar{S}$ but supply quantities q_L and q_U occur randomly with probability .5 resulting in prices P_U and P_L, respectively. Producers do not know which quantity or price will occur and are assumed to plan production based on expected price $\bar{p} = .5(p_L + p_U)$. The same total variable costs occur each year whether q_L or q_U is the outcome — the same costs that occur when price and quantity are stabilized at $\bar{p}$ and $\bar{q}$. Compared to stabilization at $\bar{p}$, consumers' loss is 2 + 3 + 4 with P_U and the gain with P_L is 9 + 10 + 11 + 12 + 13. Because each price and associated consumers surplus occurs with equal frequency, it is apparent that consumers' gains from the low price are greater than losses from their high price. Therefore consumers are better off without stabilization when supply quantity fluctuates randomly about the planned output $\bar{q}$.

The same conclusion does not hold for producers. Producers surplus is 2 + 3 + 9 + 10 + 16 - 12 - 18 when p_U occurs and is 16 + 19 - 12 when p_L occurs. If producers act as the buffer stock agency, acquiring quantity $q_U - \bar{q}$ of stocks with q_U and releasing $\bar{q} - q_L$ of stocks with q_L of production, then quantity and price would be stabilized at $\bar{q}$ and $\bar{p}$ respectively. Producers surplus would be 9 + 10 + 11 + 16. By manipulating geometrically, given the straight-line demand curve $\bar{D}$, it is apparent that stabilization increases producers surplus by 11 + 12 + .5(9 + 10 - 2 - 3 + 18 - 19) — producers could pay this amount to achieve stabilization. This outcome depends on the slope of the demand curve and in no way on the slope of the supply curve. Consumers' gain from stabilization is .5(2 + 3 + 4 - 9 - 10 - 11 - 12 - 13), which is negative — consumers lose from buffer stock operations. Net social gain, found by adding consumers' and producers' gains from stabilization, is .5(4 + 11 + 12 + 18 - 13 - 19) and is positive. With a straight-line demand curve, net social gain is .5(11 + 12) or .5(4 + 13). Producers can compensate consumers for accepting a buffer stock policy and both be better off. Unlike the previous case of shifting supply, an elastic demand will not eliminate gains from storage with fluctuating supply quantity.

When the entire supply curve shifts randomly but production costs and output can respond to price along the supply cure, the net social cost (potential benefit from buffer stocks) is .5(7 + 8 + 10 + 11), as indicated earlier. When the supply curve is stationary but production shifts randomly from q_L to q_U as above, then production costs are sunk and the net social costs of instability (potential gain from buffer stocks) is .5(4 + 13). It does not depend on the slope of the supply curve. In reality, some production costs can be changed and supply quantity adjusted as the production season progresses. It follows that reality lies somewhere between the extremes of allowing only the supply curve to vary versus allowing only the supply quantity to vary. It is of interest that the net social gain from a buffer stock policy is only half as large when only supply quantity is variable as when the entire supply curve is variable despite the same variation in quantity q_L to q_U in both cases.

It is clear that assumptions concerning the ability of producers to change output and costs in response to fluctuating prices are critical in determining the appropriate social cost to minimize in an optimal stocks policy. The issue of length of run has received insufficient attention in stock policy research. Empirical studies (see Gustafson, 1958) devising optimal stock policies have been troubled by difficulties in specifying the appropriate social cost to minimize.

IMPLEMENTING A BUFFER STOCKS POLICY

In implementing a buffer stocks policy, one must confront such issues as stock acquisition and release rules, appropriate average stock levels, reliance on the public sector or private trade, and international buffer stocks agreements. Each of these issues is examined below.

OPTIMAL STORAGE RULES

It is apparent from the foregoing analysis that the appropriate storage policy depends on the source of instability. Tweeten et al. (1971) devised an optimal storage rule for linear supply and demand curves with fluctuations only in supply. The simple rule is that stocks Q acquired or released comprise a proportion k of the difference between the actual quantity produced q and the equilibrium quantity $\bar{q}$, or $Q = k(q - \bar{q})$. If storage were costless, k would be 1.0. As a rule of thumb, k can be set equal to 1.0 minus the ratio of cost of storage to equilibrium price. Because annual storage costs are approximately 20 percent of wheat price, the storage rule for wheat is $q = .80(q - \bar{q})$. If wheat production exceeds equilibrium production by 200 million bushels, then 160 million bushels are placed in storage; if wheat production falls short of equilibrium production by 200 million bushels, 160 million bushels are taken from storage and placed on the market.

If demand rather than supply fluctuations are the source of variation, the same rule can be applied to demand quantity rather than supply quantity. If demand falls short of equilibrium $\bar{q}$ by 200 million bushels, 160 million bushels are placed in storage; if demand quantity exceeds equilibrium by 200 million bushels, 160 million bushels are removed from storage.

In earlier times when fluctuations arose mainly from supply, the above formulas were somewhat satisfactory. In recent years, fluctuations arose at least as much from variation in demand. An optimal buffer stock policy as noted earlier calls for stabilization of supply quantity when demand is the source of instability and demand quantity varies even more with than without optimal storage. An optimal stock policy calls for stabilization of the demand quantity when supply is variable, and the supply quantity fluctuates even more with optimal storage. With actual variation arising from both demand and supply, it is apparent that the quantity stabilization rule becomes very difficult to administer. However, the price is stabilized whether variation arises from either fluctuating demand or supply. Thus *price rather than quantity is the preferred instrument in storage rules*. Optimal storage using the price rule (see Just, 1975) is similar to the quantity rule for linear supply and demand curves — stocks Q are released or acquired in proportion to the difference between actual price p and equilibrium price $\bar{p}$, or $Q = k(\bar{p} - p)$. If p is wheat price in cents per bushel, k is approximately 3, and $\bar{p} = 300$ per bushel, then $q = 3(300 - p)$. If p is 400, then 300 million bushels are removed from storage and placed on the market. A problem with this approach is that disequilibrium in the form of deviation of actual from equilibrium price may become large and troublesome *before* stock adjustments are triggered. Just (1975) suggests approaches to anticipate stabilization requirements and make stock adjustments accordingly *before* disequilibrium becomes large.

While the price rule is superior to the quantity rule because price is invariant to the source of instability, other rules have been proposed, partly because they are simpler to understand, interpret, and administer in a mixed market of private and public stocks. These rules usually take the form of acquisition of stocks by a public agency when price falls to some low threshold such as the loan rate and release of stocks when price exceeds the acquisition price threshold by, say, 50 percent. Such threshold formulas have been shown by Tweeten et al. (1971) to entail nearly as low social cost as optimal rules given earlier, especially in view of the fact that supply and demand curves are not linear and because the private trade also holds stocks to dampen fluctuations within the price band that does not trigger acquisition or release of publicly controlled stocks.

It may be argued that efficiency is better served by controlling income variability than price variability. If the demand elasticity for a commodity is unitary, then changes in output caused by weather and other uncertainties are offset by changes in price, leaving farm income more stable with variable prices than with fixed prices. In view of the inelastic demand for most farm commodities and the low correlation between output of a commodity on any given farm and the national output that, in conjunction with demand, sets price, it seems unlikely that farmers have more stable net incomes with variable prices. Also buffer stock policies designed to stabilize farm incomes may increase variation in consumer food prices. Johnson (1947, p. 232) contended some years ago that, "Except for the case of perishable crops, the interests of farmers as well as the whole economy are consistent with stabilizing the price and not the value of the output."

Private Versus Government Stocks

The foregoing discussion makes a strong case for commodity stock reserves as the main approach to maintaining economic stability for American farmers and consumers and to meeting emergency food needs. But a pressing issue is whether these stocks should be privately held or supplemented by government holdings.

First, consider the case for only privately held stocks:

1. The private trade maximizes profits by acquiring and releasing stocks such that gains in prices just equal storage costs in a competitive environment. Net social cost is

minimized. Private firms acting from the profit motive will act in the public interest to optimally stabilize price and quantity. Any losses from market failure to hold enough stocks because risks are large and discount rates excessive will be more than offset by losses from government interventions in markets. Such interventions have depressed farm prices for extended periods in the past.

2. Past government holding of stocks has precluded the private trade from performing its proper role of stabilizing markets with acquisition and release policies. The private trade is not subject to untimely release of stocks motivated by narrow political concerns. A majority of farm commodities have never had a government storage policy and many observers feel that these commodities have been supplied to consumers with acceptable price stability without government interference or subsidies by relying solely on private initiative.

3. Private holdings place burdens or benefits of stock costs and returns on firms and consumers and removes the costs from taxpayers. Because many farmers are opposed to government stocks and because taxpayers wish to hold down government spending, some contend that it is politically expedient for government to rely on other devises and to forsake government reserves to stabilize prices. Large U.S. government stocks can cause foreign nations to reduce stocks, thereby shifting costs of storage from these nations to American taxpayers.

4. Public knowledge of government stock acquisition and release policies allows foreign competitors to exploit the system, selling just below release price and depriving this nation of opportunities for expanding sales and foreign exchange earnings. Government stock operations make the U.S. a residual supplier in world markets.

Now consider the case of supplementary government stocks:

1. Compelling reasoning suggests that the private trade will not hold sufficient reserves to serve the public interest because risks are too great. Unavoidable political actions by the government such as export embargoes to punish adversaries or to hold down food prices reduce returns from holding stocks below those that would prevail under a competitive, efficient market. The government can afford to hold large reserves and take large risks which private firms are too small to cope with. Holding of stocks is expensive and capital rationing either internal to the minds of those holding reserves or external in the minds of unwilling lenders of capital to those holding stocks limits holdings below levels desired by society.

2. After a period of favorable grain prices such as from 1973 to 1975, farmers and the private storage trade will hold sizable stocks, but they do not have sufficient financial reserves to do so after years of less favorable returns.

3. Private holdings can accentuate price movements. When prices are rising, the private trade may continue to hold in anticipation of higher prices, thereby driving prices higher. When prices are falling, the private trade may panic, releasing stocks and further depressing prices.

4. Some publicly, perhaps internationally, held reserves may be necessary to a humanitarian response to emergency food needs around the world. The private trade will not hold stocks to meet needs of starving people who have no buying power.

5. The government is likely to "back into" a stock operation unintentionally because, when prices fall below loan levels, alternatives to the governments acquiring stocks are few indeed. Therefore, it is best to establish a stock policy regarding acquisition, release, and appropriate stock levels before this occurs rather than after government stocks are acquired.

6. Some evidence suggests that the marketing system operates with a rachet effect, raising food prices when farm prices rise but failing to lower food prices when farm prices fall. Stocks in excess of free market levels will reduce short-term farm price gains that contribute to inflation.

7. Food will be used in foreign policy as a diplomatic tool, however unpalatable to farmers. Large stocks are necessary to use this tool most effectively.

8. Some externalities may exist in commodity stock operations. Private and social costs differ. For the government the discount rate for stock holding costs is low because events average out over many years and large geographic areas. For private firms the discount rate is high and costs of holding stock are high because random events do not average out for small and short-lived operators. The private trade may be depended upon to hold reserves for normal exigencies of the market such as droughts occurring with somewhat predictable frequency. The private trade will not hold stocks for catastrophic events such as global wars, which occur rarely.

Optimal Buffer Stock Levels

Figure 5.4, showing seasonal average wheat prices for alternative levels of expected carryout of wheat at the end of the marketing year before the new crop is harvested, highlights the dilemma facing farmers toward grain stocks. Carryout in excess of 600 million bushels means price stability, which farmers like, but low prices, which farmers dislike. With expected carryout of 1 billion bushels, a 10-million-bushel reduction in wheat carryout due to greater exports or reduced production raises wheat prices only 1¢ per bushel. Carryout of less than 600 million bushels means high prices, which farmers like, but unstable prices, which they dislike. With expected carryout of 250 million bushels, a 10-million-bushel reduction in wheat carryout raises wheat price 10¢ per bushel. Of course, consumers prefer large stocks — providing that the costs of holding stocks do not offset the advantages of low and stable wheat prices. One study (Dunn, 1975) shows that the curve bends even more sharply than shown in Figure 5.4 at about 600 million bushels, but some of the sharp bend is due to the operation of price-support loan programs in the years for which the curve was estimated. A similar curve constructed for feed grains bends at approximately 40 million tons. Thus buffer stocks of wheat and feed grains of at least 56 million tons appear to be necessary for reasonable stable prices. More recent estimates are shown by Tweeten (1983).

It is of interest that these estimates of needed reserves are broadly consistent with results from more sophisticated analysis. The Task Force of the Council for Agricultural Science and Technology (1973, p. 13) states that, "Reserves of 60 million tons might be deemed appropriate, indicating storage costs of up to $800 million annually." Optimal

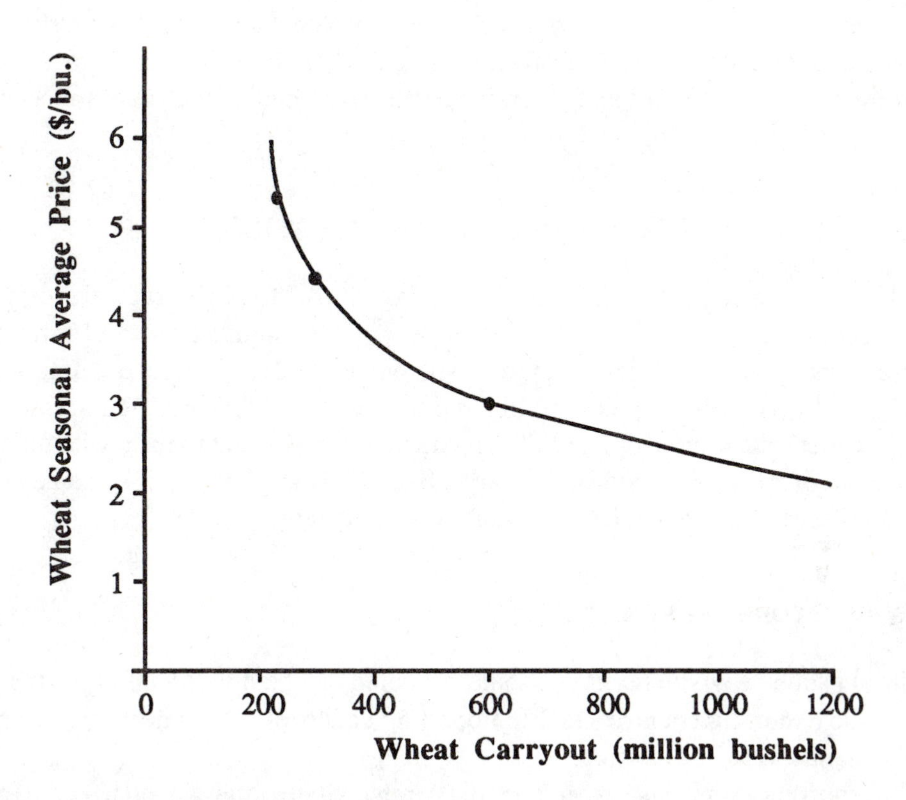

SOURCE: Anderson and Tweeten (1975, p. 4).

FIGURE 5.4. Relationship Between Wheat Price and Wheat Carryout, 1976 Price Level

reserves would be increased over time, but as suggested by the earlier theoretical analysis, the reserves should be increased in proportion to the rise in variance rather than the level of production and consumption. Trezise (1976, p. 17) estimates world reserve requirements to cover risk with 95 percent probability to be 30.5 million metric tons of wheat and 33.6 million metric tons of coarse grains by the year 1980 for a total of 64 million metric tons.

Cochrane and Danin (1976) estimate that an average reserve stock under a rule to keep grain prices within a range of 10 percent above the trend and 5 percent below the trend would require 73 million tons in 1985. Prices within a range of 20 percent above an below the price trend would require average reserve stocks of 39 million tons. Walker et al. (1976) estimated economically optimal buffer stocks required to keep wheat price with a range of $80 to $140 per metric ton and feed grain price between $70 and $120 per metric ton. With a policy so that buffer stock purchases equal buffer stock sales over a seven-year period, the mean value of buffer stocks is 18 million metric tons for wheat, 23 million metric tons for feed grains, and 41 million metric tons for combined grains.

By the late 1980s, U.S. grain buffer and pipeline stocks that are statically optimal (that would avoid a shortfall 99 out of 100 years) and economically optimal (that would minimize net social cost including carryover) were approximately 30 million metric tons of wheat and 60 million metric tons of feed grains — a total of 90 million metric tons. Actual stocks were well in excess of these desirable levels at the end of the year in most of the 1980s.

INTERNATIONAL FOOD RESERVES

It is notable that several estimates of desired world food reserves are very near estimates for U.S. reserves alone. Eaton et al. (1976) estimated world buffer stock requirements using a multi-objective linear programming model. Complete stabilization of the level of grain available for use would require a reserve of at least 172 million metric tons. A more realistic reserve of 58 million metric tons is sufficient to meet with reliability the 98 percent of the world trend supply target over the next 25 years. The same reserve size has no chance of meeting a 100 percent food security target.

WHERE TO STORE RESERVES

The question of where reserves should be stored can be divided into arguments for storage in the exporting countries and in importing countries. Arguments for storage in exporting countries are as follows:

1. Exporting countries, especially the United States and Canada, are likely to accumulate storage stocks as an effort to support grain prices, negating the need for an international reserve.

2. Storage in exporting countries is at least-cost approach — it is expensive and administratively cumbersome to deliver stocks to one country, then find they must be shipped elsewhere. The cheapest approach is to ship the commodities directly from the nation where they are produced to the nation where they are needed. Furthermore, several of the potentially food-deficit nations which require stocks are in humid tropical areas with inadequate storage facilities, causing higher costs of storage due to spoilage and pests than in temperate zones.

3. Storage in exporting countries avoids untimely release of stocks by governments which do not represent the rank and file of their people — characteristics found in several food-deficit countries. Stocks released prematurely might be very reluctantly replenished by donor countries.

4. Storing commodities in nations which are likely to be food deficit essentially destroys the international reserve system because commodities are placed under control of the local government. Opportunities for coordination, control, and financing are greater if commodities are stored in exporting countries.

Arguments for storing commodities in potential food-deficit countries are as follows:

1. This approach has been sanctioned by international agencies such as the Food and Agricultural Organization of the United Nations.

2. Commodities are readily available in emergency because they are located near consumers in food-deficit countries. Immediate supplies do not depend on political exigencies in exporting countries.

It appears unlikely that an international food reserve system will be established that provides storage potential needy countries. A more likely alternative is storage in counties which are generally net exporters with an international system controlled by several nations for finance, acquisitions, and release of the stocks rather than control either by donors alone or by recipients alone.

Many nations are committed to self-sufficiency in domestic food production. A more realistic goal is food security, defined as being assured of adequate food supplies from domestic or foreign sources. In many instances importation of some food and feed allows most efficient use of resources and highest living standards. Higher income allows countries to afford imports when local production fails. The appropriate U.S. stance is to build the reputation of being a reliable supplier and to encourage and strengthen use of the International Monetary Fund's cereal facility. The cereal facility addresses the instability problem by assuring less developed countries of financing for import of cereals when domestic crops fail or international prices soar.

OTHER OPTIONS FOR COPING WITH ECONOMIC INSTABILITY

Major alternatives to commodity stock to reduce instability include food price controls, export and import controls, trade liberalization, farm price supports, production controls, and direct payments (see Congressional Budget Office, 1976, for further exposition). Commodity programs emphasizing production controls, price supports, and direct payments are examined in greater detail in Chapter 12 and receive cursory treatment here. U.S. and world food information systems would be strengthened, with emphasis on improved crop and livestock reporting of production prospects and other data relevant to food issues.

PRICE CONTROLS

Food price controls seem unworkable except in the short run and in extreme national emergency such as major war. Price controls are counterproductive in the competitive agricultural industry, reducing output and thereby accentuating upward price pressures and shortages.

EXPORT CONTROL AND AGREEMENTS

Export embargos such as those placed on soybeans in 1973 and grain for the Soviet Union in 1975 and 1980 may restrain consumer prices but prompt importers to search for alternative supplies. The standard case is soybean expansion in Brazil, financed in part by our foreign soybean customers. The result is reduced U.S. food exports in the short and long run. Another approach is to make export agreements committing the U.S. to supply and customers to buy from us prescribed quantities of grain or soybeans. The impact of these on markets is unclear, but price volatility can increase as a result of leaving a smaller residual of "free" exports to determine price.

Among the factors that have for the most part deterred demands for export controls are: opportunities for importers of American farm products to go elsewhere for supplies, the central importance of maintaining access to world markets to earn reserves to purchase petroleum and other imports, and fear of reciprocal trade barriers.

Export controls or agreement with the Soviet Union or any other country should not be viewed as a substitute for other measures to promote stability. In years of short Soviet supplies, their import needs in excess of wheat the United States is willing to supply can be purchased in Western Europe, Canada, Australia, or Argentina. Customers normally purchasing from these countries but facing no U.S. embargoes or agreements can switch purchases to us. Or the Soviets can purchase soybeans, grain sorghum, barley, and oats rather than embargoed wheat and corn. In years of abundant Soviet supplies, their commitment to buy a prescribed minimum of grains annually can be circumvented by their selling of domestically grown wheat to other countries. Possibilities for deferred delivery and other means also reduce the effectiveness of grain agreements to stabilize markets. Export controls weaken traditional institutions for diminishing risks such as commodity futures markets, which are not prepared to handle uncertainties of the magnitude generated by government manipulations.

Comprehensive, effective export control entails substantial costs. It would require either a single public grain board to replace current private export firms or powerful controls over private firms that would make such firms essentially an arm of the state. Whether the current U.S. grain export system composed largely of private, mostly multinational, firms should be replaced by a single public corporation is an open question. While it is true that single state corporations predominate in major grain-exporting countries, the advantages over reliance on private firms is not clear. A single public corporation could have served the United States better in the seriously mismanaged grain sale to the Soviets in 1972. On the other hand, national wheat boards operating in foreign nations have frequently mismanaged sales. Under any circumstances, it is essential that the federal government monitor export sales, requiring prior approval for sizable sales. A national food policy with an adequate commodity reserve program can provide adequate stability without export controls. In other words, the cost of export controls (in foregone sales, ill-will, and the like) exceeds potential gains in the form of domestic price stability.

TRADE LIBERALIZATION

Much instability in world food prices stems from government interference in trade. Free trade would raise the price elasticity of demand for U.S. exports and dampen world price variability as noted in the next chapter. Broad-scale trade liberalization would tend to flatten the demand curve for food, hence a change in output due to weather would cause less price movement throughout the world. With assured access to world markets, the need for and cost of commodity buffer stocks would be lower. Freer trade would increase economic efficiency by encouraging nations to produce commodities in which they have a comparative advantage and is a desirable direction in multilateral negotiations with other trading nations.

PRICE SUPPORTS

In Western Europe and Japan, consumers were largely insulated by high domestic farm commodity price supports from world commodity price gyrations. In the European Economic Community high domestic food prices are maintained year after year by the variable levy system, which taxes farm commodity imports by the difference between the support price and the world price. This plan works best for food importers.

In the United States a similar program could be run along the lines of the McNary-Haugen proposal of the 1920s. All farm commodities which would not move into commercial domestic markets at high support rates would be purchase by the Commodity Credit Corporation and sold in the export market at the world price. Although high domestic prices would reduce variation in consumer food prices and hold down taxpayer costs, the procedure would severely strain another objective of policy — providing food at reasonable cost. Given the alternative of high food prices *all the time* under the above plan or high prices *some of the time* under a more market-oriented program, surely consumers would opt for the latter.

Past experience has demonstrated that voluntary production controls can stabilize crop acreages from year to year. But they are unwieldy instruments, controlling acreage rather than production. A quick response is impossible to low food inventories. Instability from volatile domestic yields and exports remains. Idled cropland constituted a valuable reserve to meet major world disasters such as wars and extended droughts in the past, but more ready reserves are required to dampen annual price volatility.

FORWARD CONTRACTS, HEDGING, AND INSURANCE

Several institutional devices discussed in Chapter 12 permit transfer of risk from one individual or group to another with greater capacity to bear risk. Months in advance of harvest, a wheat farmer can sign a forward contract with the local grain deal for a specified number of bushels of wheat to be delivered at harvest for a set price. The local grain dealer (or individual farmers) may hedge risk in commodity futures markets, thereby shifting the risk to speculators. Many farmers insure crops against natural disaster such as hail or drought through private firms and the Federal Crop Insurance program.

The important conclusion from the standpoint of public policy is that forward contracting, hedging, and insuring only nominally alleviate problems resulting from the lost value of goods and services to society because of instability — the devices only shift the burden from one group or individual to another. Such institutional devices are highly desirable and useful for pecuniary stabilization, but commodity storage and other instruments to adjust commodity stocks and flows are required for real stabilization.

SUMMARY AND CONCLUSION

Instability and uncertainty constitute a potentially huge social cost to society. The payoff is great from policies to reduce uncertainty. Some variation in prices is essential for efficiency in a dynamic economy resource and product allocations shift in response to changing weather, technology, tastes, preferences, and institutions. But prices have frequently gyrated excessively between boom and bust, with attendant problems and costs to farmers and consumers. Carefully formulated public policies can reduce uncertainty and produce benefits at the farm and retail level in excess of costs. Problems arise in determining the economically desirable limits of such programs, however.

Whether operated solely by the private trade or supplemented by government a commodity stock reserve constitutes the heart of a national food policy to stabilize markets. In the absence of commodity reserves, farmers could receive stable prices and incomes by transfer payments from nonfarmers, consumers could receive stable prices by restricting exports, or foreign customers could receive stable supplies if domestic consumers tailor their food use to absorb all the adjustments in farm output. If adequate reserves are maintained, thus making possible the maintenance of more stable total supplies from year to year, such distasteful alternatives can be avoided. Such an approach also can benefit livestock feeders, who can adjust to high grain prices or low grain prices over extended periods but who find their financial position threatened by fluctuating grain prices.

The private trade holds stocks when anticipated price gains more than offset storage costs, including a charge for risk. Fear that government action will truncate price rises injects uncertainty that leads to excessive rates of return required for socially optimal holdings of stocks by the private trade. The result is inadequate private stocks alone on the average. Inability of the private trade to obtain capital and to assume the risks involved in holding large enough stocks to meet the requirements of a national food policy also suggests the need for public involvement. Past public involvement in stock operations has been excessive, however, and some analysts contend the social cost of government intervention has exceeded the cost of the market failure being corrected. Replacing of the now massive government storage programs with annual payment of say 50¢ per bushel to holders of buffer wheat stocks would be the simplest and most direct way to correct the divergence between social and private cost. Clearer recognition also needs to be made that the role of government (if any) in buffer stocks is to reduce price volatility rather than to raise long-term farm income.

REFERENCES

Anderson, Kim and Luther Tweeten. 1975. A simplified procedure to predict seasonal average wheat and feed grain prices. *Oklahoma Current Farm Economics* 48(3):3-8.

Cochrane, Willard and Yigal Danin. 1976. Reserve stock grain models for the world, 1975-85. Pp. 77-91 in *Analysis of Grain Reserves*. ERS-634. Washington, D.C.: U.S. Department of Agriculture.

Congressional Budget Office. 1976. *U.S. Food and Agricultural Policy in the World Economy*. Washington, D.C.: Government Printing Office.

Council for Agricultural Science and Technology (CAST). 1973. *The Impact of an International Food Bank*. Ames: Department of Agronomy, Iowa State University.

Dunn, James. 1975. Another wheat price and carryover stocks equation. *Oklahoma Current Farm Economics* 48(3):8-10.

Eaton, David, W. Scott Steele, Jared Cohon, and Charles ReVelle. 1976. A method to size rural grain reserves. Pp. 39-53 in *Analysis of Grain Reserves*. ERS-634. Washington, D.C.: U.S. Department of Agriculture.

Ericksen, Milton E., Daryll E. Ray, and James W. Richardson. 1976. Farm programs and grain reserves — simulated results. Pp. 114-35 in *Analysis of Grain Reserves*. ERS-634. Washington, D.C.: U.S. Department of Agriculture.

Firch, Robert. 1977. Sources of commodity market instability in U.S. agriculture. *American Journal of Agricultural Economics* 59(1):164-69.

Gustafson, Robert L. 1958. *Carryover Levels for Grains*. Technical Bulletin No. 1178. Washington, D.C.: U.S. Department of Agriculture.

Harper, Wilmer and Luther Tweeten. 1977. Socio-psychological measures of rural quality of life. *American Journal of Agricultural Economics* 59(5):1000-05.

Hayami, Yujiro and Willis Peterson. 1972. Social returns to public information services. *American Economic Review* 62(1):119-30.

Heady, Earl. 1953. *Economics of Agricultural Production and Resource Use*. New York: Prentice Hall.

Heifner, Richard and Jitendar Mann. 1976. Market instability: Some research approaches. Pp. 118-26 in Robert Spitze, ed., *Agricultural and Food Price and Income Policy*. Agricultural Experiment Station Special Publication No. 43. Urbana-Champaign: University of Illinois.

Johnson, D.Gale. 1947. *Forward Prices for Agriculture*. Chicago: University of Chicago Press.

Johnson, Gale and Daniel Sumner. 1976. An optimization approach to grain reserves for developing countries. Pp. 56-75 in *Analysis of Grain Reserves*. ERS-634. Washington, D.C.: U.S. Department of Agriculture.

Johnson, Glenn, and Leroy Quance. 1972. *The Overproduction Trap in U.S. Agriculture.* Baltimore: Johns Hopkins University Press.

Jones, B.F. 1976. *A Grain Reserve Program.* Agricultural Experiment Station Bulletin No. 137. West Lafayette, IN: Purdue University.

Just, Richard. 1974. An investigation of the importance of risk in farmers' decisions. *American Journal of Agricultural Economics* 56(1):14-25.

———. 1975. *A Generalization of Some Issues in Stochastic Welfare Economics.* Oklahoma Agricultural Experiment Station Research Report P-712. Stillwater: Oklahoma State University.

Lin, William, G.W. Dean, and C.V. Moore. 1974. An empirical test of utility versus profit maximization in agricultural production. *American Journal of Agricultural Economics* 56(3):497-508.

Luttrell, Clifton and R. Alton Gilbert. 1976. Crop yields: Random, cyclical, or bunchy. *American Journal of Agricultural Economics* 58(3):521-31.

Officer, R.R. and A.N. Halter. 1968. Utility analysis and practical study. *American Journal of Agricultural Economics* 50(2):257-77.

Robinson, K.L. 1975. Unstable farm prices. *American Journal of Agricultural Economics* 57(5):769-77.

Tisdell, C. 1963. Uncertainty, instability, expected profit. *Econometrica* 31:243-47.

Trezise, Phillip. 1976. *Rebuilding Grain Reserves.* Washington, D.C.: Brookings Institution.

Tweeten, Luther. 1976. The demand for United States farm output. *Food Research Institute Studies* 7(3):43-69.

———. 1976. Objectives of U.S. food and agricultural policy and implications for commodity legislation. Pp. 41-63 in *Farm and Food Policy 1977,* Committee on Agriculture and Forestry United State Senate, 94th Congress, 2nd Session. Washington, D.C.: Government Printing Office.

———. 1983. Economic instability in agriculture. *American Journal of Agricultural Economics* 65:922-31.

——— and Delton Gerloff. 1977. Instability in supplies and utilization of U.S. wheat. *Oklahoma Current Farm Economics* 50:3-9.

———, Dale Kalbfleisch, and Y.C. Lu. 1971. *An Economic Analysis of Carryover Policies for the United States Wheat Industry.* Oklahoma Agricultural Experiment Station Technical Bulletin T-132. Stillwater: Oklahoma State University.

——— and James Plaxico. 1974. U.S. policies for food and agriculture in an unstable world. *American Journal of Agricultural Economics* 56(2):364-71.

——— and Dean Schreiner. 1970. Economic impact of public policy and technology on marginal farms. Chapter 3 in Center for Agricultural and Economic Development, *Benefits and Burdens of Rural Development.* Ames: Iowa State University Press.

Walker, Odell, Earl Heady, Luther Tweeten, and John Pesek. 1960. *Application of Game Theory Models to Decisions of Farm Practices and Resource Use.* Agricultural Experiment Station Research Bulletin 488. Ames: Iowa State University.

———— and A. Gene Nelson. 1977. *Agricultural Research and Education Programs Related to Decision-Making Under Uncertainty*. Research Report P-747. Stillwater: Oklahoma Agricultural Experiment Station.

Walker, Rodney, Jerry Sharples, and Forrest Holland. 1976. Grain reserves for feed grains and wheat in the world grain market. Pp. 114-35 in *Analysis of Grain Reserves*. ERS-634. Washington, D.C.: U.S. Department of Agriculture.

CHAPTER SIX

Macroeconomic Linkages

Chapter 1 listed the principal farm problems to be financial stress, instability, environment, poverty, and family farm exodus. Macroeconomic policies have contributed to each of these problems but especially to financial stress and instability. Macroeconomic policy refers to monetary policy regarding the money supply and fiscal policy regarding aggregate government spending, revenue, and the balance between them as apparent in budget deficits or surpluses.

The farming economy increasingly is exogenous, influenced by the national economy while not influencing it. Farmers and those who represent them frequently have adopted that same policy posture, reckoning that farmers are unable to influence but are influenced by macroeconomic policies. As a principal source of farm economic problems, macroeconomic policy deserves as much attention from farmers as they give to commodity program policy. Farmers and others in agriculture must be informed if they are to be a constructive force in shaping national monetary and fiscal policies.

PAST STUDIES OF FARMS IN THE NATIONAL ECONOMY

The free-enterprise system has been criticized for bunching economic activity interpersonally (case poverty), geographically (regional underemployment and poverty), and temporally (business and inflation cycles). The first two problems are dealt with in Chapter 10 and elsewhere (Tweeten and Brinkman, 1976); this chapter relates only to agriculture and the business cycle.

T.W. Schultz (1945) was one of the first to analyze systematically the relationship between agriculture and the business cycle. He postulated that an unstable nonagricultural economy giving rise to business cycles translated these problems to agriculture through a comparatively high income elasticity of demand for farm output. This parameter, coupled with a low elasticity of supply of farm output, caused the prices of farm products and farm income to rise and fall with the business cycle. The 1920s had demonstrated that the nonfarm economy could prosper while the farm economy was depressed, but the reverse was not thought to be possible; that is, the farming economy would be depressed by a downturn in the nonfarm economy. He postulated further that difficulties would not need to be so severe if cycles could be predicted. That they could not be predicted was indicated by Schultz's forecast in 1945 of an agricultural depression in 1947, which turned out to be a good year for the national economy and one of the best ever for farm prices.

Schultz's postulated relationship between the nonfarm business cycle and the agricultural economy was subsequently borne out in analysis by Hathaway (1957). Hathaway divided gross national product into ten periods of business expansion and ten periods of contraction from 1910 to 1956. Net farm income increased during nine of the ten periods of business expansion, the exception being the 1954-56 expansion. However, a consistent relationship existed between the magnitude of increase in farm income and general business activity only during periods of vigorous business expansion as measured by gross national product.

Net farm income declined in eight of ten periods of business contraction. Again a close relationship was apparent between the magnitude of farm income contraction and business contraction as measured by gross national product.

Trends noted by Hathaway to 1956 were confirmed by structural analysis in greater depth. Hathaway observed in the most recent expansion prior to 1956 that the rate of increase in prices paid and production expenses exceeded that for prices received and gross income, whereas prior to World War II the opposite held for every expansion. "As a result of greater dependence of farmers upon nonfarm produced items, in the future relatively moderate periods of business expansion may inflate farmers' cost more rapidly than either farmers' prices or income" (Hathaway, 1957, p. 55).

Gardner (1976, p. 14) found no consistent relationship between agricultural income and employment on the one hand and nonagricultural income and employment on the other hand in the postwar period. After examining six periods of recession from 1948 to 1975, he could only generalize "that real farm income falls more often than not during post-war recessions, while level of agricultural employment appears equally likely to be above or below trend during recessions." He concluded that, "Both real income and employment economic events in agriculture seem quite independent of ups and downs of the nonfarm economy." This is a very different conclusion than agricultural economists drew out of the Great Depression, e.g., T.W. Schultz's statement that "instability in farm income has its origin chiefly in business fluctuations" (see Gardner, 1976, p. 15). The low and declining income elasticity of demand for food and fiber made farm income less sensitive to nonfarm income of consumers.

One of Gardner's major contributions was to examine the relationship between farm economic conditions and rural industrialization. The economic well-being of people might be expected to be associated with nonfarm economic conditions through off-farm employment of farm people. However, Gardner (1976, p. 18) found only a very weak connection between recession and real nonfarm income of rural residents. Both Hathaway and Gardner noted a major restructuring of the impact of the business cycle on the farm economy. In the early days the impact was primarily through product markets, as emphasized by Schultz; in more recent periods the impact was coming primarily through factor markets, as recognized by Hathaway and Gardner, and through farm export markets.

Firch (1964) found declining contributions of instability in the nonfarm economy to farm income in the years following World War II, although in a later article (1977) he reported a strong relationship between farm income and the nonfarm economy in the early 1970s. The latter appeared to be primarily a statistical aberration. The myth that the farm economy had to prosper for the nonfarm economy to prosper was destroyed by realities of the 1920s, 1950s, and 1980s. The myth that the nonfarm economy had to prosper for the farming economy to prosper was destroyed by realities of 1973-76.

As indicated above, previous studies found a declining influence of variation in nonfarm income on farm product markets over time. However, growing dependence of farmers on nonfarm produced inputs causes inflation in the national economy to influence farm prices and costs more than in the past. The nonagricultural economy has become more stable than it was in the years immediately after World War I — even the recessions of 1973-75 and 1981-82, the worst in recent times, were mild compared to the Great Depression of the 1930s. The price elasticity of supply has increased, enhancing farmers' ability to respond to changing nonfarm economic conditions. Prior to 1970, the fortunes of farmers in adjusting to disequilibrium depended heavily on their ability to migrate to nonfarm jobs. The availability of jobs in turn depended markedly on the level of unemployment — the incentives in terms of higher incomes elsewhere were present given the opportunity to leave the farm. With less excess labor in agriculture and with part-time off-farm jobs more accessible to farm residents because of better roads, vehicles, and communication, farmers are no longer migrating to nonfarm opportunities in great numbers.

With off-farm earnings accounting for well over half of farm personal income, *local* nonfarm unemployment conditions strongly influence farm economic well-being. In times of nonfarm business contraction one would expect to find income of farmers reduced by an inadequate supply of local nonfarm employment opportunities. However, if the farm as well as the nonfarm economy is depressed, farmers are likely to demand more local nonfarm employment, offsetting the supply effect and accounting for the small influence of nonfarm recessions on off-farm income of farmers found by Gardner. Growing employment of farm family workers in government and other service industries and in efficient new durable goods plants not the first to be shut down in a faltering economy may reduce sensitivity of the farm economy and people to the business cycle.

Greater openness and integration of world financial capital and trade markets makes the United States, including its farmers, more sensitive to economic happenings abroad (Schuh, 1976). Flexible exchange rates, a significant share of farm output going to export markets, increased willingness of centrally planned countries to rely on imports to satisfy domestic production shortfalls, and tendencies for the European Community to export its domestic production instability increase the volatility of U.S. farm markets, especially from foreign sources relative to domestic sources. Some export instability is translated to American farmers via business cycles which tend to be worldwide. These arguments reinforce the contention that the principal impact of the nonfarm sector on farm economic conditions in the future is likely to come through purchased input markets and foreign commodity markets rather than through domestic product markets. Heavy dependence on imported oil and other inputs reinforces this tendency. Rural industries such as agriculture, mining, lumbering, and textiles competing for exports or with imports were especially disadvantaged by the overvalued dollar in the early and mid-1980s.

REVIEW OF NATIONAL INCOME AND EMPLOYMENT THEORY

It is useful to briefly review some fundamentals of national income and employment policy before analyzing the impact of policies on farmers. This analysis highlights the importance of sound monetary-fiscal policy to farmers' well-being.

Figure 6.1 shows two important reverse flows in the economy: goods and services on the one hand and payments on the other. These flows can be viewed as a hydraulic system. Unreplenished leakages cause a breakdown in the system in the form of recession or depression. Problems arise either because of leakages from the flows or from overzealous compensation for leakages.

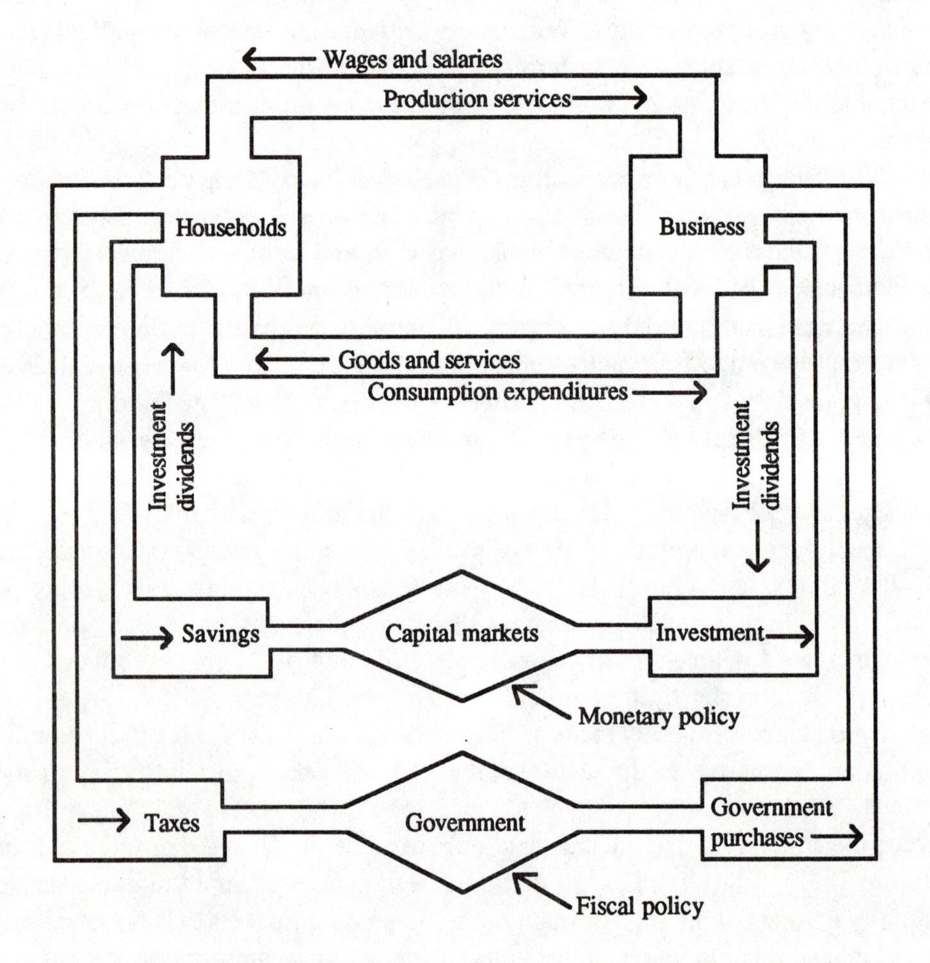

FIGURE 6.1. Flow of Goods, Services, and Payments in the National Economy

Households supply production services in the form of labor, management, and technical expertise to business. Business in turn provides goods and services to households. These flows are accompanied by money flows in opposite directions — business provides wages and salaries to households for labor services, and households provide payments to business for goods and services.

Two principal leakages from this system arise from financial capital markets and government. Taxes and savings drain off some of the funds received from households. If these leakages from households are replaced by government spending and investments by capital markets in business, the flows of real services and funds is uninterrupted and the economy continues to operate smoothly. Unfortunately, such synchronization seldom occurs and the economy is subjected to numerous undesirable shocks which public policy seeks to correct. Although advancements in the theory and application of public policy for a free enterprise economy can avoid a nationwide depression of the 1930s scope, we have been unable to avert recessions, inflation, and unemployment. "Fine tuning" of capital markets with monetary policies and of the government sector with fiscal policies remains imperfect.

When expectations of the economic outlook is unfavorable, people save more for reasons discussed later. Meanwhile under the same pessimistic economic outlook, investors reduce investment that generates future earning streams. When the economic outlook is favorable, people reduce savings no longer needed for a "rainy day" and increase consumption before the purchasing power of cash balances is eroded by higher prices. Meanwhile, investors commit capital to generate future income streams. The problem in these cases of expansion and contraction is that savings and investment are not coordinated, people substitute cash balances for goods (or vice versa), and considerable economic instability results. Because savings and investment are a chief source of instability, it is well to examine their role in greater detail.

Savings and Investment in a Keynesian Framework

The chief sources of leakage and attendant economic instability in a free-enterprise economy are savings and investment. Savings are defined as income not consumed by households or government. Motives for saving are many and include desire to (1) defer consumption to the future, (2) hold cash for economic transactions (3) provide a precautionary reserve against economic setbacks in the future, and (4) provide funds to speculate if prices decline, allowing a given cash reserve to have greater purchasing power. Savings used to increase the real capital stock constitute investment. Investment opportunities depend heavily on discovery of natural resources, new technologies, or demand expansion from population, income, and export growth.

Savings and investment are of special concern for the national economy because they are highly volatile, depending heavily on expectations (planned savings and investment activities are performed by different people and for quite different reasons that result in lack of synchronization), and they have an impact on national income out of proportion to their amount because of a multiplier effect — they determine *growth* in capital, national income, employment, and income per capita.

The critical role of savings and investment in establishing national product and employment can be illustrated with the traditional Keynesian model. A core assumption is that demand creates its own supply. The analysis applies to the short run with excess labor and capital capacity in the economy. Business is assumed to supply whatever output is demanded. Demand is comprised of consumption C, investment I, and government spending G. In equilibrium, aggregate supply equals demand, thus households and businesses are "satisfied" with flows of goods, services, payments, and employment. If demand falls short of supply, businesses do not receive compensation for goods and services produced and must cut back on production. If demand exceeds supply, businesses increase output until equilibrium is restored. Aggregate supply is equal to aggregate demand. If planned savings exceed investment, a leakage occurs in the economy and the flow of goods, services, and payments is interrupted. To the extent that business does not invest planned savings, the quantity of goods and services going to households must be cut back accordingly so that actual savings equal actual investment. Because actual savings equal investment, the national income or aggregate demand Y_d can be defined as

$$Y_d = C + I + G. \tag{6.1}$$

The phenomenon of the national product increasing by a greater amount than I or G can be demonstrated mathematically. Consumption is a linear function of national product Y_d:

$$C = C_o + cY_d, \tag{6.2}$$

where C_o is consumption when income is zero and c is the marginal propensity to consume income. If demand generates its own aggregate supply Y_s, then $Y_d = Y_s = Y$ in equilibrium. By substituting the expression for C into the demand equation 6.1 and solving for Y, national product Y is a function of I and G.

$$Y = \frac{C_o + I + G}{1 - c} \tag{6.3}$$

or

$$Y = \frac{C_o}{1 - c} + \frac{I}{1 - c} + \frac{G}{1 - c}. \tag{6.4}$$

It is clear that increasing I or G by one dollar increases Y by 1/(1 - c) dollars, the multiplier. Because c lies between 0 and 1, it follows that Y increases by more than I or G. If c is .67, then each dollar of investment or government spending raises national product by three dollars. Thus savings and investment, which depend on interest rates and highly volatile expectations, can wield a major influence on the national economy through the leverage of the multiplier shown above. It appears that G can be manipulated to bring the total demand up to that level of demand necessary to compensate for leakages in the economy associated with fluctuations in savings and investment. Unfortunately, this simplistic view of the national economy has caused much mischief.

KEYNESIAN VERSUS SUPPLY-SIDE ECONOMICS

In equations 6.3 and 6.4, 1 - c is called the marginal propensity to save s, the increase in savings associated with a unit increase in income. Therefore equation 6.4 can be written:

$$Y = \frac{C_0}{s} + \frac{I}{s} + \frac{G}{s}. \tag{6.5}$$

It follows that as the marginal propensity to save approaches zero, then each additional dollar of government outlay causes an infinite increase in national income. This finding seems to make profligacy a virtue and thrift a crime against the economy. No wonder Keynesian economics has been so widely embraced! The allure of a "free lunch" has proven irresistible; long-run federal fiscal policy has become a continuing series of short-run expedients that pump more and more deficits into an economy that responds less and less with real economic growth.

The long-run absurdity of the above Keynesian formulation becomes apparent if one recognizes that national product is maximized by consuming all that is produced, i.e., c = 1 and s = 0. The result is infinite income but only in the form of inflation! Without savings there is no investment, no real capital formation, and no real growth.

This point can be further illustrated with the Harrod-Domar growth model. In equilibrium, investment I equals savings S in (6.6). Investment is defined as net capital formation ΔK in (6.7). The average and marginal output-capital ratio g is defined in (6.8) and is a measure of the efficiency of the system in converting human and material capital into output. The average and marginal propensity to save s is defined in (6.9).

$$I = S \tag{6.6}$$

$$I = \Delta K \tag{6.7}$$

$$\frac{Y}{K} = \frac{\Delta Y}{I} = g \tag{6.8}$$

$$S = sY; \; s = \frac{S}{Y} \tag{6.9}$$

Defining r as the growth in product $\Delta Y/Y$ and rearranging terms in (6.8), it is apparent that

$$r = \frac{\Delta Y}{Y} = \frac{I}{K} = \frac{S}{K} = \frac{Y}{K}\frac{S}{Y} = gs. \tag{6.10}$$

Product or real income grows at a rate gs. If g is .2 and s is .3, then output grows at a rate .2(.3) = .06 or 6 percent per year. A high propensity to save and invest coupled with efficient use of resources (high g) lead to high income, a conclusion quite opposite from that of the Keynesian model. The government sector can be introduced as forced savings and investment if planned savings are unsynchronized with investment so that national growth is too high, low, or unstable. But forced savings and subsequent investment by government must pay attention to the lessons of equation 6.10 to contribute most effectively to real growth — the savings must be invested efficiently where payoffs are high. *Supply side economics* stresses the importance for economic growth of incentives for savings and for allocation of investment capital and other resources where returns are highest. Nearly all economists subscribe to *Keynesian economics* which prescribes government deficit spending to revitalize a depressed economy operating at less than full employment. But most economists shun *neoKeynesian economics* which holds that government deficits are appropriate and necessary in a full-employment economy. Reaganomics featuring large full-employment deficits in the 1980s is an example of neoKeynesian economics.

Quantity Theory of Money

Disenchantment with the Keynesian model led to revival of neoclassical economic theory coupled with the quantity theory of money to explain national income and employment. One shortcoming of Keynesian analysis is that it gives little or no insight into inflation as long as excess capacity exists so that additional income will come forth in real goods and services rather than as inflated values of the "old" real volume of goods and services. Keynesian economics concentrates on aggregate investment and ignores whether investment is in uses providing returns in excess of costs.

To better understand inflation, we examine an identity called the quantity theory of money. In equation 6.11, M is money supply, V is the velocity or rate of circulation of money, P is the national price level, and Q is real output of goods and services. Although (6.11) is an identity, it provides very valuable insights into inflation defined as an increase in P. (P is actually an average of all prices in the economy, weighted by quantities of goods and services; thus an increase in P does not necessarily mean that all prices have increased, only that a weighted average of prices has increased.)

$$MV = PQ. \tag{6.11}$$

Taking V as a constant k, then the relationship between the money supply, inflation, and output can be expressed as:

$$M = \frac{PQ}{k}; \; P = \frac{kM}{Q}. \tag{6.12}$$

Nominal national product PQ is proportional to the money supply. This proposition suggests manipulation of monetary policy to adjust savings-investment flows to proper levels. The money supply can be expanded to the point where unused capacity and unemployment are dissipated. Rearranging terms in (6.12), the general price level P is a function of real output of goods and services Q and the money supply.

Equation 6.12 implies that with a fixed real national output and money supply, federal deficit spending (with deficits financed by borrowing), higher food prices, "inflationary" union wage settlements, monopoly pricing, higher oil import prices, and higher food costs do not directly cause inflation. With the money supply and real output fixed, increases in some wages and prices are offset by decreases in other wages and prices.

Government spending often diverts resources from private uses to less productive public programs. Thus M is increased and Q is decreased, causing P to rise. Higher wage settlements and petroleum import prices are inflationary in part because of a pervasive and important element in the economy — downward inflexibility of prices, especially wages.

Those who view monetary policy as paramount in determining the general price level and the course of the economy are called *monetarists*. Money does matter and inflation is "always and everywhere" a monetary phenomenon in the long run. But because the velocity of money is important, varies widely, cannot be predicted reliably, and cannot be controlled by public policy, management of the economy and the general price level by controlling the money supply remains as much an art as a science.

One final point should be observed before moving on. Some argue that the way to reduce inflation is to increase Q, the denominator in equation 6.12. If M is held constant and fiscal policy is used to expand output through public service employment programs, then it is necessary that increased output not come from resources that are bid away from more productive uses. And if the expansion in Q comes from expanding the money supply, then M must increase less than Q. Public policy has had little success meeting these requirements when unemployment rates are less than 6 percent. Thus neoclassical microeconomic theory becomes important in explaining macroeconomic performance.

RATIONAL EXPECTATIONS AND THE INFLATION CYCLE

Depression-born Keynesian theory that "demand creates its own supply" seems as discredited today as Say's law that "supply creates its own demand" was by the Great Depression. Disenchantment with theories sanctioning policies that stagnate supply to stimulate demand and that substitute perennial short-run pump-priming for a coherent long-run policy motivates a further look at the causes, repercussions, and remedies for an unstable macroeconomy which has caused the farming industry so much grief.

Central to the reexamination of theory is the Phillips curve, which reveals a tradeoff between inflation and unemployment. The traditional Phillips curve showed that a society willing to tolerate high inflation could have low unemployment rates. The curve sanctioned government deficits and expansionary monetary policies as a positive sum "game," i.e., with net gains for society. Neither Keynesian economics nor the traditional Phillips curve could account for stagflation — inflation and unemployment rates near double-digit levels in some years. Keynesian analysis and the traditional Phillips curve apply only to short-run periods of excess capacity. Expectations account for macroeconomic trajectory. Expectations are self-fulfilling; expecting a business downtown causes one; expecting a boom causes one. Federal macroeconomic policy was designed to stabilize the economy in the face of fickle expectations but public decision makers are not especially adept at running the economy. The result is that the old business cycle is now overshadowed by its analog, the inflation cycle.

Inflation is best viewed as an integral part of a cycle of two phases — expansion and stabilization (see Morley, 1971). Suppose initially that the economy is in a recession at point A in Figure 6.2. Expansionary policies are pursued to decrease unemployment by moving from A to C along the short-run Phillips curve SR_1. The inflation cycle begins with the expansion phase as government deficit spending and an increase in money supply creates excess demand. In the expansion phase, excess demand for goods and services pressing a limited supply generates higher product prices which raise profits and investment in the system. As long as nominal wage rates hold steady and product prices increase, the marginal value product of labor moves upward — employment and output increase.

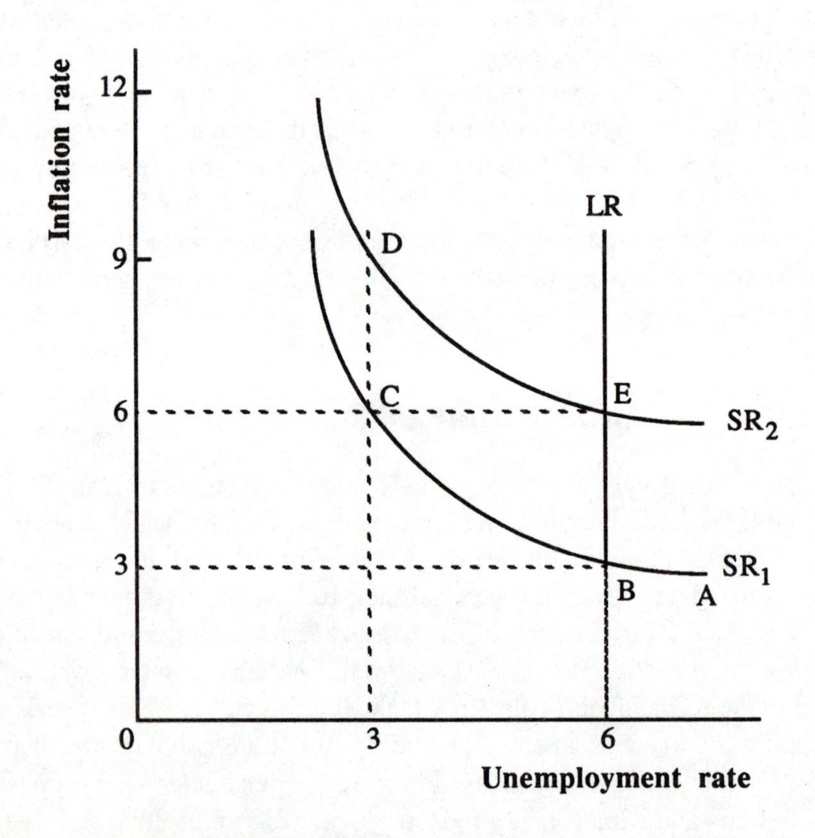

FIGURE 6.2. Hypothetical Phillips Curve, Showing Short-Run (SR) and Long-Run (LR) Relationship of Inflation to Unemployment

If the originating monetary-fiscal stimulus is removed, commodity prices return, albeit with trauma for the economy, to initial levels with movement along the traditional Phillips curve back to the natural level B where aggregate labor supply and demand intersect. If the initial monetary-fiscal stimulus continues at C, employment returns to a new equilibrium E characterized by the same unemployment but higher inflation rate than at B, as explained below.

Labor does not initially perceive its decline in purchasing power. In time it does and is willing to supply less labor at the old nominal wage. Wages increase not only to maintain real wages but also to recoup past real wage losses and to anticipate future increases in cost of living. Labor is in a position to impose such demands if fewer wages are set by pure competition and more are administered or negotiated wages or salaries established in industries with labor bargaining power. Especially in imperfectly competitive industries, management can push some of the higher wage costs to consumers through higher prices, but such opportunities are limited in part because prices have already increased in the expansion phase and in past because of foreign competition.

Thus the *stabilization* phase sets in. Real labor wages to the firm at C are higher than at A or B, and employment is reduced below the long-term equilibrium. Profits, investment, and output fall in this stabilization phase. Termination of the monetary-fiscal stimulus and subsequent movement from C to A with eventual stabilization at the equilibrium point B causes severe hardships in the form of high unemployment, idle plants, and reduced real incomes as inflationary expectations must be destroyed. An alternative is to continue the monetary-fiscal stimulus, but rising labor costs result in rising unemployment as labor raises wage demands to maintain purchasing power, as explained above. Eventually, employment and output return to the natural or equilibrium level E on a new short-run Phillips curve, SR_2, if the monetary-fiscal stimulus is continued that brought the economy from A to C.

If natural unemployment at E is greater than "full" employment as perceived possible or desirable by those who make monetary and fiscal policies, policies again initiated to reach the unattainable full employment recreate excess demand and the cycle is repeated. Additional monetary-fiscal stimulus is required to move from E to D. Again, when labor anticipates the inflation, movement is back to E if the new monetary-fiscal stimulus is removed and is horizontal to approximately 9 percent inflation and 6 percent natural unemployment if the stimulus continues. Several conclusions follow from this explanation of the inflation cycle.

To generalize the forgoing explanation of the inflation cycle, the concept "national output of goods and services" can be substituted for employment and "input prices" can be substituted for wage rates. The long-run relation between a graph of input price inflation on unutilized capacity is a vertical line, LR, in Figure 6.2. Because employment and output expansion in the expansion phase is offset by contraction in the stabilization phase, no net gain in employment or output accrues from the inflation cycle. Because monetary-fiscal stimulus does not cause much inflation when unemployment is greater than 6 percent in Figure 6.2, the actual long-run Phillips curve is ABE. Federal deficits and rapid expansion in monetary supply can be justified when unemployment is above the equilibrium rate. The equilibrium rate has not been estimated with accuracy, partly because it changes over time with public policy and shifting composition of the labor force.

The traditional Phillips curve applies only to given inflationary expectations. If excess demand creates no expectation that the inflation generated thereby is permanent because decision-makers feel assured that the expansion phase will be immediately reversed by strong measures to erase inflation, it is possible to move along the traditional Phillips curve. But if inflation expectations shift from an anticipated permanent inflation rate of (say) 3 percent to 6 percent, then the Phillips inflation-unemployment curve shifts vertically to a new and higher level. A concerted attempt to achieve high employment associated with high inflation will continue to shift short-run Phillips curves upward until the vertical long-run Phillips curve is traced.

The *rational expectations* hypothesis holds that decision makers use all information it pays to use and that they learn from the past so they do not make systematic errors. They learn to omit the processes that give rise to the short-run Phillips curve and follow only the long-run curve. Thus monetary-fiscal policy loses its impact as LR in Figure 6.2 tends to become the short-run as well as the long-run Phillips curve.

Marginal workers made up disproportionately of minorities, youth, women, and the poor gain from inflation as marginal workers are called to work in the expansion phase and if social welfare benefits increase at a faster rate than inflation. Unfortunately, the gains achieved in the expansion phase are followed by losses in the stabilization phase.

Inflation may be so extended and internalized that the cost of ending it with stabilization is deemed greater than the cost of allowing it to continue. To avoid the stabilization phase, new monetary-fiscal stimulus is added to the initial stimulus. This action only postpones the inflation cycle. New inflation is superimposed on the old rate; the cost of foregoing the traumatic stabilization phase is to substitute a natural or equilibrium inflation rate of (say) 6 percent for the "old" rate of 3 percent but each at 6 percent unemployment. Continually foregoing the stabilization phase eventually leads to unacceptable inflation rates. We do not know how long this takes. The sobering conclusion is that only by accelerating inflation from year to year in a situation where expectations of laborers and other suppliers of resources do not anticipate the acceleration can unemployment be reduced "permanently" below the natural level indicated by intersection of the aggregate supply and demand curves for labor.

Many of the real changes discussed above would not occur if inflation were anticipated. There would be inflation but less pronounced "business" cycles. Several limitations are apparent, however. The equilibrium unemployment rate is unchanged. A more serious problem is that inflation cannot be anticipated by all participants in the economy. Even indexing of wages, interest, and rents to adjust to inflation are not effective ways to avoid many of the problems of inflation.

Substantial resources are used in adjusting to macroeconomic policy changes. Resources are required to shift funds from financial assets to real assets during the expansion phase and back again during the stabilization phase. Policymakers are pressured to manage wages and prices, thereby distorting efficient allocations, and to fund low benefit-cost ratio projects of all types to stimulate the economy. Such spending invariably attracts a political clientele which makes it difficult to phase out inefficient and inequitable programs when the need for them has passed. Labor and management devote resources to activities such as changing price tags, training for temporary employment, performing hiring and layoff functions, building sophisticated forecasting devices, and the like that would be unnecessary in a more stable economy.

IMPACTS ON AGRICULTURE OF UNFAVORABLE MACROECONOMIC POLICIES

Sound monetary-fiscal policy will be discussed later but the following depicts the four outcomes possible from a two-way classification of unsound monetary and fiscal policies divided into restrictive and expansionary modes:

	Fiscal Policy	
Monetary Policy	Restrictive	Expansionary
Restrictive	A	B (Reaganomics)
Expansionary	C (Carternomics)	D

An overly restrictive monetary policy combined with a restrictive fiscal policy (combination A) brings recession or depression. Such a policy has not been followed in the U.S. since the Great Depression and is unlikely to be followed again. Also, an expansionary fiscal policy combined with an expansionary monetary policy (combination D) leads to hyperinflation which is intolerable in most advanced industrial economies. The unsound practice of an expansionary monetary policy and restrictive fiscal policy was followed during the administration of President Jimmy Carter and a restrictive monetary policy coupled with an expansionary fiscal policy was followed during the administration of President Ronald Reagan. "Carternomics" and "Reaganomics" describe the macroeconomic policies of the U.S. during these administrations.[1] These policies also are not sustainable but can last longer than the extremes listed above.

CARTERNOMICS

Macroeconomic policy characterized by a relatively tight fiscal policy (modest federal budget surpluses or deficits) and expansionary monetary policy (rapid increase in money supply) is classified for convenience as "Carternomics" after Jimmy Carter who was U.S. President in the 1976-80 period when the policy was followed. The expansionary monetary policy traces in no small part to OPEC oil price increases. If the money supply is restrained as energy prices rise, other prices and wages will fall to minimize increases in the

---

[1]The labels are based on who was president when the policies occurred and are used for convenience. It is well to note that both sets of policies had considerable bipartisan support and were not products of one person.

general price level (inflation). But because prices and wages are inflexible downward in industrial economies, higher energy prices are associated with worker layoffs, idled industry capacity, and recession. To avoid this outcome, monetary authorities increased the money supply to accommodate higher energy prices until inflation reached 13 percent in 1979 and 12 percent in 1980. The result was *cash-flow, wealth, instability,* and *cost-price* impacts on the U.S. farming sector in the 1976-80 period (see Tweeten, 1981; 1983; 1986).

The principal benefit to farmers of Carternomics in the expansion phase was real wealth gains and the principal hardships were cash-flow stress and instability. The cash-flow theory of the farm problem was discussed in Chapter 4 and is reviewed only briefly here. In a well-functioning market a durable asset such as land receives a price equal to the expected future earnings discounted at a rate α — the desired real rate of return. If land earnings or rents R_t are expected to increase or decrease at a constant percentage rate, the rent in year t is

$$R_t = R_0 e^{(i + i')t} \qquad (6.13)$$

where subscript zero refers to the initial year, e is the base of natural logarithms, i is the expected inflation rate or rate of increase in nominal rents, and i' is the expected future rate of increase in real rents. Integrating over all future rents, the discounted value is the present land price P_0, i.e.,

$$P_0 = \frac{R_0}{\alpha - i'} \qquad (6.14)$$

or

$$\frac{R_0}{P_0} = \alpha - i' \qquad \text{(current rate of return).}$$

The ratio of rent to land price is the current rate of return on land investment. If land earnings are expected to remain constant in real terms, the current rate of return is also the real rate of return to land.

Because land rent and land prices are related to each other by a constant α - i' each year, it follows that current land price is equal to

$$P_t = P_0 e^{(i + i')t} \qquad (6.15)$$

hence, the rate of increase in land price or rate of capital gain is i + i', the same as the rate of increase in rent. Total nominal return on land is current return plus capital gain or summed from (6.14) and (6.15) is

$$(\alpha - i') + (i + i') = \alpha + i. \qquad (6.16)$$

Adjusting for inflation by subtracting i from the nominal rate of return in (6.16), the real rate of return is α.

Bonds are frequently used to finance farmland mortgages and pay a fixed nominal return each year. That nominal return declines in real terms by the inflation rate, hence i' = -i in equation 6.14. If follows that the current return or market interest rate on a bond is $R_0/P_0 = \alpha + i$ if α is the desired real rate of return.

The cash-flow problem associated with inflation can now be identified. The cash-flow shortfall on land as a percent of land price may be defined as the current interest rate (α + i) less the current rate of return on land (α - i') or i + i' based on equation 6.14. Assume i' is zero for simplicity. Then if the inflation rate is 10 percent, earnings from land will fall short of the interest on a fully indebted acre by 10 percent of the land price. The shortfall as a proportion of land price on a full indebted acre is the inflation rate. If the desired real rate of return α = 5 percent and inflation is 10 percent, then the current rate of return on land is 5 percent and the interest rate is 15 percent. Under such conditions, it is normal for the current earnings from three acres to be required to pay the interest on one acre. The real return on land is 5 percent when capital gain is included, thus markets are in equilibrium. Inflation raises immediate costs to farmers but defers returns — the cash-flow problem with Carternomics.

Under an expansionary monetary policy that brings unanticipated inflation, farmers benefit from real wealth gains. A land owner-operator who purchased land when inflation was expected to average 5 percent in the future might be able to lock into a long-term mortgage interest rate of 10 percent. If inflation rises to 10 percent (i = 10 percent) and α is 5 percent, then the total return on land is 15 percent and the cost of interest only 10 percent, giving the farmer a real wealth windfall gain of 5 percent. When interest rates on intermediate and long-term loans are fixed, unanticipated inflation reduces the real rate of interest. Debtors gain and creditors lose from lower real interest rates. Because farmers on the whole are net debtors, they benefit at the expense of creditors by lower real interest costs and by asset appreciation with unanticipated inflation. Variable interest mortgages reduce opportunities for such windfall gains, however.

As noted in equation 6.14, the value of current assets is inversely proportional to the real interest rate. A lower real interest rate raises present value and, along with lower real interest costs, brings real wealth gains under Carternomics. After averaging 2 - 3 percent in the 1950s and 1960s, real farm mortgage interest rates averaged zero in the 1970s.

Inflation does not impact all prices equally. Real price effects result from changes in the general price level. One hypothesis is that the food, marketing, and production components of agriculture are a "flexible-price" sector, reacting mostly by changes in price rather than in quantity to changes in the general price level. In contrast the input supply sector is presumed to be a "fixed-price" sector, adjusting mostly by changes in quantity so as to maintain a stable sector price as the general price level changes. If this hypothesis holds, farm prices received react more quickly than prices paid to inflation, making farmers better off by inflation. Also if inflation is of demand-pull origin and consumers whose cash balances are expanded by expanding money supply use them to purchase food, then farmers also are favored by inflation and expanding money supply.

An alternative hypothesis is that the negotiated and administered price structure in the marketing and input supply sectors may quickly pass inflated costs to farmers. Because of a weak competitive position, farmers not only cannot raise prices received but they may actually see prices received decline as marketing margins are expanded by inflation in the imperfectly competitive agribusiness industries. Hence prices received might be depressed and prices paid raised by general inflation, making farmers worse off.

These conflicting arguments must be resolved empirically. Although empirical estimates are not all consistent among studies, well-specified models indicate that an increase in the general price level first increases prices paid more than prices received so farmers are worse off after the first year (Tweeten, 1983). The second year, however, prices received increase more than prices paid. This cycle continues to dampen and by the end of four years has been completed with no *net* real impact in domestic terms of trade. The contribution of inflation to the farm cost-price squeeze is transitory, but inflation does add to the perennial problem of farming — instability.

Other oil importing countries expanded money supply in the 1970s to reduce the shock of higher oil prices. The expansion in worldwide money supply was attended by recycling of petrodollars from oil exporters through Western banks to developing countries. Some of the loaned proceeds were used to purchase farm exports which contributed to higher commodity and land prices of U.S. farmers. The important point, however, is that world money supply and credit were expanded at rates that could not be sustained.

The *expansion phase* of the inflation cycle characterized by Carternomics eventually brought high inflation rates which monetary policy corrected by reducing the money supply. The second or *stabilization phase* of the inflation cycle originated in the U.S. when the Federal Reserve Bank switched in late 1979 from a policy of controlling interest rates to a policy of controlling money supply. This monetary restraint, in response to public demands for less inflation, induced a recession which was severe in 1981 and 1982 and which reduced the inflation rate to 4 percent in 1982 where it remained into 1986.

Reaganomics

President Ronald Reagan was elected in 1980 and was ideologically committed to lower tax rates, to a strong national defense, and to reduced welfare spending. As the intellectual foundation for his economic policy, he was strongly attracted to the "Laffer Curve" concept of economist Arthur Laffer who maintained that lower tax rates would increase the tax take — which could be used to finance increased military spending.

The concept was erroneously labeled "supply-side" economics but in fact was mostly discredited neoKeynesian economics. The latter holds that advanced industrial market economies are chronically prone to unemployment which can only be alleviated by full-employment (structural) federal deficits. In contrast to neoKeynesian and Laffer Curve economics, mainstream economics holds that lowering the tax rate reduces the tax take, and that federal deficits are appropriate to bring a nation out of recession but that a balanced or surplus budget is appropriate in a full-employment economy.

President Reagan, who had conceived his economic policy before the 1981 recession was apparent, successfully steered his economic policy legislation through Congress to form the Economic Recovery Tax Act of 1981. The resulting large deficit was timed perfectly to lift the nation out of recession. The economy approached full employment in late 1983. (Seven percent unemployment in 1983 was "full employment" because less unemployment induced by macroeconomic policy would cause inflation. Lower unemployment can be sustained only by structural economic policies such as manpower training and wage supplement programs and removal of institutional restraints to employment as discussed in Chapter 10.)

If the federal government would have moved towards a balanced budget in 1983, the script would have followed mainstream economic fiscal policy prescriptions to deal with the inflation cycle. But no action was taken by President Reagan and Congress to balance the federal budget in 1983; deficits continued to rise in a full employment economy.

The result was embarkation on an economic policy in uncharted waters. The large demand for funds to finance the deficit coupled with strong private demand for funds to finance a full-employment economy in the face of limited savings and tight money raised the real interest rate (market rate less inflation) to unprecedented levels — 8-10 percent versus historic rates of 2-3 percent. Although many economists predicted that high real interest rates would truncate the recovery, in fact, the consumer and service industry sectors continued a boom lasted longer than typical recoveries. But agriculture and other "traded goods" industries were distressed by high real interest rates and real foreign exchange rates.

The major impacts of Reaganomics came through international linkages. Lower tax rates added only modestly to domestic private savings. Federal deficits were negative savings which brought overall savings rates to historically low levels. But the high real interest rates attracted savings in the form of financial investment from abroad to keep interest rates from rising even further. The emergence of an efficient global capital market mobilized worldwide savings to finance American debt, allowing Reaganomics to be sustained much longer than would have been possible in earlier decades. The strong foreign demand for dollars to invest in U.S. financial markets relative to a limited supply of dollars abroad raised the real value of the dollar 40 percent from 1980 to 1984. The high dollar made imports cheap and abundant to hold down inflation and maintain prosperity for consumers and service industries. However, the high dollar diminished U.S. export shares in world markets.

Reaganomics intensified the financial crisis in developing countries. They cut imports and expanded exports to service debt. This along with slow recovery of developed countries from the 1981-82 recession contributed to low demand for U.S. exports. In the European Economic Community (EC), high rigid price supports coupled with no production controls brought large grain surpluses which were exported. The EC sharply expanded its farm export share at the expense of the U.S. share. The policy probably would have required export subsidies in excess of politically tolerable levels to the EC in the absence of the overvalued dollar. Hence Reaganomics had diverse and often subtle impacts on U.S. agriculture.

Federal deficits and trade deficits were related and reached similar magnitudes. Given the low elasticity of domestic savings with respect to interest rates, savings to finance the deficit had to be imported. The dollars from abroad providing these savings had to be earned by trade deficits. The process was coordinated by interest and exchange rates. To interrupt the process by trade barriers would have forced U.S. interest rates to levels truncating economic expansion. Charles Dickens' *A Tale of Two Cities* might describe "A Tale of Two Economies" in the U.S. For consumers and service industries it was the best of times; for traded goods sectors it was the worst of times because the latter found it difficult to compete with imports or to compete for exports. Imports made it possible for the nation to consume much more than it produced. Reaganomics was particularly devastating to the farming industry because it (1) uses twice as much capital per worker as other industries and high real interest rates raised the cost of capital, (2) is a net debtor, and (3) depends much more heavily on exports than do other industries on the average.

Figure 6.3 summarized key linkage in Reaganomics. The first key linkage is between federal deficits and real interest rates. Tweeten (1985) calculated that each $100 billion increase in federal deficits added 2 percentage points to real interest rates. The impact remains positive but diminishes over time. An estimate by Barclay and Tweeten (1988) using a different methodology did not differ statistically from Tweeten's earlier results. Other, conflicting, findings are summarized by Belongia and Stone (1985, p. 11).

High real interest rates have direct effects on farm input prices and expenses and indirect effects on farm exports through international markets (Figure 6.3). Turning first to the latter, a key linkage is between real interest rates and the exchange rate. Each 1 percent increase in real interest rate increases the value of the dollar approximately 10 percent, other things equal. Again the impact is dampened with time.

Another key linkage is between exchange rates and exports. Estimates of U.S. wheat export responses to the dollar exchange rate range from -.5 in the short run (Dunmore and Longmire, 1984) to -1.5 percent in the long run (Chambers and Just, 1981). Thus federal deficits bring high real interest rates which bring a high value of the dollar which reduces U.S. imports causing low commodity prices and land earnings. An alternative is government intervention and high government costs to maintain farm income.

The direct effect of high real interest rates is to raise farm real interest expenses. Interest expenses constitute the largest single farm expense; each 1 percentage point increase in interest rate raised farm costs $2 billion and reduced net income a like amount in the mid-1980s, other things equal. The tripling of real interest rates from normal levels helped reduce farm real estate values by one-third and contributed greatly to the $250 billion farm capital loss from 1981 to 1987. These direct and indirect affects caused the financial crisis in the 1980s (see Figure 6.3).

Like Carternomics, Reaganomics is unsustainable because large federal deficits of mid-1980s proportions accumulated for 55 years would require the entire GNP just to pay interest on the national debt. Continued trade deficits of 1980s proportions would cause the dollar to continue to fall and bring other changes as foreigners lost confidence in the dollar. Carternomics and Reaganomics are merely phases of a more general cycle.

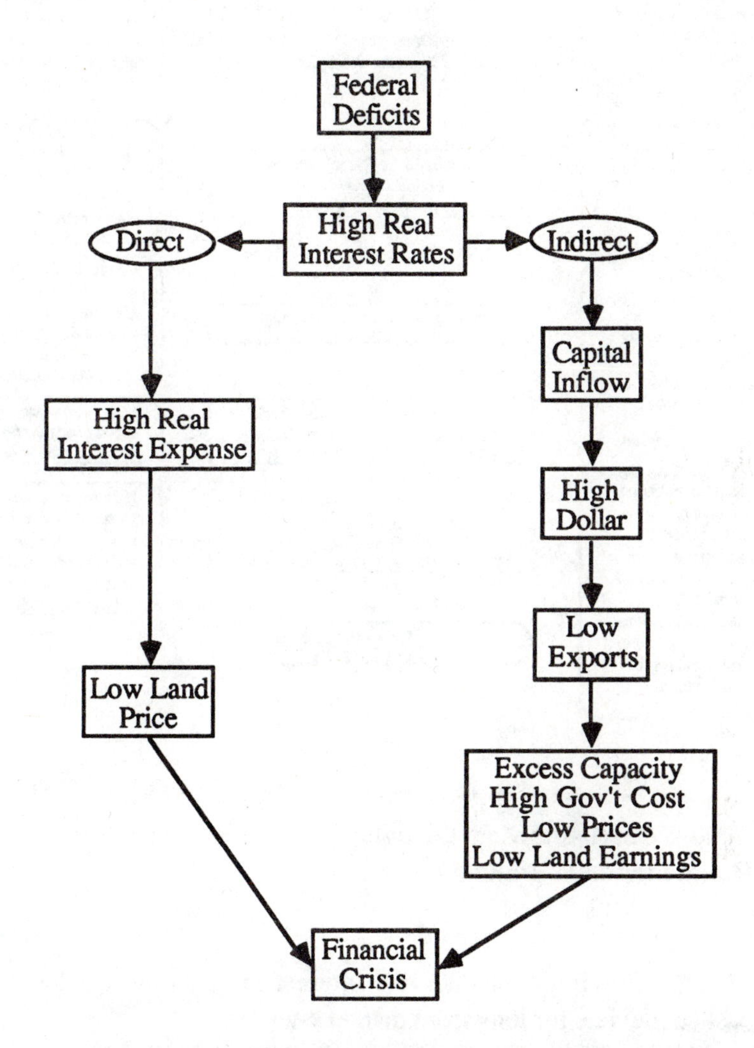

FIGURE 6.3. Flow Chart Showing Impact of Reaganomics on the Farming Economy

A General Theory of Macroeconomic Policy Linkages to the Farming Industry

Principal elements of a favorable macroeconomic policy include a monetary policy of increasing the money supply at the rate of growth of real output plus (say) 3 percentage points, and a fiscal policy of a government deficit in recession but a balanced or surplus budget in a full-employment economy. The average deficit could increase over time at a rate no faster than the rate of gain in real GNP. In addition, a "structure" policy would work to eliminate barriers to competition and commerce as a means to reduce the natural

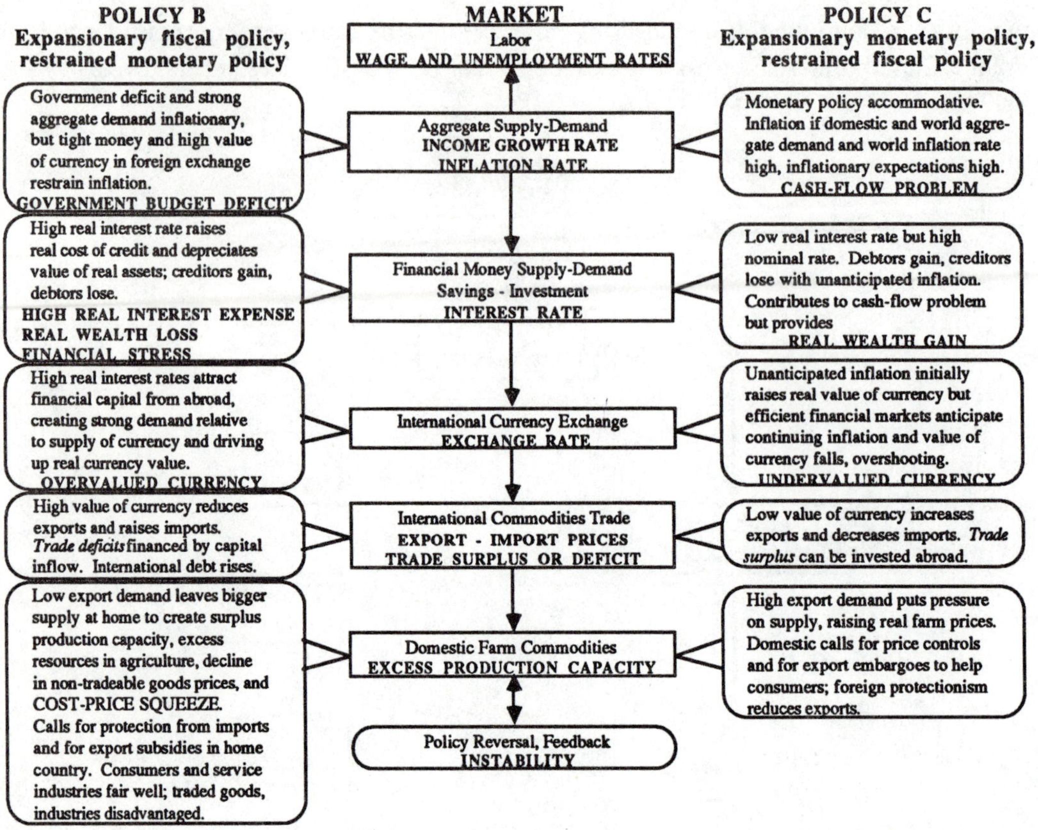

POLICY B
Expansionary fiscal policy, restrained monetary policy

MARKET

POLICY C
Expansionary monetary policy, restrained fiscal policy

Labor
WAGE AND UNEMPLOYMENT RATES

Government deficit and strong aggregate demand inflationary, but tight money and high value of currency in foreign exchange restrain inflation.
GOVERNMENT BUDGET DEFICIT

Aggregate Supply-Demand
INCOME GROWTH RATE
INFLATION RATE

Monetary policy accommodative. Inflation if domestic and world aggregate demand and world inflation rate high, inflationary expectations high.
CASH-FLOW PROBLEM

High real interest rate raises real cost of credit and depreciates value of real assets; creditors gain, debtors lose.
HIGH REAL INTEREST EXPENSE
REAL WEALTH LOSS
FINANCIAL STRESS

Financial Money Supply-Demand
Savings - Investment
INTEREST RATE

Low real interest rate but high nominal rate. Debtors gain, creditors lose with unanticipated inflation. Contributes to cash-flow problem but provides
REAL WEALTH GAIN

High real interest rates attract financial capital from abroad, creating strong demand relative to supply of currency and driving up real currency value.
OVERVALUED CURRENCY

International Currency Exchange
EXCHANGE RATE

Unanticipated inflation initially raises real value of currency but efficient financial markets anticipate continuing inflation and value of currency falls, overshooting.
UNDERVALUED CURRENCY

High value of currency reduces exports and raises imports. *Trade deficits* financed by capital inflow. International debt rises.

International Commodities Trade
EXPORT - IMPORT PRICES
TRADE SURPLUS OR DEFICIT

Low value of currency increases exports and decreases imports. *Trade surplus* can be invested abroad.

Low export demand leaves bigger supply at home to create surplus production capacity, excess resources in agriculture, decline in non-tradeable goods prices, and COST-PRICE SQUEEZE. Calls for protection from imports and for export subsidies in home country. Consumers and service industries fair well; traded goods, industries disadvantaged.

Domestic Farm Commodities
EXCESS PRODUCTION CAPACITY

High export demand puts pressure on supply, raising real farm prices. Domestic calls for price controls and for export embargoes to help consumers; foreign protectionism reduces exports.

Policy Reversal, Feedback
INSTABILITY

FIGURE 6.4. Flow Chart Showing Behavior of Markets and Farm Problems Under Alternative Macroeconomic Policies

unemployment rate. Foreign exchange and interest rates would be determined by the market or at least not deviate for long from market rates.

These principles of sound policy are rarely followed. Two key elements, fiscal policy and monetary policy, instead of being moderate, usually are overly expansionary or overly restrained as noted earlier in the combinations: restrained monetary and expansionary fiscal policy B (e.g., Reaganomics), expansionary monetary and restrained fiscal policy C (e.g., Carternomics), and restrained monetary policy and fiscal policy D. The objective is two-fold: (1) to summarize how macroeconomic policies influence agriculture for good or ill, and (2) to lay out rudiments of policies, markets, variables, and their consequences that can be used in modeling of macroeconomic linkages. Figure 6.4 is a generalization which will not hold in all situations. The figure identifies agricultural problems of *cash-flow, real wealth loss, cost-price stress,* and *instability* caused by unsound macroeconomic policies. Problems are listed in capital letters in the *policy* columns. Key linkage variables are in capital letters in the *market* column squares.

1. Aggregate Demand and Supply. Monetary and fiscal policy work through aggregate demand and supply to determine national income and the general price level. The income growth rate and inflation rate may be used to monitor performance. Policies B and C stimulate aggregate demand relative to supply, especially if they are unanticipated. Under policy B demand is stimulated directly and under policy C it is stimulated indirectly as greater money supply expands cash balances in the hands of individuals and firms. Faced with disequilibrium in the form of too much cash relative to goods and services, participants enter the market and expand aggregate demand by using cash balances to purchase goods and services. Expectations for favorable economic outcomes initially are improved, stimulating the economy during the expansionary phase. When expected favorable real factor returns and product prices do not materialize, however, losses in the stabilization phase will offset gains in the expansionary phase as noted earlier. Unlike the short-term, the *long-term growth rate* is not very sensitive to money supply and government deficits. Policies B or C tend to increase farm income in the expansionary phase and to retard farm income in the stabilization phase, hence, the policies contribute to farm economic instability but do little to raise farm income or employment over a longer period of time.

2. Labor. The labor market determining wage and employment rates was discussed earlier regarding the Phillips curve. Unstable macroeconomic policies cause labor to seek institutional means to increase job security but which reduce employment. Pitfalls emerge in attempting to short-circuit the development processes by forcing high wages through legislation or allowing high concentration of economic power in organized labor. This can price output out of international competition and make wages even less flexible in response to changing economic circumstances. It can slow employment growth. It can slow outmovement of surplus farm labor to uses having higher value. It can make it more difficult for farmers to diversify and utilize off-farm employment as a means of financial survival under unfavorable macroeconomic policies.

3. Domestic Financial Market and the Interest Rate. The interest rate is the result of forces of demand and supply working in financial markets. A high interest rate encourages savings and increases the supply of financial capital which speeds economic growth but retards consumption, investment, and demand for capital which slows growth. In actual markets the short-run supply of loanable funds is determined not only by savings but also by money supply and other factors. And the demand for loanable funds is influenced not only by expected rates of return on investment but also by factors such as the need for liquidity. Rates vary by risk involved. Risk-aversion dominates so rates of return are higher in risky investments than in secure investments. High real interest rates increase instability by raising costs of holding buffer stocks.

Under expansionary fiscal and tight monetary policy B, the demand for savings to finance government deficits combined with a strong private demand for funds relative to a limited supply of funds (due in part to monetary restraint) causes real interest rates to be high. The immediate impact is high real interest expenses to firms and individuals as noted earlier under the discussion of Reaganomics.

Under tight fiscal policy and expansionary monetary policy C, as noted earlier the result is a lower real interest rate which raises asset present value and, along with lower interest costs, brings real wealth gains. Policies B and C are cyclical, that is, they reverse results between phases. For example, the second phase of policy C is *real wealth loss* and *financial stress* problems of expansionary fiscal policy B. With the present value of farmland inversely proportional to the real interest rate, higher real interest rates under policy B reduce farm asset values and bring real capital losses. Along with higher real interest expenses and a transitory *cost-price* squeeze, the result is financial stress as noted in Figure 6.4.

4. International Finance and Currency Markets. International currency markets are large, fluid, and efficient. Millions of dollars can be transferred electronically among banks in milliseconds at low cost per unit, hence worldwide markets are closely integrated by arbitrage. High real interest rates associated with a strong demand for currency to invest in a country relative to the international supply of the currency raises the value of the currency in foreign exchange markets under expansionary fiscal policy B. Low real interest rates under expansionary monetary policy C work in the opposite direction to reduce the value of currency. Initially, inflation tends to overvalue currency. Exchange market participants quickly perceive the impact of inflation and adjust the value of currency accordingly, often overreacting, overshooting, and causing unduly low real rates.

The real exchange rate is the nominal rate adjusted for differences in inflation rates between countries. For example, if the exchange rate in India changes from 10 rupees per dollar to 20 rupees per dollar while inflation in India is 100 percent more than in the United States, the real rupee exchange rate remains unchanged with respect to the dollar. The real rather than nominal exchange rate determines trade.

An overvalued currency causes numerous distortions. Imports are implicitly subsidized, causing uneconomic substitution of imports for domestically produced goods and services. Consumers are benefited relative to producers. A market-determined flexible exchange rate is optimal but is highly unstable. A second-best solution is a pegged rate allowed to float periodically and adjusted between floats for relative inflation rates compared to trading partners.

5. International Trade Markets. It makes sense for a country such as the U.S. in the 1980s which placed a high value on current consumption to run a trade deficit with a country such as Japan which placed a high value on savings and deferred consumption. Ordinarily, however, a wealthy country such as the U.S. with much capital will export capital. An overvalued currency makes imports cheap and exports dear, shifting a country towards a negative balance of trade. If the agricultural exporting country goes from equilibrium to an overvalued currency, export demand will fall and reduce the agricultural trade surplus. A balance of payments deficit will need to be offset by foreign grants or loans. Expansionary fiscal policy B is likely to accelerate tendencies for a country to be a net debtor. But again, a distortion is apparent — international loans and transfers ordinarily should be used to finance high payoff investments in infrastructure and other nonmarket goods rather than to subsidize consumption in a country living beyond its means as apparent in a trade deficit in excess of justifiable foreign investment inflow.

An exchange rate equating the demand with supply of exchange is preferred, but an error in undervaluing exchange has certain advantages to overvaluing exchange (see Figure 6.4). An undervalued exchange rate raises exports and employment but may reduce living standards because imports are dear. An overvalued rate reduces exports and employment but living standards of the employed temporarily are improved by lower import prices. The former policy can be sustained longer than the latter because financial crisis comes more quickly with overvalued exchange and trade deficits. Some nations pursue a neomercantilist policy of a chronically undervalued currency and foreign exchange accumulation to promote employment. Such policies need to be discouraged in international forums because they unsettle world trade and capital markets.

6. Domestic Commodity Markets. With expansionary fiscal policy B, overvalued currency and attendant low export demand reduce overall demand for farm products of the country following the policy, creating excess production capacity and excess resources at current prices. In the absence of commodity price supports the result is a *cost-price squeeze,* defined as falling product prices relative to prices paid by producers. Producers resist falling prices and resource adjustments, instead calling for export subsidies, commodity import quotas, or other market interventions. Thus improper macroeconomic policies beget improper (protectionist) trade policies.

With expansionary monetary policy C and a functioning exchange market, currency may become undervalued and exports expand. Domestic producers may be hard pressed to meet foreign and domestic demand for agricultural products. The rise in real domestic prices reduces real income of consumers and brings calls for food price controls, export restrictions, and other market interventions.

Expansionary fiscal policy B allows a nation to live beyond its means for a time until money and credit run short. Mounting federal debt and trade deficits cause loss of confidence by the domestic and foreign lenders, bringing a financial crisis. The second phase, austerity, under policy B may be brought on in a developing country by the International Monetary Fund or other sources called on to rescue the country from financial collapse. Expansionary monetary policy C also is two-phased and unsustainable. High inflation rates lose both political support and economic stimulus after the expansionary phase. The stabilization phase follows, offsetting gains made in the expansion phase.

In summary, *instability* may be the single greatest problem caused by unsound macroeconomic policies. Price signals are distorted, long-term plans for investment in durables are thwarted; capital flight is encouraged. A second problem caused by unsound macroeconomic policies is microeconomic price distortions and market interventions both nationally and internationally.

7. *Interaction and Feedback.* In Figure 6.4, markets are treated as a single causal chain emanating from monetary-fiscal policy. In fact, interaction and feedback characterize the political and economic relationships. Economic instability begets political instability and vice versa. Financial stress caused by high real interest rates and falling asset values is reinforced by the cost-price squeeze induced by reduced export demand. Government interventions to protect groups from macroeconomic policy distortions in turn distort markets and weaken the ability of the price system to bring needed allocations for economic growth. Distortions, though detrimental to society as a whole, benefit some groups. Once in place, interventions are difficult to dismantle even after the original purpose for them has long disappeared.

The tangled web of reactions and counter-reactions could be traced further, but it is apparent that one unfortunate macroeconomic policy such as deficit spending can lead to a complex set of distortions, not only causing misallocation of resources but also shifting the distribution of income and wealth. Attempts to shield an economy from nontransitory shocks eventually build economic pressures for change to crisis proportions. The time to stop a "hangover" is "the night before." Early acceptance of market realities with a series of small adjustments avoids the large crisis of "the morning after."

THE TRADE, FEDERAL BUDGET DEFICIT, AND FOREIGN DEBT ACCUMULATION LINKAGE

To conclude this chapter and provide a transition to the following chapter on trade, the linkage between the trade and federal budget deficits is formalized.

Gross national product (GNP) can be defined in terms of income disposed as

$$GNP = C + FT + ST + S \tag{6.17}$$

where C is consumption expenditures, FT is federal tax payments, ST is state and local tax payments, and S is gross private (individual and corporate) domestic savings. Alternatively, GNP can be defined as expenditures on final product

$$GNP = C + FE + SE + I + (X - M) \tag{6.18}$$

where FE is federal expenditures, SE is state and local government expenditures, I is gross private domestic investment, X is exports, and M is imports. Subtracting (6.18) from (6.17), the result is

$$(S - I) + (ST - SE) + (FT - FE) = (X - M) \tag{6.19}$$

	(S - I) Private Savings Surplus	+	(ST - SE) State and Local Gov't Surplus	+	(FT - FE) Federal Budget Surplus	=	(X - M) International Trade Surplus
1980	41.4	+	26.8	+	(-61.3)	=	9.5
1982	109.8	+	35.1	+	(-145.9)	=	.3
1986	(-3.9)	+	60.8	+	(-204.0)	=	(-125.7)

(Data from Council of Economic Advisors, 1987)

Equation 6.19 is interpreted with the help of specific numbers for 1980 (a somewhat normal employment year with a modest budget deficit), 1982 (a recession year), and 1986 (a full-employment budget deficit year). The federal budget deficit can be financed from net domestic private savings surpluses (individual and corporate savings less investment), from state and local government surpluses (revenue less expenditures), or from foreign savings inflow supplied dollars by our international trade deficit. In 1980, the federal deficit was financed domestically because private savings surplus plus the state-local government surplus exceeded the deficit, hence a trade deficit was unneeded to provide foreigners with dollar "savings" to finance the U.S. deficit.

With recession in 1982, individuals and corporations saved for a "rainy day" and cut investment in a pessimistic business climate. Hence, the private savings surplus was large. A federal deficit was appropriate in such circumstances to stimulate the economy. The budget deficit did not lead to a trade deficit because domestic net private savings ($109.8 billion) plus domestic state-local government surplus ($35.1 billion) approximately equalled the deficit. (The relationship is not exact in equation 6.18 because of time lags, drawing on foreign currency reserves, and earnings from international capital investments.)

With full employment in 1986, private investment exceeded private savings so the net domestic funds available to finance the $204 billion deficit totalled only $60.8 - 3.9 = $56.9 billion. The $125.7 billion trade deficit created by high real interest and exchange rates supplied dollar savings to foreigners to finance the U.S. debt but fell short by $21.4 billion. Hence it was necessary to draw on dollar reserves.

The important conclusion from the above analysis is that in a full-employment economy where private savings surplus and state and local government surplus sum to near zero, the federal budget deficit and the trade deficit tend to be roughly equal to each other and to the additional foreign debt incurred by the U.S. The domestic private savings surplus is not very responsive to higher interest rates so foreign savings must be attracted. The associated high real interest and exchange rates generating the trade deficit to finance the budget deficit damage agriculture and other U.S. trading sectors. Use of trade barriers to end the trade deficit while continuing the budget deficit would force interest rates to levels that would crowd out private investment and bring recession, also damaging to agriculture and the nation.

SUMMARY AND CONCLUSIONS

Former President Harry Truman once said, "Those who can't stand the heat should stay out of the kitchen." Technology and normal vagaries of nature creating uncertainties from weather, pests, disease, and productivity advances require change which might be viewed as normal "heat" farmers and markets can bear without government interventions. Change induced by price and other incentives giving rise to sustainable gains in productivity might be viewed as tolerable to farmers and a reasonable price to pay for greater economic efficiency and real national income growth.

In the case of monetary, fiscal, and trade policy, however, governments play a major role in creating unnecessary uncertainty and change in agriculture. Change induced by improper macroeconomic policies is only cyclical because neither high inflation nor large budget and trade deficits are sustainable. Long-term economic efficiency and growth are not enhanced. Even if farm and national income averaged over the cycle is as high as farm and national income without the cycle, the changes induced by the cycle are socially traumatic. Most farmers lost in the period of high real interest and exchange rates under Reaganomics were not retrieved when real interest and exchange rates returned to more normal levels. The cycles probably reduced long-term farm and national income because resources were wasted in the adjustment process.

One alternative to avoid such change is with a centrally planned and tightly controlled agricultural and national economy. Americans are likely to continue to reject such a policy because it sacrifices too much economic efficiency and freedom to make decisions. A more realistic alternative is to pursue a sound macroeconomic policy. Macroeconomic tools are as yet too crude to fine-tune an economy to steady growth without inflation, and some macroeconomic variability is inevitable in a free enterprise economy. That is part of the "heat in the kitchen" which farmers will have to live with.

In the global, internationalized economy characterized by integrated trade and financial markets, macroeconomic shocks from one country are felt around the world. An attempt by one country to stimulate or restrain its economy can fail if other countries do not cooperate. In an integrated world economy troubled by inflation or slow growth, less macroeconomic stimulus and hence less distortion is required to correct the situation if countries coordinate macroeconomic policies. An international institutional framework is much needed to provide that coordination.

REFERENCES

Barclay, Tom and Luther Tweeten. 1988. Macroeconomic policy impact on United States agriculture. *Agricultural Economics* (Journal of the International Association of Agricultural Economists) 1:291-307.

Belongia, Michael and Courtenay Stone. November 1985. Would lower federal deficits increase U.S. farm exports? *Review* (Federal Reserve Bank of St. Louis) 67:5-19.

Boyne, David H. 1964. *Changes in the Real Wealth Position of Farm Operators*. Technical Bulletin No. 294. East Lansing: Michigan Agricultural Experiment Station.

Bradley, Edward, Jay Andersen, and Warren Trock. 1986. Outlook for U.S. agriculture under alternative macroeconomics policy scenarios. WREP 102. Laramie: Department of Agricultural Economics, University of Wyoming.

Chambers, Robert and Richard Just. 1981. Effects of exchange rates changes on U.S. agriculture. *American Journal of Agricultural Economics* 63:32-46.

Council of Economic Advisors. 1987. *Economic Report of the President*. Washington, D.C.: U.S. Government Printing Office.

Dunmore, John and J. Longmire. January 1984. Sources of recent changes in U.S. agricultural exports. Staff Report No. AGES831219. Washington, D.C.: International Economics Division, ERS, USDA.

Firch, Robert. 1977. Sources of commodity market instability in U.S. agriculture. *American Journal of Agricultural Economics* 59(1):1964-69.

Firch, Robert. 1964. Stability of farm income in a stabilizing economy. *Journal of Farm Economics* 46:323-40.

Gardner, Bruce. 1976. The effects of recession on the rural-farm economy. *Southern Journal of Agricultural Economics* 8:13-22.

Hathaway, Dale. 1957. Agriculture in the business cycle. Pp. 51-76 in *Policy for Commercial Agriculture,* Joint Economic Committee Print 97226, 85th Congress, 1st Session. Washington, D.C.: Government Printing Office.

Henneberry, David, Shida Henneberry, and Luther Tweeten. 1987. The strength of the dollar: An analysis of trade-weighted foreign exchange rate indexes. *Agribusiness* 3:189-206.

Morley, Samuel. 1971. *The Economics of Inflation.* Hinsdale, IL: Dryden Press.

Schuh, G. Edward. December 1976. The new macroeconomics of agricultue. *American Journal of Agricultural Economics* 58:802-811.

Schultz, Theodore W. 1945. *Agriculture in an Unstable Economy.* New York: McGraw-Hill.

Tweeten, Luther. October 1986. Macroeconomic policy as a source of agricultural change. *Agrekon* 25:5-10.

Tweeten, Luther. 1985. Farm financial stress, structure of agriculture, and public policy. Pp. 83-112 in B.L. Gardner, ed., *U.S. Agriculture Policy: The 1985 Farm Legislation.* Washington, D.C.: American Enterprise Institute.

Tweeten, Luther. July 1983. Impact of federal fiscal-monetary policies on farm structure. *Southern Journal of Agricultural Economics* 15:61-68.

Tweeten, Luther. June 1981. Farmland pricing and cash flow in an inflationary economy. Research Report P-811. Stillwater: Agriculture Experiment Station, Oklahoma State University.

Tweeten, Luther and George Brinkman. 1976. *Micropolitan Development.* Ames: Iowa State University Press.

CHAPTER SEVEN

Foreign Trade and Aid in American Farm Products

Since colonial times, foreign markets for American farm products have been a vital source of our nation's farm prosperity and economic growth. Periods of boom and bust in domestic agriculture have been closely tied respectively to export expansion and contraction. Figure 7.1 shows the close association between exports and farm prices in recent years. The relationship would be even closer without government intervention to support the farm economy when exports dropped in the 1980s.

Despite their importance, foreign markets have often been manipulated by U.S. policymakers with seeming disregard for other countries or the long-term domestic consequences of our policies. This chapter documents the advantages of foreign trade, the importance of such trade to the United States, and the progress made over the years to remove trade barriers. The chapter concludes with a section on foreign aid, emphasizing ways to improve the use of food aid.

THE MAGNITUDE OF U.S. AGRICULTURAL EXPORTS

The United States accounts for approximately one-fifth of all agricultural commodities entering free world trade, hence it has a major stake in keeping trade channels open. In 1985, U.S. agricultural exports totaled $30 billion, or 14 percent of all U.S. exports although the farming sector accounted for only 2 percent of the nation's economy. Farm product exports of $30 billion in 1985 comprised one-fifth of farm receipts. Because less than 10 percent of world farm output finds its way into world trade, the United States is absolutely and relatively more heavily involved in world commerce in farm products than are other countries on the average. World trade has increased faster than agricultural output, indicating that economic development and world trade progress together.

Most U.S. exports are unprocessed wheat, soybeans, and feed grains. Processing is minimal despite the fact that the U.S. food processing industry is the most advanced in the world. Trade distortions are widespread in farm commodities but processed farm products are some of the most heavily subsidized. Much raw farm output is produced with greater capital intensity than processed products, hence the U.S. has comparative advantage in the

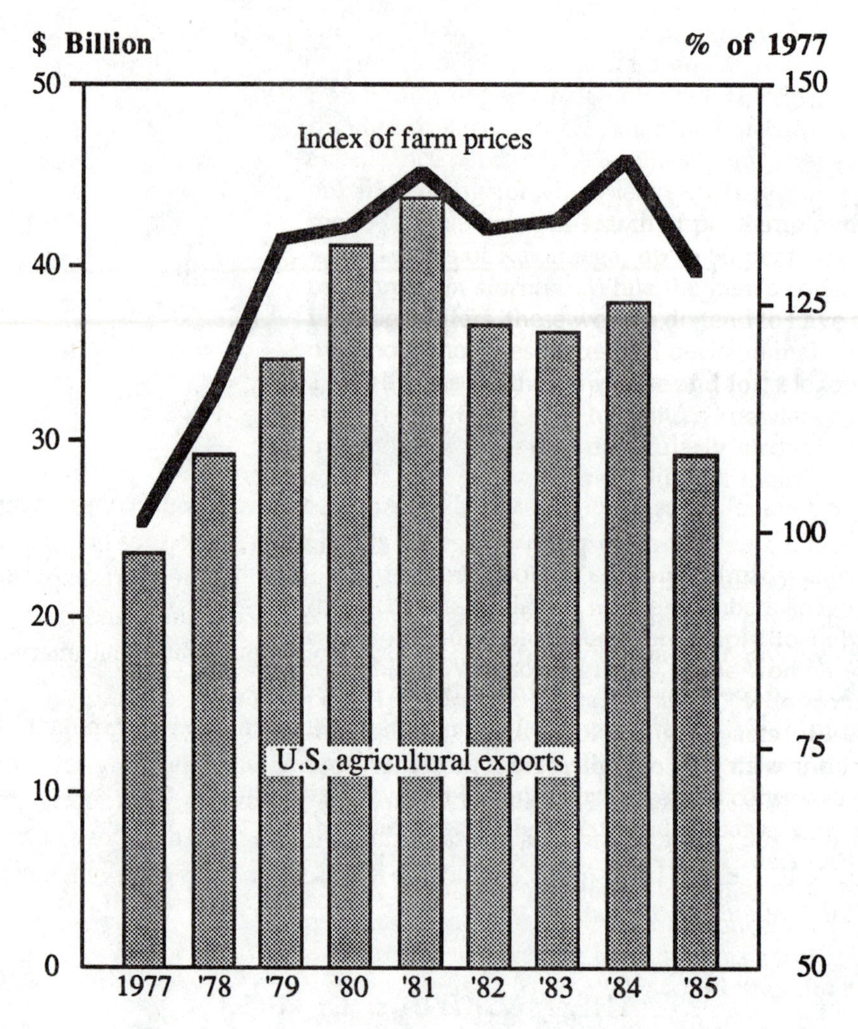

SOURCE: U.S. Department of Agriculture (1987).

FIGURE 7.1. U.S. Agricultural Exports and Farm Prices

former. These factors help to explain why this nation is not more prominent in supplying high-value-added processed farm commodities in the world market.

The share of production exported varies considerably by commodity. High proportions of wheat (including flour equivalent), almonds, cattle hides, and soybeans are exported. The principal exports from the U.S. are soybeans, feed grains, and wheat but considerable portions of cotton, tobacco, and rice also are exported. Japan is leading dollar markets for U.S. agricultural exports. The European Economic Community comprises the largest single market bloc for U.S. farm exports.

Developed countries traditionally have been the leading outlet for U.S. exports but less developed countries are becoming the leading outlet (Figure 7.2). Centrally planned countries are a smaller and highly variable outlet for U.S. farm exports.

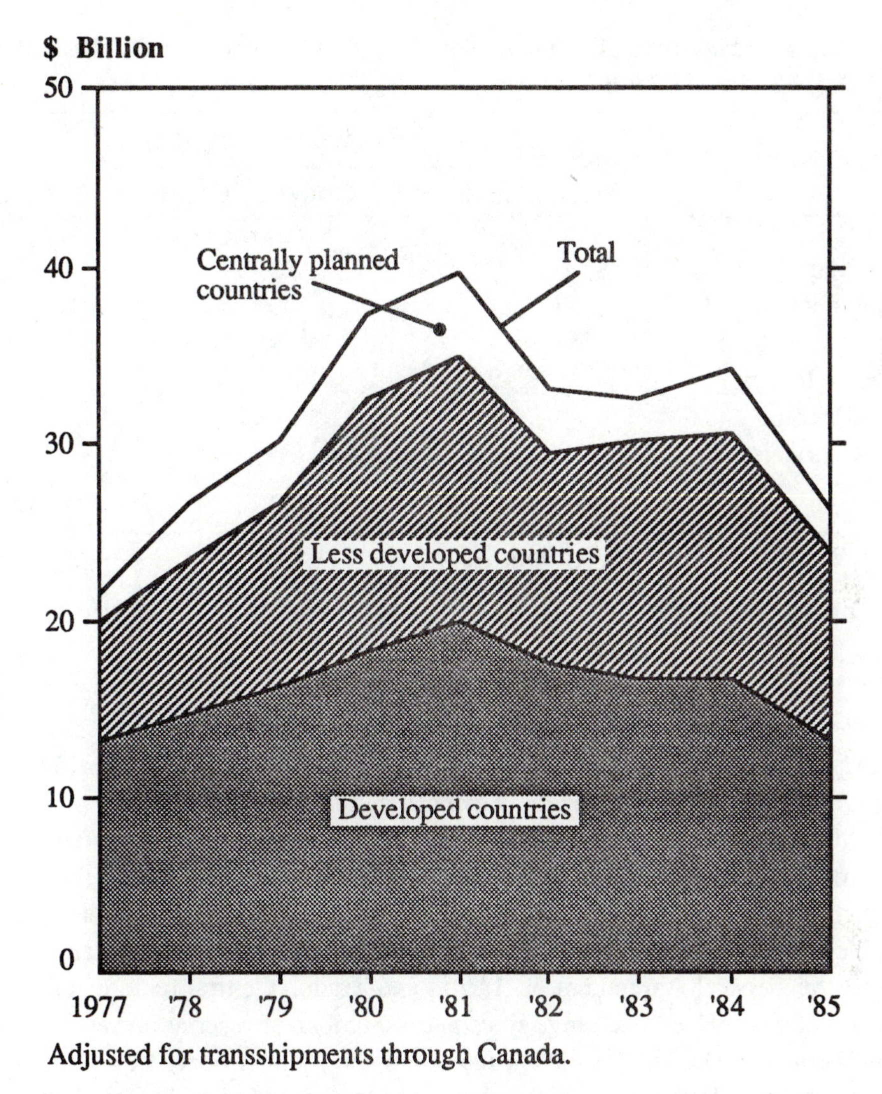

Adjusted for transshipments through Canada.

SOURCE: U.S. Department of Agriculture (1987).

FIGURE 7.2. U.S. Agricultural Exports to Major Areas

As countries increase per capita income, they tend to shift from rice or sorghum to wheat imports, then from wheat to feed grains for livestock and meat, and finally to soybean imports as livestock rations are enriched. That trend is apparent in Table 7.1 — food grain imports from the U.S. increased most rapidly in developing countries and least rapidly in developed countries. Oilseed imports increased most rapidly in industrial countries.

TABLE 7.1. Percentage Distribution of Growth in U.S. Net Imports from 1961-63 to 1981-83

Group	Growth in Imports of		
	Food Grains	Coarse Grains	Oilseeds
		(percent)	
Industrial Market	3	23	44
Planned Economies	34	28	17
Developing Countries	63	49	39
Middle Income	19	35	19
Low Income	41	10	20
Oil Exporters	3	4	--
	100	100	100

SOURCE: World Food Institute (various issues).

EMERGING DEVELOPMENTS IN WORLD TRADE

Comparatively recent institutional and other developments of worldwide scope highlight the importance of viewing the American agricultural economy in a world context.

1. Most notable of the institutional changes is the shift from fixed to flexible exchange rates. The rules that governed trade relations among nations in the post World War II period were largely established by the Bretton-Woods Conference of 1944: the remarkable effort that established the World Bank, International Monetary Fund, and, ultimately, the General Agreement on Tariffs and Trade. Central features of the system included reliance on fixed exchange rates and a number of reserve currencies, the most important being the U.S. dollar. With the dollar the major reserve currency, the world tolerated a persistent deficit in the U.S. balance of payments to provide trade liquidity.

As inflation accelerated in the U.S. economy during the late 1960s and early 1970s, the U.S. dollar became increasingly overvalued in relation to currencies of its major trading partners, and the deficit in the balance of payments grew. In August 1971, the dollar was devalued in relation to gold by 8 percent and again in February 1973 by another 10 percent. In the process the United States closed its gold window. De facto generalized floating among the industrialized countries was adopted in March 1973. In 1976, a set of rules reflecting the status quo was ratified by the United States and several other nations.

These successive devaluations of the U.S. dollar and the shift to floating exchange rates ended a rather long period of discrimination in economic policy against the agricultural sector. An overvalued currency is in effect an implicit export tax that, depending on the elasticities of export demand and domestic factor supplies, falls on the exporting sector (Schuh, 1976). With fixed exchange rates, surpluses or deficits of foreign exchange would build for extended periods. To correct for a chronic surplus of foreign exchange, a nation could pursue stimulative monetary and fiscal policies raising income and prices which in turn would encourage imports and discourage exports. Opposite policies would be pursued to correct chronic balance-of-payment deficits, but repressive monetary-fiscal policies to reduce imports, encourage exports, and attract foreign capital would produce undesirable side effects such as reduced income, raised unemployment, and inflated interest rates. It would not correct fundamental changes that effect comparative advantage including different rates of technology growth, depletion of natural resources, and changes in tastes and preferences among nations.

Flexible exchange rates determine terms of trade by market forces of supply and demand. If a nation accumulates excess foreign exchange, the value of its currency rises relative to that of other nations, decreasing the price of its imports from other countries and increasing the price of exports. The opposite results occur when a nation sustains an extended deficit in its international accounts. The exchange rate adjustments can correct trade imbalances and free domestic macroeconomic policies to pursue other objectives. In short, flexible exchange rates make domestic policies less dependent on balance-of-payments considerations. Flexible exchange rates reduce discipline on the part of public officials to pursue sound monetary-fiscal policies.

Flexible exchange rates can create short-run uncertainty in commodity markets — the economic attractiveness of a commodity to foreign buyers can change quickly due to unpredictable changes in the value of the dollar relative to other currencies. The instability is exacerbated by a highly integrated world financial capital market capable in seconds of moving billions of dollars electronically around the world in response to minor shifts in exchange and interest rates.

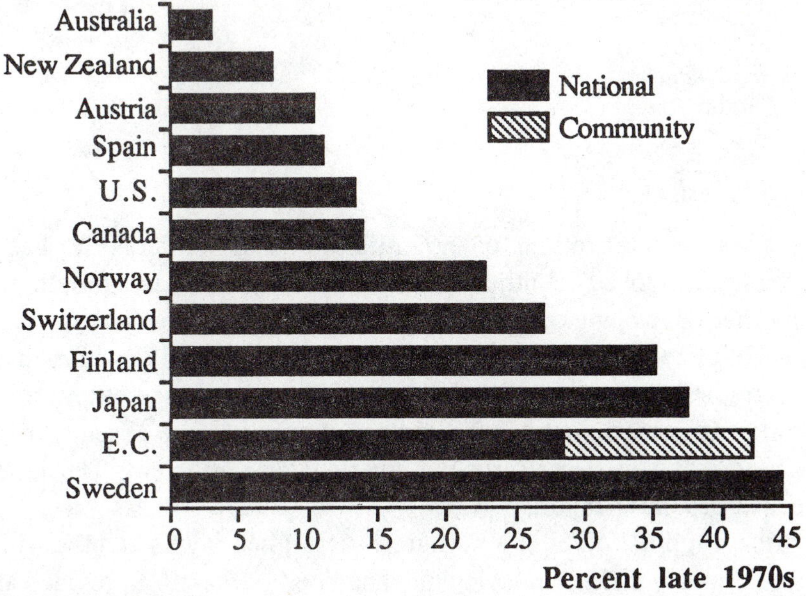

SOURCE: Blandford (1986, p. 14).

FIGURE 7.3. Government Support as a Percent of Agricultural Value Added in Major Industrial Countries

2. The growth and strengthening of worldwide liquid financial markets, the shift to flexible exchange rates, and growth in world trade now integrate world capital and trade markets. It is said that agriculture is part of a global market, that it is internationalized. In the U.S. such statements are more prominent when exports are a large share of utilization such as after World War I and in the early 1980s than when export shares drop.

3. Many of the trade problems attributed to international trade policies in fact stem from commodity program policies. Figure 7.3 shows relative economic support of agriculture by government in selected industrialized countries and regions. Government support as a percent of value added by agriculture ranged from less than 5 percent in Australia to over 40 percent in the European Community and Sweden. The U.S. and Canada tied with a 13 percent support rate.

An alternative estimate of government commodity program intervention is the producer subsidy equivalent (PSE). It measures revenue that would be needed to compensate producers if existing government programs (import quotas, variable levies, some input subsidies, price supports, etc.) were eliminated. PSEs as a percent of domestic value of production plus direct payments in 1982 through 1984 were estimated to be (U.S. Department of Agriculture, April 1987):

Developed Countries	Weighted Average PSE
Japan	72
European Community	33
Canada	22
United States	22
Australia	9
Developing Countries	
South Korea	64
Mexico	40
Taiwan	18
India	8
Brazil	7
Nigeria	-9
Argentina	-22

It is apparent that generalizations of agricultural subsidy patterns are hazardous for developing versus developed countries. Nigeria and Argentina tax agriculture but South Korea, a borderline developing country, heavily subsidizes its agriculture.

The level of price supports is an inadequate indicator of how a country distorts world trade. For example, price supports and payments used to reduce output in the United States distort world prices much less than price supports in the absence of production controls characterizing western European and Japanese policies. The latter policies markedly reduce U.S. farm exports.

U.S. price supports fixed by Congress at higher levels than supply-demand conditions would justify under the Agriculture and Food Act of 1981 provided an umbrella under which competing exporters found shelter to undersell U.S. farm products in world markets. High rigid price supports without production controls helped the European Community to shift from a net importer of 20 million metric tons of grain per year in the early 1970s to a net exporter of 15 million metric tons in the mid-1980s. Japan would have imported much rice and enjoyed higher living standards and more urban living space in recent years in the absence of policies supporting domestic rice prices at several times the world level. Sugar commodity policies have made the U.S. virtually self-sufficient in sugar but at a far greater cost to U.S. consumers and foreign producers than gain to U.S. producers.

Developing countries also practice domestic commodity policies distorting trade. The World Bank (1986) reported that in the late 1970s and early 1980s domestic farm prices for grains and cotton were below world price levels (appropriately adjusted) in 32 developing countries and were above world prices in 14 developing countries. These comparisons underestimate the discrimination against agriculture because they do not account for distorted exchange rates and high industrial tariffs which raise farm input prices. Developing countries frequently do not have the economic resources to sustain a policy discriminating against domestic agriculture and encouraging food imports; when exchange reserves and foreign borrowing capacity are exhausted the policies are reversed with attendant reduction in imports of U.S. farm products. This makes U.S. export markets less stable.

4. As noted earlier, macroeconomic policies frequently overshadow trade policies in governing imports and exports. U.S. macroeconomic policies played a major role in creating high real interest rates in the 1980s. This caused financial crises in several heavily indebted developing countries and motivated the push for a more favorable trade balance (higher exports to and lower imports from the U.S. and elsewhere) to service debt. High real interest rates in the U.S. raised real interest rates worldwide, siphoning off world savings to service U.S. debt rather than to be invested in productive capital at home. High U.S. real interest rates created high interest rates worldwide, mobilizing foreign savings to finance U.S. consumption that otherwise would have provided capital to raise income and living standards abroad. The higher foreign income growth would have raised U.S. farm exports.

5. A related element explaining changing trade patterns is economic growth featuring an accumulation of capital relative to labor. This cheapens capital relative to labor, and is attended by growth of human capital-intensive industry such as services relative to more raw labor intensive industries such as manufacturing. As economic growth progresses it is quite normal for labor intensive manufacturing of shoes, textiles, clothing, steel, and automobiles in developed countries to give way to imports from newly industrialized countries. Developed countries shift to high-tech manufacturing using robots and to service industries such as entertainment, finance, insurance, higher education, health care, and science. Capital will flow from where it is abundant (in developed countries) to where it is scarce (in developing countries). Hence, the balance of merchandise trade normally will be negative for developed countries but the trade deficit will be offset from export of services and from earnings on capital. Thus developed countries especially have strong incentive to remove impediments to trade in services. Respect for patents, trademarks, and copyright laws, and for open financial markets are examples.

Those who competently build, operate, and manage the highly complex machines and institutions of an advanced developed economy have substantial human resources and command high economic rewards. Those left behind by inadequate investment in human resources in developing countries tend to have high reservation wages and are often left unemployed or underemployed. Measures to preserve jobs tend to interfere with trade.

6. From time to time, agricultural groups in the United States have viewed trade as a threat and have attempted to cut foreign aid improving agricultural productivity in developing countries. Such aid, it is charged, reduces U.S. farm exports. Empirical data indicate that countries expanding their agricultural production at the most rapid rate also expanded imports of U.S. farm products at the most rapid rate (Paarlberg, 1987). That increased domestic farm production would increase imports from the U.S. may seem implausible.

On average, agricultural production accounts for only 36 percent of gross domestic product in the poorest countries and 10 percent in the upper middle-income countries where import demand for U.S. farm products has increased. Still, demand for grains and soybean meal from the U.S. may increase faster than farm output because the income elasticity of demand for wheat and meat exceeds 1.0. Another explanation is that countries which are good at increasing farm output are equally good at increasing nonagricultural output and overall income.

The arithmetic is compelling. If the overall income elasticity of demand for food is .6, if real income per capita is growing 4.0 percent per year, and if population is growing 2.6 percent per year, the growth in total demand for farm output is 2.6 + .6(4) = 5.0 percent annually. If food production is increasing at a brisk 4. percent per year, imports as a percent of output grow 1 percent per year. If only 10 percent of consumption is from imports, imports grow at a hefty 10 percent per year.

Recognition that investments in technology and infrastructure to improve agricultural and industrial sectors also improve the well-being of people in developing countries may be motivation enough for the U.S. and multinational agencies to help such countries. That the process might increase U.S. farm exports is a side benefit. Neither farmers nor other sectors of the U.S. economy benefit from the poverty that pervades Haiti, Mozambique, Angola, or Ethiopia. The U.S. benefits much more from growing nations such as South Korea, Kenya, Singapore, and Taiwan.

Growth of exports follows a pattern of stages. The poorest countries exhibit little increase in agricultural output or imports. Middle-income countries increase agricultural and industrial output fastest; agricultural output though increasing rapidly cannot keep up with the rapid increase in demand from population and income growth. As countries approach developed status, birth rates decline, population growth slows, and indigenous institutionalized agricultural research expands productivity to meet food demand with only modest increases in conventional resources. Demand for U.S. farm imports slacks off as nations mature economically.

In conclusion, recognition of the role played by domestic commodity programs, macroeconomic policies, economic growth, and other forces in trade helps to divert attention from counter-productive trade sanctions to the root causes of shifting trade patterns. For example to correct a trade deficit, a more nearly balanced federal budget with attendant lower real interest and exchange rates sustained over a period of years is likely to be more constructive than a 25 percent surtax on imports from other nations. Just as it is a myth that unfair trade and commercial policies are the principal cause of U.S. trade deficits, so it is also a myth that removal of such barriers alone will correct trade deficits. An attack on unbalanced trade relationships worldwide must begin with domestic commodity program, macroeconomic, and structural growth policies.

REASONS FOR TRADE

Freer trade is valued because it raises domestic and foreign incomes and stabilizes international markets. At the individual or personal level, the advantages of free trade are even more obvious. An individual who trades with no one would be a Robinson Crusoe. He would spend long hours providing minimal stone age necessities of life — primitive shelter, food, and clothing. There would be no time nor resources for disease prevention or for self-fulfillment in the form of education, culture, or entertainment. Life would be short, miserable, and brutal.

COMPARATIVE ADVANTAGE AND THE GAINS FROM TRADE

The fundamental proposition that participants benefit from exchange of goods and services underlies foreign trade. Figure 7.4 illustrates the concept of comparative advantage and the benefits of trade. The farm resources of the United States and Brazil will produce wheat and sugar in the combinations indicated by the respective production possibility curves P. In isolation, the highest societal indifference curve that can be reached with the given resources is I_0 in each country. The terms-of-trade line (not shown) in isolation, indicating the ratio of the price of sugar to the price of wheat, is tangent to the product transformation curves and the societal indifference curves I_0 at A. The slope of this price line is considerably steeper for the United States than for Brazil. The relatively high price for sugar in the United States and wheat in Brazil in Figure 7.4 reflects differences in production capabilities rather than in consumer preferences.

In isolation, quantities of sugar and wheat produced and consumed in the countries are W_i and S_i. The resources of the United States are better suited to produce wheat than sugar, and those of Brazil are better suited to produce sugar than wheat. Hence, even in isolation, with the same preferences reflected in similar indifference curves in each country, the United States consumes relatively much more wheat than sugar. The reverse is true for Brazil.

The Unites States is said to have a *comparative advantage* in production of wheat, Brazil in sugar. That is, with resources available to each country, the U.S. produces a higher ratio of wheat to sugar than does Brazil. A nation is said to have an *absolute* advantage when it can produce a commodity at a lower cost (measured in hours of labor, or, more properly, in a weighted value-sum of inputs) per unit than can another country.

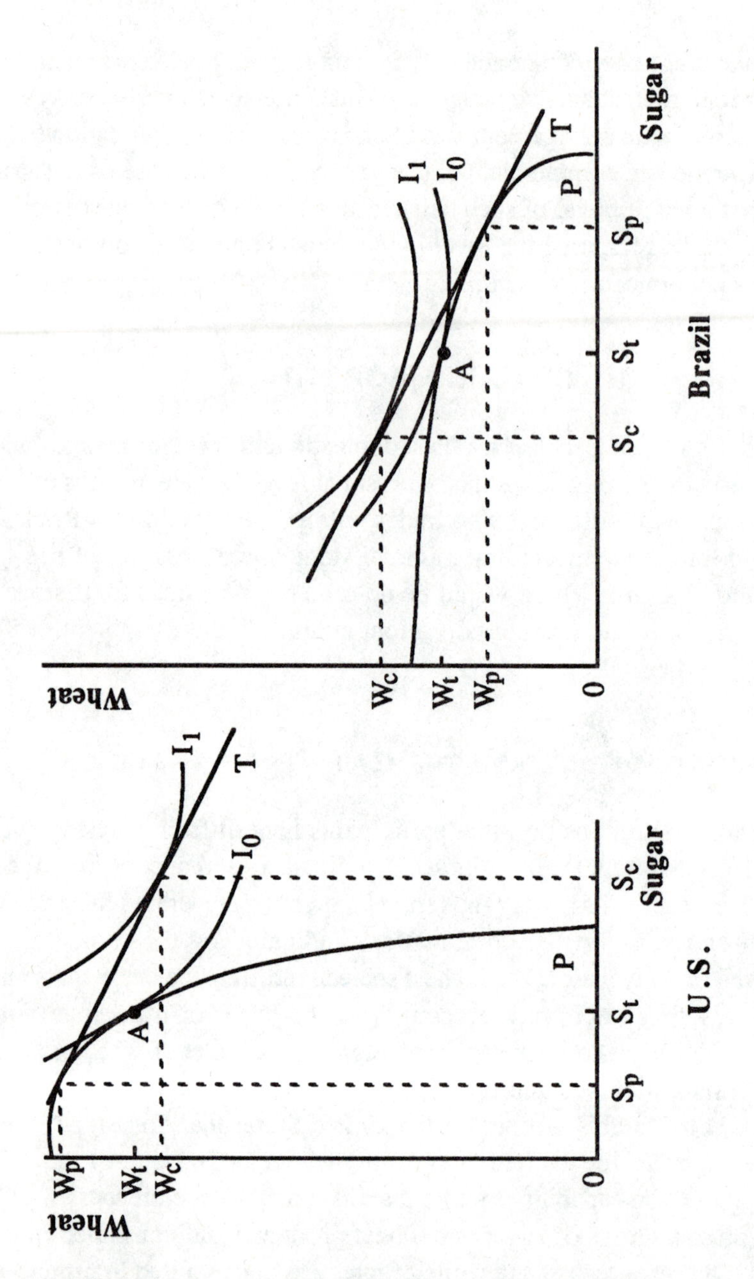

FIGURE 7.4. Production Possibility Curves and Indifference Curves for Wheat and Sugar in the United States and Brazil

The classical economists pointed out that one nation (say the United States in Figure 7.4) could have an absolute advantage in production of both wealth and sugar, and yet trade would be advantageous. It is only necessary to possess a *comparative* advantage.

The modern theory of comparative advantage is illustrated with the two production possibility curves P in Figure 7.4. The curve for Brazil is rotated 180 degrees and placed on top of the curve for the United States. By making the two curves tangent at various points, different combinations of total sugar and wheat will be produced efficiently by the two countries. It is apparent that, because of the nature of the curve, the United States will tend to specialize in wheat and Brazil in sugar. Theory must also take into account the nature of consumer preferences and demand.

Trade results in greater specialization in production of wheat (W_p) in the United States and sugar (S_p) in Brazil. However, more wheat (W_c) is consumed in Brazil and more sugar (S_c) in the United States after trading than in isolation. The quantity $W_p - W_c$ is a net wheat export from the United States and a net import into Brazil. The quantity $S_p - S_c$ is the net sugar export from Brazil and a net import to the United States. This trade enables each country to move from a lower indifference curve (I_0) in the absence of trade to a higher indifference curve (I_1) through greater specialization in production of what it does best. The new terms of trade line T would represent the same price ratio for both countries in the absence of trade barriers. It follows that tangency of the same price line to the product transformation curves and indifference curves in each country indicates equal marginal rates of substitution in consumption and production (see Chapter 2). The price line in reality does not have the same slope for each country because of transport costs and institutional impediments to trade such as duties, quotas, export subsidies, and domestic price supports.

Comparative Profits

Recognition that T in Figure 7.4 may differ among nations because of demand, transport, and institutional circumstances leads to rejection of comparative advantage based only on relative production costs as a basis for trade. The concept of comparative profits, which takes into account consumer preferences, comparative production possibilities, and barriers to trade through costs and returns data, provides a more complete explanation of the basis for trade. A nation or region will emphasize production of those goods in which its profits are greatest per unit of fixed resources. Country A may have higher profits in all potentially exportable commodities than does country B. But if the profit, which is the return to fixed resources, is highest in wheat among all commodities produced in A and is highest is sugar among all commodities produced in B when exposed to the world market, then A will export wheat to B and import sugar from B.

Supply, Demand, and Trade

Figure 7.5 shows trade between the United States and foreign countries, given the domestic demand and supply curves in the United States (a), and in foreign countries taken as an aggregate entity in (c). The supply of U.S. exports is the amount by which domestic supply exceeds domestic demand $S_{US} - D_{US}$ at all possible prices, hence is the excess supply curve S_e in (b). The demand for U.S. exports is the amount by which foreign demand exceeds foreign supply $D_f - S_f$ at all possible prices, and hence is the excess

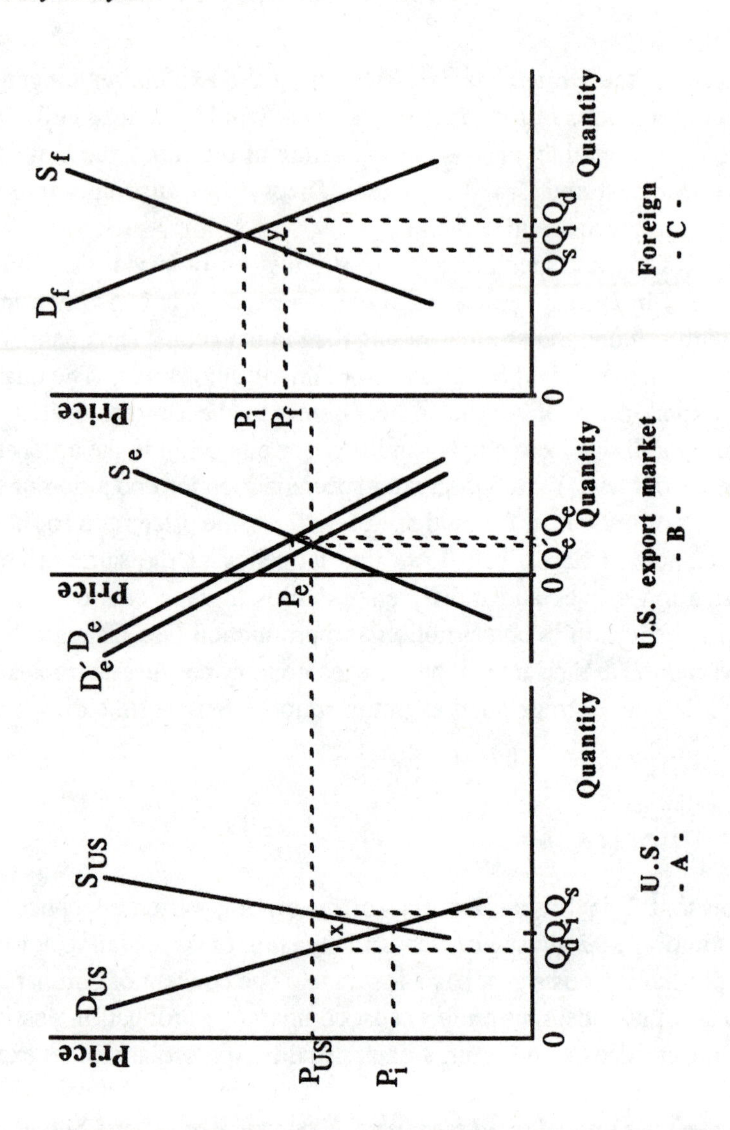

FIGURE 7.5. Domestic Demand and Supply Curves (a), U.S. Export Demand and Supply Curves (b), and Foreign Demand and Supply Curves (c)

demand curve D_e in (b). Export supply and demand intersect at an equilibrium price P_e, and Q_e is the quantity exported by the United States. This does not allow for transport costs P_f - P_{US}. Transport costs lower the United States' export demand by the amount of the cost — from D_e to D_e'. The equilibrium price is then P_{US} in the United States and P_f in the foreign market. Other barriers to trade such as foreign tariffs have an effect similar to transport costs in reducing the demand for United States exports. The exports from the United States are Q_s - Q_d in (a); the same quantity as Q_e' in (b), and foreign imports Q_d - Q_s in (c).

Export markets raise to P_{US} the domestic price of the commodity in the United States from the equilibrium price P_i in market isolation. The price P_f in the foreign market is lower than the market-isolated price P_i. With trade, the gain in producers surplus exceeds the loss in consumers surplus in the United States by the net social gain represented by the triangle *x* in Figure 7.5. In the foreign market, the gain in consumers surplus exceeds the loss in producers surplus by the net social gain represented by the triangle *y*. Thus the public (made up of consumers and producers) in each trading area realizes a positive net gain from trade, but U.S. consumers are disadvantaged and foreign producers are disadvantaged in the example in Figure 7.5. These groups may resist freer trade. The losses focused on a few well-organized foreign producers may motivate such producers to press for barriers to curtail trade and cut off the widely dispersed consumer gains, though the consumers' gains in total outweigh producers' losses. This is perhaps the single most important explanation for seemingly irrational trade barriers. A tariff equal to the price difference in isolation, P_t (foreign) - P_t (U.S.), stops trade even in the absence of transport costs.

ARGUMENTS FOR TRADE BARRIERS

Several reasons have been given to justify trade barriers. These include efforts to protect or promote national security, infant industry, balance of payments, countervailing power, and employment.

Protection is said to be required for some items to maintain necessary domestic production when foreign supplies are cut off by drought, by political decree in peacetime, or by enemy attacks on shipping in wartime. Rather than depend on cheaper foreign sources, the reasoning goes, it is better to maintain domestic production as security, even at high cost. One weakness of this conclusion is that the pattern of wars has changed. Brush-fire wars of the Korean and Vietnam types do not cut off foreign supplies, and a major war using nuclear weapons would most likely be of short duration. In peacetime, diversification of supply sources and opportunities for substitution of one commodity for another limit serious supply shortages.

The Japanese continue a costly price support program to maintain self-sufficiency in rice. The program not only raises rice price to consumers to several times the world price, it also raises land costs to levels that encourage families to live in cramped quarters and realize a lower quality of life than necessary. Most additional rice output is from pesticides and nitrogen fertilizer derived from petroleum imports more vulnerable to cutoff than food imports in an unstable world. Food supplies have been readily available in world markets to any country with buying power since World War II. Developing countries which impoverish themselves and exhaust foreign exchange to become self-sufficient are highly vulnerable to hunger when domestic harvests fail as they too often do. Without income as foreign exchange to purchase in the export market, they are at the mercy of charitable food donors. A policy of food security emphasizing efficient economic growth and ability to access world food markets as necessary makes more sense than food self-sufficiency. The *cereal facility* of the International Monetary Fund insures access to world food markets for developing countries. This ready access to world food supplies frees developing countries to pursue a policy of *food security* and comparative advantage rather than waste resources on an ill-advised policy of self-sufficiency.

The protect-infant-industry argument has validity where a domestic industry needs to grow to achieve external and internal economies of scale. Initially, the industry is not competitive in world markets, but, with time, economies of scale and maturity of know-how reduce costs to competitive levels, eliminating the need for protection. This argument has particular appeal in developing nations, but it has little validity for dairy, beef, sugar, textile, steel, and chemical industries in the United States.

Another rationale for trade barriers in the past has been to improve balance of payments. This argument would have greater validity if exports were independent of imports, but efforts to curtail imports, such as the notorious Smoot-Hawley Tariff, lead to countermeasures by foreign countries to protect their trading position. Unless there is a real need, well recognized by other nations, to protect a balance-of-payment situation (and then only after domestic policies such as currency devaluation and curbs on inflation have been pursued), increasing trade barriers are not likely to be tolerated without reciprocal action on the part of foreign nations.

Proponents of trade barriers have reasoned that unilateral reduction of trade barriers may not be in the interests of the United States and that countervailing trade barriers may be necessary. Out of such thinking has grown a theory of the second best i.e., what kind and level of trade barriers are optimal for country A facing a world of existing and mounting institutional barriers to free trade? That issue is addressed in the next section.

Major support for trade barriers comes from domestic industries and labor which want their price and income position protected from foreign competition. The fight for trade barriers is seldom waged in the name of maintaining or increasing income of the protected industry. Though this is the real reason, "good" reasons are officially stated such as protection against dumping (goods sold here below the production cost in the exporting country), contributions to national defense, and balance of payments. The latter has little validity with the shift to flexible exchange rates.

Workers press for trade barriers to protect jobs. The presumption is that workers are incapable of adjusting to alternative employment. To honor that presumption would be to cut off virtually all economic change and progress. Programs of counseling, training, job search, and other mobility assistance can ease labor adjustment pains.

Seldom is the macro-micro inconsistency more apparent than in negotiations over trade barriers. Economists have repeatedly demonstrated — and history has supported — the contention that freer trade generates greater economic progress with but few exceptions. A major impediment to movement toward freer trade is that in reality it is not a Pareto optimum or Pareto better situation, because someone is made worse off. Removal of trade barriers is consistent with the new welfare economics, which stresses greater efficiency irrespective of the distribution of the efficiency gains (see Chapter 2). Given resources are able to produce more output — greater efficiency means that gainers can compensate the losers and make them no worse off than before the change. The problem is that compensation is seldom made. The gains are often widely dispersed over millions of consumers. The losses on the other hand are often rather narrowly focused on a few producers.

In the arena of pressure groups and power politics, millions of indifferent gainers are no match for the intense opposition generated by a few determined big losers. A small dollar loss to a group with high marginal utility of money, coupled with a large dollar gain to a group with low marginal utility of money, *may represent* a sizable *net* welfare loss from liberalized trade arrangements. The value judgment of most economists is that this is not usually the case, and they continue to press for freer trade. Commitment is almost universal among economists to the proposition that the United States and the world have far more to gain than to lose from a reduction of trade barriers.

TOWARD A COHERENT TRADE POLICY

As background to suggesting a trade policy, Figure 7.6 is used to illustrate further the impact on consumers, producers, taxpayers, and society of free trade and trade distortions. The hypothetical demand and supply curves in the U.S. are d and s and in the rest of the world (ROW) are D and S respectively. In isolation, price is 0 in the U.S. and P in ROW; quantity is q in the U.S. and Q in ROW.

The world demand curve for U.S. exports is the excess demand curve ED formed by subtracting the demand quantity from the supply quantity at each price along D and S. The supply curve for U.S. exports is the excess supply curve ES found by subtracting the supply quantity from the demand quantity at all prices along d and s. Assuming no transportation costs or trade barriers, the equilibrium world price is $p_e = P_e$ and quantity is $q_e = q_s - q_d$ (U.S. exports) $= Q_d - Q_s$ (ROW imports). Compared to isolation, the gains from free trade are as follows:

	U.S.	ROW
Gain to consumers	-1 - 2	a + b + c + d + e + f
Gain to producers	1 + 2 + 3 + 4 + 5 + 6	-a - b
Gain to society	3 + 4 + 5 + 6	c + d + e + f

U.S. producers gain and consumers lose; the opposite holds in the rest of the world.

Both the U.S. and ROW benefit from free trade. If producers in ROW are strategically placed and well organized relative to consumers as is often the case, they may succeed in fully or partially curtailing trade despite positive gains from trade for society.

We now introduce trade imperfections. A quota limiting imports or exports to zero or a tariff equal to P - 0 would erase the gains from trade and make the world worse off. However, one country may gain by taxing exports or imports.

Suppose the U.S. wishes to impose an optimal tax on exports. The marginal revenue in the foreign market is MR. Because MR is negative in the elastic portion of the demand curve, the optimal strategy would not be for the U.S. to export more than the free trade quantity q_e. The optimal export is quantity $q_e' = q_s' - q_d' = Q_d' - Q_s'$ where MR intersects ES. Price is p_d in the U.S. market and P_d in the ROW market with a tax of $P_d - p_d$ per unit

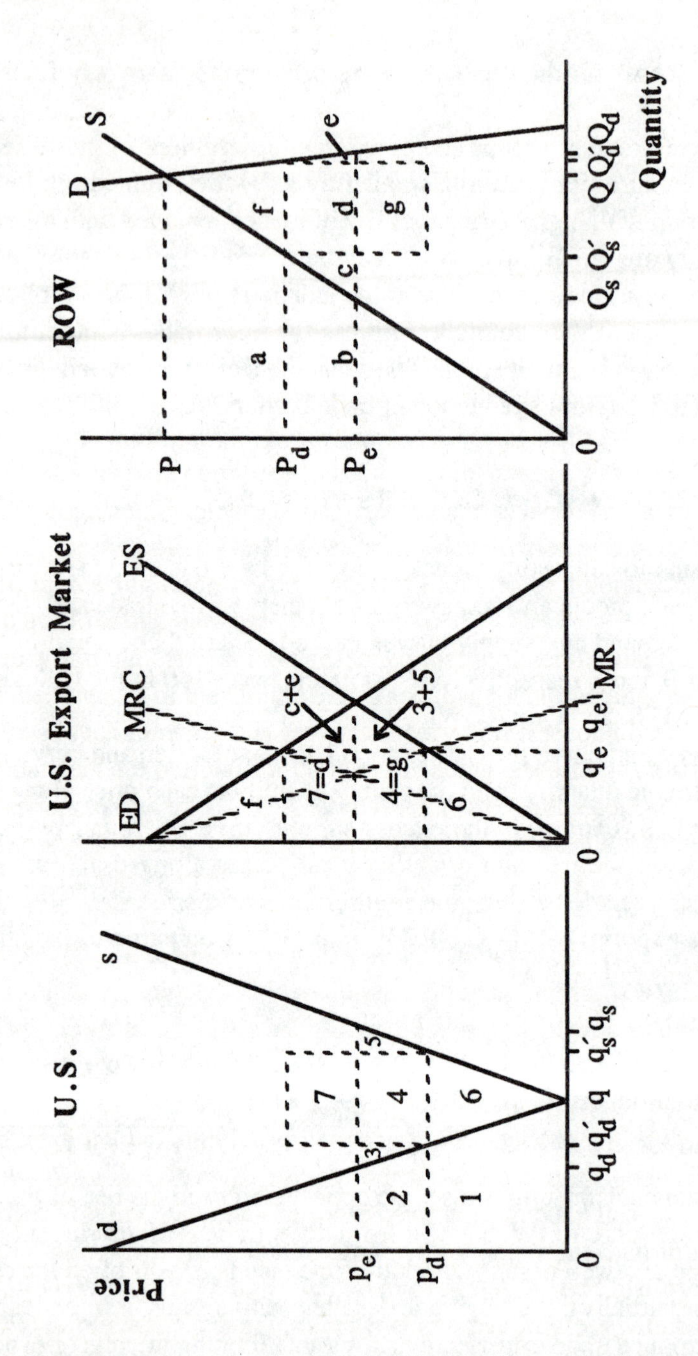

FIGURE 7.6. Illustration of Benefits from Free Trade

bringing revenue of 4 + 7 in Figure 7.6. The summary of gains to the U.S. and ROW from monopoly selling by the U.S. compared to free trade is as follows:

	U.S.	ROW
Gain to consumers	2	-b - c - d - e
Gain to producers	-2 - 3 - 4 - 5	b
Gain to taxpayers	4 + 7	—
Gain to society	7 - 3 - 5	-c - d - e
Gain to world (7=d)	-3 - 5	-c - e.

U.S. consumers gain and producers lose from the export tax. Producer losses exceed consumer gains but tax revenues of 4 + 7 more than offset net loss so in theory producers could be compensated. ROW is a net loser; national income falls by c + d + e and the big loss is borne by ROW consumers. Part of the loss, d, is a transfer 7 to the U.S. leaving c + e as the deadweight loss. ROW loses more than the U.S. gains; the world deadweight (real income) loss from the U.S. export tax is 3 + 5 + c + e.

An alternative trade distortion is for ROW to impose a tariff on imports from the U.S. The optional tariff to maximize net revenue is where the marginal resource cost MRC intersects ED at price P_d in ROW, p_d in the U.S., and with import quantity $q_e' = Q_d' - Q_s'$. Results are as follows:

	U.S.	ROW
Gain to consumers	2	-b - c - d - e
Gain to producers	-2 - 3 - 4 - 5	b
Gain to taxpayers	—	d + g
Gain to society	-3 - 4 - 5	-c - e + g
Gain to world (4=g)	-3 - 5	-c - e.

The tariff of $P_d - p_d$ per unit brings revenue d + g to taxpayers of ROW. Producers in ROW gain b while consumers lose b + c + d + e. Consumers and producers experience the same redistribution as with the export tax imposed by the U.S. However, the import tariff brings sufficient revenue to compensate consumers in ROW for losses. Deadweight loss to the world from the case of monopsony buying by ROW equals that from monopoly selling by the U.S. (This outcome arose from the special way the curves in Figure 7.6 were constructed; deadweight losses would occur with other constructions although not necessarily of the magnitudes shown in Figure 7.6.)

Any number of combinations of U.S. export taxes and ROW import tariffs could produce similar outcomes for prices and quantities. Suppose the U.S. export tax were $p_e - p_d$ and the ROW tariff were $P_d - P_e$. The outcomes would be the same as with distortions shown above except tax revenue would be only 4 in the U.S. and d in ROW. Neither the U.S. nor ROW would be able to compensate losers from tax revenues, and deadweight or national income loss would be 3 + 5 in the U.S. and c + e in ROW.

U.S. producers controlling production and exports could obtain the gains (area 7) accruing to taxpayers in Figure 7.6. However, if ROW retaliates to remove 7, producers will be worse off than with free trade. Whether gains from an export cartel are feasible depends not just on retaliation but also on the elasticity of demand for exports. The elasticity of receipts with respect to export price is 1 + E where E is the price elasticity of export demand. If export demand is elastic, 1 + E is negative and raising price or reducing quantity reduces export receipts.

It is apparent that whether it would pay the U.S. to restrict exports and raise export price to increase export receipts and farm income depends heavily on the elasticity of export demand. The following analysis indicates that for major U.S. export commodities export revenues may be raised by restricting exports in the short run but are lowered by restricting exports in the longer run. Laypersons may be surprised by the high absolute magnitudes of elasticity of demand for export because *domestic* demand and supply are frequently inelastic. The formula for export demand E_x provides insights into reasons for that high elasticity and may be written (Tweeten, 1967):

$$E_x = \sum_{i=1}^{n} \left[E_{di} \, E_{pdi} \left(\frac{Q_{di}}{X} \right) - E_{si} \, E_{psi} \left(\frac{Q_{si}}{X} \right) \right].$$

The formula ordinarily utilizes data for each country involved in exports and imports of a commodity. However, average but reasonably realistic values over all countries for wheat are assumed for illustrative purposes in this example where

E_{di} = price elasticity of domestic demand in country i with respect to domestic price in country i, assumed to average -.15 in both the short and long run over all countries for wheat.

E_{pdi} = demand price transmission elasticity, defined as the percentage change in domestic price to consumers in country i associated with a 1 percent change in world price. The value would be 1.0 in a perfect market. In the example, the U.S. price is assumed to be the world price, and the transmission elasticity is only .17 in both the long and short run for wheat, implying major impediments to free trade.

Q_{di}/X = ratio of consumption in country i (a total of 484 million metric tons of wheat in 1987/88) to U.S. exports (33 million metric tons of wheat in 1987/88), or a ratio of 14.7.

E_{si} = price elasticity of domestic supply in country i with respect to domestic supply price in country i, assumed to average .11 in the short run and .66 in the long run for wheat over all countries.

E_{psi} = supply price transmission elasticity, defined as the percentage change in price to producers in country i associated with a 1 percent change in world price. In a perfect market, the value would be 1.0. In the example, the U.S. price is assumed to be the world price and the transmission elasticity is assumed to be only .15 in both the short run and long run for wheat, implying major market distortions impeding the flow of price signals to producers.

Q_{si}/X = ratio of production in country i (a total of 451 mmt of wheat in 1987/88) to U.S. exports (33 mmt of wheat in 1987/88), or an average ratio of 13.7.

The simple illustration using judgment consensus estimates from previous studies is instructive in that with reasonably realistic but modest estimates of domestic demand and supply elasticities for the world and with extremely low but again reasonably realistic price transmission elasticities from previous estimates, the calculated U.S. elasticity of wheat export demand E_x is a sizable -.60 in the short run and -1.73 in the long run. The low price transmission elasticity indicates that on average only one-fifth of the price change in world markets is passed to domestic markets. Worldwide open markets could raise the transmission elasticities to 1.0 and quintuple the elasticity of U.S. export demand for wheat. Disaggregate data from Tyers and Anderson (see World Bank, 1986, p. 131) support this conclusion. World price variation also would be sharply dampened.

Gardiner and Dixit (1986) reviewed a large number of U.S. export demand elasticity estimates from diverse sources which used various methods of parameter estimation. Their results are summarized as follows:

	U.S. Export Elasticity of Demand	
	Short Run	**Long Run**
Wheat	-.60	-1.71
Coarse grains	-.73	-2.00
Corn	-.35	-.63
Sorghum	-1.57	-2.36
Soybeans	-.76	-1.13
Soybean meal	-.47	NA
Soybean oil	-.59	NA
Rice	-.57	-7.00
Cotton	-.40	-4.60

These average estimates from many previous studies (dropping the highest and lowest estimates to avoid distorted outliers in cases such as wheat and cotton which had a large number of estimates) indicated that export restrictions or a price hike would lose revenue for major U.S. exports except in the short run. The averages tend to underestimate elasticities for rising prices and overestimate elasticities for falling prices. Thus imposition of an export tax (even if legal) would not benefit the U.S. except perhaps in the short run.

An alternative is export subsidies which potentially could raise revenue and sales. However, extensive use of such measures brings charges by American consumers that they are unfairly paying more than foreigners for U.S. farm products if a two-price plan is used, charges by taxpayers that they should not have to subsidize U.S. farmers or foreigners and charges by foreigners of unfair competition if either a two-price plan or subsidies are used to dump produce abroad. Retaliation may be swift. The United States could contend that it is not subsidizing exports but only retaliating against subsidized foreign exports. Again, however, export subsidies are a dubious long-term strategy because political support for them erodes or a trade war erupts which merely transfers income from exporters to importers.

Given the above discussion, the appropriate trade policy for the U.S. is as follows:

1. The first best policy is worldwide free trade or as close to such a policy as possible in a highly imperfect world. Losers need to be compensated through retraining and other adjustment efforts among workers who must adapt to changing conditions. Exports have contributed to instability of U.S. food demand and prices (Tweeten, 1983). Rather than to pursue isolation to seek stability, a wise move is to raise the level and

stability of national income with freer trade, a policy that also would raise U.S. farm income given strong evidence from numerous studies indicating the nation has a comparative advantage in farm products (CAST, 1988; Office of Technology Assessment, 1986; Vollrath, 1987). Trade liberalization is consistent with a higher, more stable, and more equitably distributed income called for by the social welfare function discussed in Chapter 2.

2. The second best policy in a world of trade imperfections is for the U.S. to maintain trade barriers but only to use in successful confrontations and negotiations removing world trade barriers and moving the world to free trade as in (1) above. Data from Tyers and Anderson (World Bank, 1986) indicate the shock of moving to free trade would be sharply reduced if other trading nations also adopted open trading.

3. The third most ideal policy but *first best policy* in a world of trade barriers which cannot be eliminated by negotiation, threat, or intimidation is unilateral removal of trade barriers even if other nations do not remove their barriers to trade. If other countries are "shooting themselves in the foot" by trade barriers, export subsidies, and the like, we need not feel compelled to do likewise.

4. The fourth best or worst policy is for the U.S. to join the world in erecting sustained barriers to trade. The larger and longer-maintained the barriers, the more damaging the policy to economies throughout the world.

If only the developed countries liberalized markets, developing countries might be worse off. Tyers and Anderson (World Bank, 1986, p. 131) estimated that worldwide trade liberalization would raise efficiency (real income) by $41 billion. Gains in national income from liberalization as measured by deadweight losses avoided have been estimated to be $15 billion to the European Community alone, $4.1 billion to Japan alone, and $4.4 billion to the U.S. alone — for a total gain of $25 billion in these three countries or regions alone (World Bank, 1986, p. 121) The U.S. economy as a whole would benefit from freer trade but producers would not necessarily be better off, especially if only the U.S. liberalized trade. Except for sugar and dairy products, U.S. and world prices of major traded commodities are similar. Hence better world prices tend to mean better U.S. prices.

The impact on world price and quantity from trade liberalization as estimated by Tyers and Anderson (World Bank, 1986, p. 129) is as follows:

Liberalization from:	**Wheat**	**Coarse Grains**	**Beef and Lamb**	**All Dairy Products**
U.S Alone				
% change world price	1	-3	0	5
% change world trade vol.	0	14	14	50
All Market Economies				
% change world price	9	4	16	67
% change world trade vol.	6	30	235	190

Price and export gains to American producers would be substantial if all market economies liberalized as noted above. U.S. producers would benefit from the estimated 6 percent increase in world wheat trade and 30 percent increase in world coarse grain trade. The 190 percent increase in world trade in dairy products with liberalization would not necessarily be good news for U.S. dairy farmers. Some of the increased trade would be imports into the U.S. which would depress markets.

Variability of world price would also decline with worldwide trade liberalization. The estimated coefficient of variation (in percent) in world price would fall from 45 to 10 for wheat, from 19 to 8 for coarse grain, from 6 to 3 for beef and lamb, from 16 to 4 for dairy products, and from 20 to 4 for sugar (World Bank, 1986, p. 131).

In conclusion, trade liberalization would bring a higher, more stable, and more equitably distributed income among countries. The more broad-based the liberalization, the more favorable the outcomes.

TOWARD FREER TRADE: HISTORICAL PERSPECTIVE

After World War I, American farmers and industry were committed to the protectionist policies under which they had prospered prior to the war. Protectionism had little impact on agriculture in the immediate prewar period because farmers were net exporters, the world economy was relatively strong, and domestic prices were determined to a large extent by world markets. The drop in U.S. farm prices in 1921, following recovery of world agricultural production, brought new demands by U.S. farmers for protection from foreign competition. These demands were met to a considerable degree. Industry requested similar protection, which it received in the Tariff Act of 1922.

Unlike the prewar situation, Europe was now a debtor rather than creditor to the United States. Europe reacted by increasing its own level of protection, especially for agricultural products. U.S. agriculture continued to be depressed in the 1920s. In 1929 a sharp drop in farm and industry exports led to demands for even greater protection. Again the demands were met in the Smoot-Hawley Tariff of June 1930, creating the highest import duties in the century. Many trading nations reacted by increasing their levels of protection. In 1932, for example, the United Kingdom, then our largest customer for farm commodities, created legislation and Commonwealth agreements which required outsiders to pay a 10 percent duty on farm products but allowed most agricultural products of Commonwealth countries to enter duty free.

Partly as a result of protectionist measures and counter measures, the value of U.S. agricultural exports in 1930 fell to one-third of the average export value in 1925-29. Increased protectionism was clearly a disaster, but farm interests were reluctant to recognize that the United States could not expand exports without at the same time allowing expansion of imports.

THE TRADE EXPANSION ACT

The Trade Expansion Act of 1934, the reciprocal trade agreement program, represented a major change in policy. Under the act, the President was empowered to reduce Smoot-Hawley Tariff rates up to 50 percent in return for reciprocal tariff reductions from foreign countries. Negotiations were largely bilateral, but the benefits of agreements were more widely dispersed through frequent applications of the Most Favored Nation principle. This treatment applied, as a rule, to all countries the lowest rate of duty or other import charge granted to any country. As noted in Table 7.2, negotiations in the 1934-47 period reduced duties one-third.

TABLE 7.2. Duty Reductions Since 1934 Under the U.S. Trade Agreements Program

GATT Conference	Proportion of Dutiable Imports Subjected to Reductions	Average Cut in Reduced Tariffs	Average Cut in All Duties	Remaining Duties as a Proportion of 1930 Tariffs
		(percent)		
Pre-GATT, 1934-1947	63.9	44.0	33.2	66.8
First Round, Geneva, 1947	53.6	35.0	21.1	52.7
Second Round, Annecy, 1949	5.6	35.1	1.9	51.7
Third Round, Torquay, 1950-1951	11.7	26.0	3.0	50.1
Fourth Round, Geneva, 1955-1956	16.0	15.6	3.5	48.9
Dillon Round, Geneva, 1961-1962	20.0	12.0	2.4	47.7
Kennedy Round, Geneva, 1965-1967	79.2	45.5	36.0	30.5
Tokyo Round, 1974-1979	n.a.	n.a.	29.6	21.2

SOURCE: Lavergne (1981).

GATT

The General Agreement on Tariffs and Trade (GATT) originated in 1947 almost by coincidence a part of an ambitious plan for an International Trade Organization (ITO) to regulate trade relations and promote free trade among countries. GATT was a makeshift arrangement to begin multilateral trade negotiations while the ITO charter was being ratified. Congress never ratified the ITO despite the support of President Harry Truman, but the GATT continued under Executive Agreement only — Congressional approval was not required.

The GATT Secretariat, or administrative agency, began very small and weak but grew in influence and size to 300 employees by 1987. By 1987, 93 countries accounting for four-fifths of world trade were members of GATT and another 31 abided by its rules (CBO, 1987, p. 15). Only the Soviet Union, Taiwan, and the Peoples Republic of China were not members in 1987 but the latter appeared destined for membership.

Part I of the General Agreement on Tariffs and Trade by 1987 contained critical components:

1. GATT provides so called Most-Favored Nation (MFN) treatment; that is, tariffs on imports must be applied equally to all members. A decrease in tariffs negotiated with one member must be extended to all members. Tariffs are to be negotiated downward over time but if any trade barriers are used tariffs are favored over other interventions. Unconditional MFN treatment precludes bilateral and preferential agreements that favor one or a group of countries. Variances have been generous for bilateral and multilateral trade agreements, including free trade zones such as the common market.

2. GATT Article II binds member countries to tariff concessions; that is, tariffs can be lowered but they cannot be raised. Any rescission feature of trade agreements causing harm to other countries must be compensated. Disagreements must be settled by consultation with affected parties.

Part II of GATT contains a number of provisions, many relating to elimination of nontariff barriers.

1. Members are prohibited from circumventing tariff concessions by employing nontariff barriers to offset lower tariffs. The same article requires taxes and regulations to be applied at least as favorably to imported as to domestically produced goods.

2. All laws and regulations regarding trade are to be applied in a transparent manner which requires public disclosure and impartial administration of trade laws. For example, an unwritten decision by an organization of wholesalers to accept only domestically produced goods violates rules of GATT. Customs fee formalities or marks of origin that discriminate against foreign goods are prohibited. The Japanese in particular inhibit trade through custom and tradition, favoring domestically produced goods and services over imports.

3. Dumping, selling abroad below cost or the price charged domestic consumers, must be proved and injury to domestic producers must be shown before antidumping duties can be imposed. Such countervailing duties to offset dumping or foreign government export subsidies are not to exceed the dumping margin or the export subsidy.

4. GATT calls for removal of quantitative restrictions (QR) or quotas on trade but allows them to be used to safeguard balance of payments and to temporarily relieve domestic industry suffering injury and adjustment problems from trade. Quantitative restrictions are to be applied on a nondiscriminatory basis. Numerous countries use balance of payment arguments to justify trade barriers. Payment shortfalls are better dealt with by devaluation of currency.

5. Export subsidies are discouraged and are to be eliminated on nonprimary products. Export subsidies for primary products are not to be used by a country to gain more than an equitable share of world trade in the product. This provision especially has been violated by the EC in agricultural products.

Many exceptions in addition to those noted above are allowed. Trade barriers are permitted for national security purposes. Variances are especially generous to developing countries.

GATT has no policy power and relies on passive enforcement of rules. If consultations between affected countries fail, disputes can be submitted to third parties for arbitration. GATT Council rulings require a unanimous vote, hence can be vetoed by the offending parties. Moral suasion and sanctioning of retaliation have had modest success in stopping improper behavior, including some such action by the United States. The most effective enforcement tool is retaliation sanctioned by GATT and taken by countries damaged by trade barriers. Such retaliation tends to favor countries with strong bargaining positions. Small countries with weak bargaining positions have little scope for retaliation.

The first GATT sponsored multinational trade negotiation convened in 1947 (Table 7.2). Although much of the negotiation remained bilateral, the interplay among trading partners resulted in sizable tariff cuts. The seven GATT rounds and earlier negotiations brought tariff duties to only 3.1 percent of all imports by 1980, an almost negligible amount by standards of earlier decades.

The last completed GATT negotiation, the Tokyo Round was significant in taking place in a world trading system characterized by the breakdown of the fixed exchange rate system which had held since the Bretton Woods agreement. Oil prices had tripled in 1973 and protectionism was on the rise and was fed by instability and shifting trading patterns. Nontariff barriers were replacing tariff barriers to trade. A most important contribution of the Tokyo Round was to hold back protectionism and make progress in dealing with nontariff barriers which had replaced tariffs as the principal form of protectionism.

To be sure, the earlier Kennedy Round had for the first time addressed nontariff barriers, reaching agreements on an Antidumping Code and eliminating the U.S. system of American Selling Prices. The latter applied a tariff rate for selected imports to a price set artificially high (to equal the price of a competing good produced domestically) instead of to the imports' actual invoice price. The Tokyo Round established codes of conduct regarding antidumping rules, subsidies, and countervailing measures; government procurement practices, customs valuation; technical standards, and import licensing. The Tokyo Round grappled with but by no means resolved the issue of export subsidies.

Almost all governments subsidize domestic producers and products to some degree. Examples include education, research, concessional access to government held natural resources including forests, police protection, road construction and maintenance, and the like. When a product is exported, at what point do such "subsidies" become unfair trading practices? Export subsidies on nonagricultural products were prohibited and a code was established to differentiate between a domestic subsidy acceptable under GATT for promotion of social and economic policy objectives and an export subsidy which was not.

Export subsidies were permitted on farm products except to gain more than an equitable share of world trade.

The Uruguay Round

The eighth or Uruguay Round under GATT in Geneva is scheduled to end in 1991. Negotiations are the most difficult ever attempted under GATT because many of the more easily settled issues such as tariffs on manufactured goods had already been largely resolved, leaving the most intractable issues unsettled.

Many exceptions to GATT principles had been tolerated and needed to be addressed. Exceptions included (Congressional Budget Office, 1987, pp. xi, xii):

— The Multifiber Agreement governing trade in textiles and apparel, and placing limits on such imports into the U.S. from developing countries.

— Voluntary export restraints such as limits on automobile imports into the U.S. from Japan.

— Agricultural import quotas and agricultural subsidies.

— Barriers to trade in services and failure to protect property rights through copyrights, patents, and trademarks.

— Free trade areas, such as the European Community, which does not extend Most-Favored-Nation treatment to countries outside the area. A free trade area protects itself from world competition by erecting import barriers while pursuing free trade among nations within the area.

— Preferential treatment for developing countries,

— Nontariff barriers to trade, and

— Retaliatory trade actions.

Because Japan, the EC, and to a lesser extent the U.S. have been major impediments to free trade, 13 countries formed the *Cairns group* under leadership of Australia to apply pressure in the Uruguay Round for freer agricultural trade in the Uruguay Round. The Cairns group wants GATT rules revised to provide compensation to third countries impaired by unfair trade policies of rival exporters — notably the U.S. and EC. Allied with the U.S., the so called Cairns group can be a strong negotiating tool to mediate the protectionist policies of Japan and the EC.

With a comparative advantage in agricultural products, the U.S. has vigorously promoted joint negotiation for farm and nonfarm products under GATT. The attempt to link more open access to foreign markets for U.S. farm products to more open access of foreign industrial products into the U.S. has not succeeded, however. Much agricultural trade is carried out either outside of GATT or in violation of GATT. Principal reasons are:

— Nations contend that self-sufficiency in agricultural products is a matter of national security and national sovereignty, hence non-negotiable.

— Most major agricultural trade participants (e.g., western Europe and Japan) have well-placed, well-organized farm interests over-presented in the political process. Such groups are skilled at special pleading and manipulating public opinion to serve farm interests. These developed nations subsidize their farmers with domestic commodity programs maintained in the name of transferring income to farmers and preserving the family farm.

The success of GATT in reducing tariff barriers on nonagricultural products is only exceeded by its lack of success in reducing agricultural trade barriers. The U.S. shares much of the blame for failure of GATT to address agricultural trade barriers. The U.S. has often been at the forefront in obtaining special trade exemptions for agriculture, a notable example being the breaching of the prohibition against import quotas when contracting parties agree to import quotas to sustain domestic price supports under Section 22 of the Agricultural Adjustment Act of 1933. The Multifiber Arrangement permits a multilateral system of import quotas on textiles and apparel trade. The U.S. has restricted imports of subsidized pork and lumber form Canada, inviting retaliation by Canada to subsidized U.S. grain exports. It subsidizes some exports directly by various low-interest or guaranteed loan credit arrangements and by payment-in-kind (humanitarian food assistance is permitted). Quotas or other restrictions are placed on U.S. imports of sugar, dairy products, and beef.

The root cause of GATT breakdown on agricultural trade, however, is worldwide tendencies for self-sufficiency, farm fundamentalism, and overrepresentation of farming interests in the political process. Significant agricultural trade reform would require adherence to GATT article 11 prohibiting import quotas, and to article 16 prohibiting export subsidies. In the Uruguay Round of GATT negotiations the Reagan Administration and the Cairns group proposed that nations subsidizing farmers be confined to *decoupled* supports, essentially to direct farm payments unrelated to production incentives. Producers would not be rewarded by payments for producing more. Payments would not interfere with access of consumers to low (world) food prices. While such decoupled of payments is unlikely to be successful and would not provide pure transfers, they would be a major improvement over current policies.

The thrust of trade liberalization under GATT has been confined mostly to trade in manufactured goods, excluding textiles and apparel, and to developed nations without severe balance-of-payments problems. GATT does not cover trade in services, intellectual property, and financial investment. Patents, copyrights, and trademarks are not covered but need to be. GATT is ambiguous regarding subsidies. It, for example, recognizes the right of a country to subsidize firms to help economically disadvantaged geographic areas, to facilitate economic restructuring, to maintain employment, and to promote other "important" objectives of social and economic policy. These issues need clarification.

Nontariff barriers which have replaced tariff barriers as the principle restraint to trade pose new problems for GATT. To deal with NTBs, techniques need to be developed to quantify and report NTBs perhaps on a tariff equivalent basis. Appropriate rules will need to be developed to delineate what is "fair trade" in this context and sanctions devised to deal with these violations. Much effort in the Uruguay Round must go into making nontariff trade barriers more transparent, developing set of rules that gradually reduce such barriers on a multilateral most-favored-nation basis, and streamlining and strengthening enforcement procedures. GATT is unlikely to become a primary and strong enforcer, however; for some time its major role will be to facilitate dispute settlements and sanction retaliations by member countries designed to bring noncompliers into line.

CONCLUDING COMMENTS

Successful trade negotiations need to recognize the relationship between macroeconomic policies and trade. Unstable macroeconomic policies breed protectionism. For example, an overvalued dollar making imports into the U.S. cheap and export dear causes Congress to pursue policies to protect U.S. industries from imports and to subsidize exports. An undervalued dollar causes our competitors to protect their industries that compete with farm imports from the U.S. and to subsidized their products that compete with U.S. products in third-country export markets. Coordination of macroeconomic policies among nations to achieve steady, sustainable economic growth with price stability is complementary with an open world trade policy.

As noted in Table 7.3, tariffs were once the major source of funds to run the federal government, accounting for nearly all federal revenue in 1791. Duties as a proportion of all imports reached 57 percent in 1830 and erratically declined thereafter. However, the Smoot-Hawley Tariff raised average tariff rates on U.S. dutiable imports to almost 60 percent by 1932 (Table 7.3).

The Reciprocal Trade Agreements Act of 1934 recognized the folly of high tariffs and shifted most authority over tariffs from Congress to the President. Congress gave the President flexibility to cut tariff rates on individual commodities up to 50 percent in other nations reciprocated but no tariff was to be cut if it threatened serious injury to a domestic industry. By the end of World War II, bilateral negotiations with major U.S. trading partners (with agreements extended on a Most-Favored-Nation basis) had reduced rates about one-third from Smoot-Hawley levels (Table 7.3). Because tariff barriers have been reduced to very modest levels in the 1980s, it follows that the remaining challenge is reducing nontariff barriers such as quotas — including those on imports of manufactured goods from developing countries.

The major new markets for U.S. agricultural products lie in industrializing developing countries. Obvious, massive gains are possible by opening U.S. markets to labor-intensive products of light manufacturing industries such as shoes, textiles, and apparel and of lower-skilled service industries such as construction, shipping, personal services, and printing. Large numbers of U.S. workers would be displaced, creating political fallout few politicians wish to face. Developing nations make less economic progress and buy less of our farm products because we will not accept the labor intensive industrial and service industry products which would earn dollars. A similar impasse is apparent in sugar. Developing countries cannot buy our farm and industrial products because trade barriers will not allow us to buy the sugar they produce with comparative advantage. Negotiations to end this socially and economically disastrous situation for less developed countries are complicated because GATT is dominated by developed countries.

TABLE 7.3. U.S. Tariff Rates Through 1984

				Ratio of Calculated Duties:		
Year	Imports	Percent Duty-Free	Calculated Duties	Total Imports	Dutiable Imports	Federal Revenue
	($ mil.)	(%)	($ mil.)		(percent)	
1791	n.a.	n.a.	4	n.a.	n.a.	99.5
1800	91	n.a.	9	9.9	n.a.	83.7
1810	85	n.a.	9	10.6	n.a.	91.5
1820	74	n.a.	15	20.3	n.a.	83.9
1830	50	8.0	28	57.3	61.7	88.2
1840	86	48.8	15	17.6	34.4	69.3
1850	164	9.8	40	24.5	27.1	91.0
1860	336	20.2	53	15.7	19.7	94.9
1870	426	4.7	192	44.9	47.1	47.3
1880	628	33.1	183	29.1	43.5	55.9
1890	766	33.7	227	29.6	44.6	57.0
1900	831	44.2	229	27.6	49.5	41.1
1910	1,547	49.2	327	21.1	41.6	49.4
1915	1,648	49.2	206	12.5	33.5	30.1
1920	5,102	61.1	326	6.4	16.4	4.8
1925	4,176	64.9	552	13.2	37.6	14.5
1930	3,114	66.8	462	14.8	44.7	14.1
1932	1,325	66.9	260	19.6	59.1	16.3
1935	2,039	59.1	357	17.5	42.9	9.0
1940	2,541	64.9	318	12.5	35.6	5.9
1945	4,098	67.1	381	9.3	28.2	0.7
1950	8,743	54.5	522	6.0	13.1	1.0
1955	11,337	53.3	633	5.6	12.0	0.9
1960	14,650	39.5	1,084	7.4	12.2	1.2
1965	21,283	34.9	1,643	7.7	11.9	1.2
1970	39,756	34.9	2,584	6.5	9.9	1.2
1975	96,516	32.2	3,780	3.9	5.8	1.3
1980	244,007	43.8	7,535	3.1	5.7	1.4
1984	322,990	31.9	12,042	3.7	5.5	1.4

SOURCE: Lande and VanGrasstek (1986).

The United Nations Conference on Trade and Development (UNCTAD) is on organization of mostly developing countries committed to improve export outlets and prices for products of member countries. Emphasis has been on commodity agreements and other price fixing schemes to assure markets and (at least) minimum prices. The efforts have largely failed. Many of the proposals of UNCTAD have been unworkable schemes to transfer wealth from wealthy to poor nations. UNCTAD lacks bargaining power to impose its demands on other countries.

Developed and less developed countries would do well to enter a dialogue of potential mutual benefit. The less developed countries of UNCTAD have not but need to join with developed countries in serious negotiations in which the less developed countries place their considerable trade barriers on the bargaining table with those of developed countries. Reciprocal barrier reductions will accomplish much; commodity agreements on the other hand have had a mixed record of success at best and offer little hope for improving terms of trade. Declining terms of trade in agricultural products in which developing countries have comparative advantage is not the result of a conspiracy against them and will not be solved by price fixing. Instead, developing countries increasingly must recognize that development proceeds as nations increase productivity through human and material capital formation. Diversification into new industries is vital. The development process begins with improvement in agricultural productivity and then moves to light (labor-intensive) industry, and eventually to high-skilled service industry. That process is cut off by trade barriers. And barriers will be removed only through mutual reductions in barriers by developed and less developed countries.

FOREIGN ECONOMIC ASSISTANCE PROGRAMS

With World War II, foreign economic assistance became a tool in American foreign policy and a major factor in world affairs. On March 11, 1941, Congress passed the Lend-Lease Act providing an economic weapon for Great Britain and a few other nations' war efforts. From 1941 through 1945, the United States provided $50.2 billion in aid under this act (Benedict and Bauer, 1960, p. 28). Of this amount, $6.5 billion were spent on agricultural commodities.

Lend-Lease assistance was entirely a wartime arrangement, with no important carryover of policy from it. Current economic assistance programs have their antecedents in the creation in 1942 of the Institute of Inter-American Affairs, a government agency responsible for providing U.S. technical assistance to Latin American countries (U.S. Department of State, 1966).

By the end of World War II, the United States had become involved in relief and rehabilitation in a large number of countries in Europe and Asia. Soon afterward the Truman doctrine led to large-scale military aid to Greece and Turkey. As the Cold War developed and the Marshall Plan was introduced to promote and accelerate war recovery in Western Europe, the volume of U.S. foreign assistance rose to unprecedented heights. The idea of aid for development of poor countries emerged as a logical extension of economic aid under the Marshall Plan and the United Nations Relief and Rehabilitation Administration program for reconstruction of war-damaged economies.

The main features of what was to become the European Recovery Program, or Marshall Plan, were outlined in 1947. The first appropriation under the program was made the same year, and in 1948 a four-year program was established. Most of the assistance was given in the form of U.S. commodities, including large shipments of wheat and other foodstuffs. No direct payment was made to the United States for these commodities. When received by European governments, the commodities were sold through regular trade channels. The funds obtained were placed in a trust account held in the name of the recipient country and could be released only on U.S. approval. The funds were generally released for long-term development projects in the countries. The economic assistance provided under the Marshall Plan totaled $13.2 billion.

The apparent success of the sales for nonconvertible currencies rather than outright grants or loans as a means of economic assistance in the European Recovery Program undoubtedly was of great significance in structuring the later Public Law 480 to aid developing countries. The Marshall Plan proved to be very successful in transforming the war-damaged economies of the European countries into highly productive economies with great potential for sustained growth. Hope for results in the poor countries of Asia, Africa, and Latin America, similar to those obtained in Europe, undoubtedly was a major force behind the sizable foreign aid appropriations after 1952. Barriers in culture, institutions, and human capital made for much less success from development funds directed to poor countries, leaving many Americans disillusioned with foreign aid.

The Mutual Security Act of 1951 included all current assistance programs except the Export-Import Bank activities. The objective of the act was "to maintain the security and promote the foreign policy of the United States by authorizing military, economic, and technical assistance to friendly countries to strengthen the mutual security and individual and collective defenses of the free world" (Brown, 1953, p. 509). It is apparent that the foreign assistance authorized under the Mutual Security Act of 1959 primarily was aimed at strengthening the position of the United States toward the Communist Bloc, while economic development of the recipient countries per se was, at best, a secondary objective.

The portion of the foreign aid funds allocated to direct military assistance was very large during the early 1950s. Military assistance accounted for 5 percent of the total foreign aid allocation in 1948-49, 32 percent in 1951, 16 percent in fiscal year 1967, and 40 percent in 1986 (Benedict and Bauer, 1960, p. 38; U.S. Senate, 1967, p. 334; Nelson and Folta, 1986).

Official economic development assistance, excluding military assistance, constituted .24 percent of U.S. GNP in 1984, the lowest among industrialized market economies (Simon, 1987, p. 9). The Scandinavian countries and the Netherlands spend nearly 1 percent of their GNP on foreign economic assistance. Even if the 40 percent of U.S. foreign assistance that is military is included, the U.S. ranks low among major industrial nations in aid *effort* as measured by proportionate spending. But because the U.S. economy is so large, it provides more total assistance than any other country.

Attention now turns to individual components of aid. The $14.5 billion spent by the U.S. on foreign aid in fiscal 1986 can be divided as follows:

		$ Billion		Percent
Development Aid		**3.86**		**26.7**
Bilateral	2.62		18.1	
Multilateral	1.24		8.9	
PL 480 Food Aid		**1.24**		**8.6**
Title I	.58		3.9	
Title II	.54		3.9	
Title III	.12		.8	
Economic Support Fund		**3.54**		**24.5**
Military		**5.82**		**40.2**
		14.46		**100.0**

Development aid totalling $3.86 billion in 1986 was the largest component of economic assistance (excluding military) and the majority of this was bilateral — country to country. Most of the bilateral aid was through the Agency for International Development (AID) which helps to develop the capacities of developing countries' people and institutions through agriculture, nutrition, family planning, health, and education. Emphasis in AID has varied over time from industrialization and modernization of infrastructure to basic needs (Carter administration) to influencing national policies (Reagan administration) through greater emphasis on the private sector and price incentives. The Agency for International Development (AID) was established in 1961 to carry out the functions of the Foreign Assistance Act. The AID has responsibility for carrying out nonmilitary U.S. foreign assistance programs and for continuous supervision and a general direction of all assistance programs under the Foreign Assistance Act. It also helps to allocate some of the local currency funds made available under PL 480. Multilateral development assistance is channeled with aid from other developed countries through the World Bank, Inter-American Bank, Asian Development Bank, African Development Bank, and other agencies such as the International Fund for Agricultural Development.

In 1987, strong efforts were underway to stop development assistance that aided technology, production, or transportation of foreign farm commodities that compete with U.S. farm commodities. Although certain types of foreign agricultural development depresses U.S. farm exports, on the whole it has been strongly contended that greater rates of growth in farm output in developing countries have been associated with greater rates of growth in U.S. farm exports to those same countries as noted earlier.

The Mutual Security Acts mentioned earlier had opened the way for disposal of agricultural surpluses, and commodity stocks were mounting during 1953 and 1954. This situation, along with the real need for food and fiber in a large number of developing nations, pressed the U.S. Congress for a more extensive program of surplus food disposal to developing nations. As an outcome of these pressures, the Agricultural Trade Development and Assistance Act, popularly known as Public Law 480, or PL 480, was enacted in 1954. Enactment of this law laid the foundation for what came to be the most extensive international transfer of agricultural commodities, on an aid basis, that the world has even experienced.

Food aid totalled $1.24 billion, nearly one-tenth of U.S. foreign assistance in 1986. Much of it moved through Public Law 480 (PL 480), also known as the Food for Peace program. PL 480 has led a troubled existence, torn by attempts jointly to dispose of surpluses, promote economic development abroad, revive the U.S. merchant fleet, and build foreign markets.

Title I of PL 480 provides food on easy dollar credit terms to developing countries. Commodities are sold by recipient governments to their citizens. As a condition for concessional sales, the recipient government agrees to use the proceeds to undertake needed development measures such as financing agricultural extension and research.

Title II of PL 480 was about the same size as Title I in 1986 and provided food as a grant distributed mostly by private voluntary organizations. The program has been large in famines such as have occurred in Africa from time to time. Title II also supports school lunch and infant and child-bearing-women programs as well as Food for Work programs where public works construction laborers are paid with food.

Title III entailed outlays of only $116 million in 1986. It changed Title I loans to grants for countries agreeing to undertake larger scale development activities than feasible under Title I.

The Economic Support Fund (ESF) promotes U.S. security and political interests. Nearly 60 percent of ESF funds went to Israel and Egypt in 1986. The $3.6 billion of aid to Israel and $2.5 billion to Egypt (BFW, 1987, p. 9) and $5.8 billion of military assistance left relatively little of the $14.5 billion in U.S. economic assistance for the billions of people in other developing countries in 1986.

NEW DIRECTIONS FOR FOOD AID

Our food aid programs have been widely criticized, and the Food for Peace Act of 1966 and subsequent revisions were designed to correct some of this criticism. One charge is that food aid has depressed farm prices in recipient countries and hence has discouraged private investment in agriculture. A related charge is that food aid has been more concerned with disposal of U.S. farm surpluses than with economic development in recipient countries. Furthermore, it is said that food aid has encouraged complacency in the recipient governments, thereby retarding needed public investment in research facilities, roads, and other infrastructure essential for a more productive agriculture.

To blunt these and other criticisms, PL 480 has shifted emphasis from sales for "soft" foreign currencies to sales on long-term dollar credit. An attempt is made to allocate PL 480 and dollar credit to countries and uses most conducive to development of agriculture. A third change in PL 480 emphasizes aid in commodities needed by aid recipients, not just commodities in surplus in the United States. This permits greater flexibility in aid programs and the ability to tailor commodity shipments more nearly to the needs of each country. This third revision also blunts misconceptions in the United States and foreign countries that food aid costs the U.S. Treasury nothing, because surplus capacity exits in agriculture. It also helps to blunt another misconception held by many Americans that surplus food is equal in its economic benefits, dollar for dollar, to aid in any other form. A final revision is greater reliance on private charitable organizations to distribute PL 480 emergency food assistance. This not only reduces U.S. government costs but also diminishes opportunities for U.S. food aid to be siphoned off to benefit corrupt government officials in recipient countries.

Shortcomings of PL 480 remain, however. PL 480 is not always allocated to most productive uses and it continues to interfere with local production incentives. Even greater effort is needed to direct PL 480 proceeds to high-payoff much-needed activities such as adaptive agricultural research knowing that such allocation may reduce commercial U.S. farm exports.

Finally, three-fourths of PL 480 must be shipped in U.S. flag vessels. A purpose is to subsidize and hence preserve the U.S merchant marine for national security reasons. Two shortcomings of such policy are apparent. First, large numbers of U.S. owned cargo vessels fly Liberian, Panamanian, and other flags to avoid union wage scales but would be available in national emergency. Second, if the U.S. merchant marine must be subsidized, a more appropriate approach is a direct cash subsidy unrelated to and hence not compromising U.S. food aid.

A Proposal to Raise the Efficiency of Foreign Aid

Because desire to dispose of surpluses was a primary motivation for U.S. food aid programs, and foreign economic development was often a secondary consideration, it may be concluded that in fact the alternative to food aid might have been no aid. Still, principles of economics apply in getting the maximum benefits from a given amount of federal funds. Of special concern here is the optimum amount of (1) food and nonfood foreign aid for maximum well-being of recipients from a given total aid appropriation, and (2) the maximum contribution to U.S. net farm income from a total government outlay for farm price and income supports through voluntary production controls.

Analysis indicates that on the average the United States has provided a proper balance of domestic farm production controls versus food exports to handle excess farm production capacity (Tweeten, 1966; White et al., 1974). The allocation has been less than optimum at the margin, however. An augmented market mechanism has been proposed that uses the equilibrating forces of the market to bring a more nearly optimum allocation.

The proposal is that foreign aid be provided in cash (or "script" usable only in the United States to purchase items most conducive to economic development if restrictions are deemed essential for balance-of-payment or other reasons). Allowable purchases might be restricted to the set of items that would contribute most to economic growth in developing nations. Foreign aid recipients would receive on each dollar's worth (market value) of food purchased in this country a discount equal to the cost to the U.S. Department of Agriculture of inducing farmers not to produce a dollar's worth of output with voluntary acreage-diversion programs.

Responsible governments might be free to use the aid money as desired. In other instances, the array of permitted purchase would be narrowed to whatever items would contribute most to economic development and eventual "weaning" of the recipient country from foreign aid. The proposal would give responsible governments greater flexibility than currently to chose the mix of assistance most nearly meeting development needs.

PLACE OF FOOD AID IN DEVELOPMENT ASSISTANCE

In conclusion it is important to note that food aid is a useful component of development assistance but its appropriate place is often not recognized. Assistance to agricultural development in less developed countries (LDCs) need not interfere with, indeed it is likely to assist, developed country (DC) exports.

The following set of priorities are suggested for DCs to promote development in LDCs:

1. ***Open Markets to LDC Exports.*** If countries do not import, they can't export, and the world loses. Opening up DC markets would do more to promote development in LDCs than do all current aid programs combined.

2. ***Hold Emergency Food Reserves.*** Adequate food reserves are unlikely to be held by LDCs to meet emergency food needs. Nor is it economic for them to hold such reserves. The food security facility of IMF remains in a formative stage but has promise. Financial insurance to be able to purchase food in export markets when supplies are short is much cheaper for LDCs' food security than are food stocks held in the country or attempts at food self-sufficiency. LDCs must have a place to turn for food supplies in the world when nature and pestilence combine to provide a short crop. Food reserves can be held by food exporting developed countries.

3. ***Basic Research in DCs.*** LDCs cannot afford basic research but it is the single brightest hope for dispelling the Malthusian specter in the long run. Biotechnology offers vast promise, but LDCs cannot afford the luxury of investing in costly research with such uncertain payoffs and large spillover of benefits to other countries.

4. ***Adaptive Research in LDCs.*** It is critical for the LDCs to have local research capacities to adapt research from elsewhere. Failure of LDCs to attract the brightest and best scientists available to them and maintain continuing support for their efforts is a major oversight which needs to be corrected mainly by the LDCs but with assistance of the DCs. Earnings from PL 480 imports can help pay for research and infrastructure investments.

5. ***Improve Infrastructure in LDCs.*** The market alone will not provide adequate infrastructure in the form of roads, bridges, port facilities, and schools. Public sector involvement is essential. Adequate infrastructure assists the market to work better.

DCs can help to fund public infrastructure local markets will not fund. But roads should not be built where LDCs will not maintain them.

6. ***Human Resource Development.*** DCs can help LDCs improve educational and vocational training facilities and services. Improved health care and family planning is possible in part from outside help in technology and funding from DCs. Such help in human resources is already apparent in the lower infant mortality and death rates in LDCs.

7. ***Macroeconomic, Trade, and Commodity Program Policies.*** Unfavorable monetary-fiscal policies in the United States and high rigid commodity price supports in the European Community and Japan have had serious unfavorable repercussions for LDCs in recent decades. One result is unstable world prices and markets as well as unfair competition in agricultural commodities. Ending or sharply revising such policies in developed countries would assist LDCs.

The magnitude of foreign trade and advantages of freer trade demonstrated in this chapter point to the need for this country to strive to keep trade channels as free a possible from institutional barriers. This has been attempted, particularly during the period extending from the Reciprocal Trade Agreements Act of 1934 through rounds of trade negotiations under GATT. While considerable progress has been made, many trade barriers remain. Neo-mercantilism remains among commodity groups in the United States and among foreign nations and trading blocs. Substantial efforts will be needed in the future to avoid rising levels of trade protection, especially as nations gain greater wealth to subsidize farmers.

REFERENCES

Andersen, Per. 1969. The role of food, feed, and fiber in foreign economic assistance. Ph.D. thesis. Stillwater: Oklahoma State University.

Benedict, Murray R. and Elizabeth K. Bauer. 1960. *Farm Surpluses, U.S. Burdens or World Assets?* Berkeley: University of California Press.

Blandford, David. 1986. Developed country agriculture: A clash of protectionism. Proceedings of conference on *U.S. Agriculture and World Markets*. Washington, D.C.: American Enterprise Institute.

Brown, William, Jr. 1953. *American Foreign Assistance*. Washington, D.C.: Brookings Institution.

Congressional Budget Office. June 1987. *The GATT Negotiations and U.S. Trade Policy*. Washington, D.C.: U.S. Government Printing Office.

Council for Agricultural Science and Technology (CAST). 1988. Long-term viability of U.S. agriculture. Ames, Iowa.

Gardiner, Walter and Praveen Dixit. 1986. Price elasticity of export demand. ERS Staff Report No. AGES860408. Washington, D.C.: ERS, USDA.

Lande, Stephen L. and Craig Van Grasstek. 1986. *The Trade and Tariff Act of 1984: Trade Policy in the Reagan Administration*. New York: Lexington Books.

Lavergne, Phillipe Real. 1981. The political economy of U.S. tariffs. Ph.D. thesis. Toronto: University of Toronto.

Nelson, Paul and Sam Folta. December 1986. Foreign aid. Background Paper No. 92. Washington, D.C.: Bread for the World.

Office of Technology Assessment. 1986. *A Review of U.S. Competitiveness in Agricultural Trade*. Washington, D.C.: U.S. Congress, OTA.

Paarlberg, Robert. 1987. Agriculture and the developing world: Partners or competitors? Chapter 8 in Randall Purcell and Elizabeth Morrison, eds., *U.S. Agriculture and Third World Development*. Boulder, Colorado: Lynne Rienner.

Schuh, G. Edward. 1976. The new macroeconomics of agriculture. *American Journal of Agricultural Economics* 58(5):802-11.

Simon, Arthur. August 1987. Basic facts. Background Paper No. 99. Washington, D.C.: Bread for the World.

Tweeten, Luther. 1966. A proposed allocative mechanism for U.S. food aid. *Journal of Farm Economics* 48:803-10.

Tweeten, Luther. 1967. The demand for United States farm output. *Food Research Institute Studies* 7:343-69.

Tweeten, Luther. December 1983. Economic instability in agriculture. *American Journal of Agricultural Economics* 65(5):922-931.

U.S. Department of Agriculture. April 1987. Government intervention in agriculture: Measurement, evolution, and implications for trade negotiations. FAER-39. Washington, D.C.: ERS, USDA.

U.S. Department of Agriculture. 1987. Handbook of agricultural charts. Washington, D.C.: ERS, USDA.

U.S. Department of State. 1966. *The AID Story*. Washington, D.C.: Government Printing Office.

U.S. Senate. 1967. *Foreign Assistance Act of 1967*. Hearings before the Committee on Foreign Relations, 90th Congress, 1st Session. Washington, D.C.: Government Printing Office.

Vollrath, Thomas. 1987. Revealed comparative advantage for wheat. Staff Report AGES861030. Washington, D.C.: ERS, USDA.

White, Fred, Luther Tweeten, and Per Andersen. 1974. Allocation of agricultural production capacity among commercial markets, food aid, and production control. *Southern Journal of Agricultural Economics* 6(2):129-35.

World Bank. 1986. *World Development Report 1986*. New York: Oxford University Press.

World Food Institute. 1986. World food trade and U.S. agriculture, 1960-1985. Ames: Iowa State University.

CHAPTER EIGHT

Agribusiness Conduct, Structure, and Performance

The agribusiness sector is frequently blamed for an alleged cost-price squeeze defined as a chronic tendency for prices received by farmers for crops and livestock to be low relative to prices paid by farmers for production inputs. We observed in Chapter 4 that rates of return on farm resources are better measures of economic health than is the ratio of prices received to prices paid by farmers in an industry characterized by dynamic changes in productivity.

At issue is whether farmers are exploited by the agribusiness sector comprised of input supply firms and product marketing firms. We test the proposition stated by Tony Smith (1986) that "a competitive [farming] sector sandwiched between two oligopolistic [agribusiness farm input supply and product marketing] sectors will inevitably experience disadvantaged terms of trade." The proposition is evaluated at a conceptual and empirical level.

CONCEPTUAL FRAMEWORK

Before appraising the impact of the agribusiness sector on the farming industry within the conceptual framework of workable competition, it is well to review what neoclassical economic theory tells us about a competitive farm sector facing a monopoly selling inputs to farmers and a monopsony purchasing farm output. The markets in which farmers purchase inputs and sell commodities are generally characterized as some form of imperfect competition. The actions of a single firm can influence the market price. Whereas the sales of any one farmer have no apparent impact on total output and price in the farming industry, the sales of one firm supplying farm inputs or buying farm output often can perceptibly influence industry output and price. An industry characterized by *oligopoly* has just a few firms and the actions of one firm are felt by the other firms, hence pricing and output decisions are interdependent. An industry in which there is just one firm that can adjust industry price and output at will and can keep other firms from entering the industry is called *monopoly*. A market characterized by one firm which is the sole seller of a product or resource is called monopoly, and a market characterized by a single firm which is the sole buyer of a product or resource is called *monopsony*.

Most industries which sell farm inputs and market farm products are oligopolistic. There are no neat analytical models to predict the behavior of oligopolistic industries. These industries are best analyzed in the framework of market structure, conduct, and performance presented later. Nevertheless, the rigorous models of monopoly and monopsony suggest hypotheses and illustrate *tendencies* characterizing pricing and output under imperfect competition and hence are the point of departure in this section. The implications of imperfect competition for farm prices and quantities, both for resources and products, are examined.

IMPACT ON FARM SECTOR IF MARKETING SECTOR BEHAVES AS A MONOPSONIST

Assume for the sake of analysis that one giant monopsonistic firm buys all farm output. The analysis also applies to a single farm commodity market facing only one buyer because it is isolated from a large number of buyers by high transportation costs. A single processor may buy the entire area output of fruits and vegetables, for example. The possible effect on farm pricing and output is apparent in Figure 8.1. The demand for the

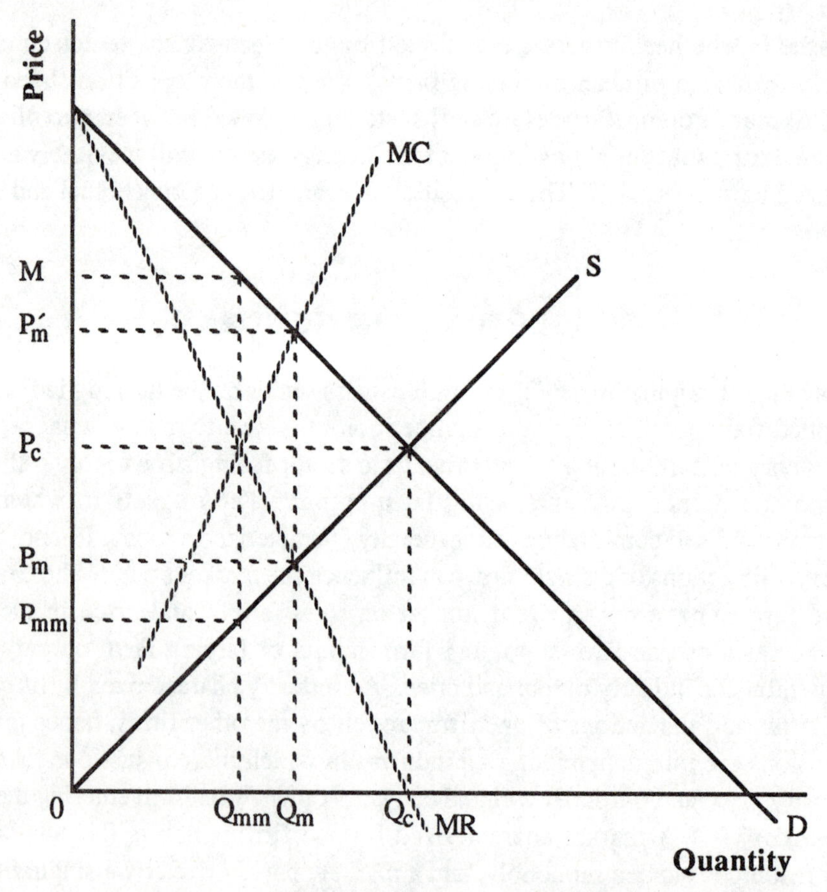

FIGURE 8.1. Hypothetical Supply and Demand Curves

commodity by the single buyer is D and the supply is S. Under competitive conditions in buying and selling, the equilibrium price is P_c and the quantity is Q_c. If the single buyer could perfectly discriminate among producers, paying only the supply price for each unit of production, the marginal cost curve would be S. But if the monopsonist must pay one price for the entire quantity purchased, then the marginal resource cost is MC. The monopsonist firm which buys from many farmers and which sells in a competitive market maximizes profits at that quantity where demand equals MC. The quantity purchased from farmers is Q_m and the price paid farmers is P_m. The effect is a lower price, lower quantity, and lower income to farmers than under competitive conditions.

If the monopsonist firm is also the single monopoly seller of the processed commodity, the firm's profit is maximized at the point where marginal revenue MR is equal to MC. The result of monopoly and monopsony in the firm buying the commodity is to reduce the quantity to Q_{mm} and the farm price to P_{mm}, hence to reduce farm income even further than in the previous case. Net social cost is the area bounded by D, S, and M-P_{mm} at Q_{mm}.

Next consider the situation where farmers form a cohesive bargaining group to sell their output as a monopoly, given the demand D for their output Q. The marginal revenue for farmers is MR. If they were to maximize profit, they would market that quantity Q_m where MR intersects S. They would receive a price P_m' for the commodity and would make a greater profit than at the competitive equilibrium price P_c and quantity Q_c.

Finally, assume that the demand for farm production is D and that farmers have a bargaining group which pools production to sell as a monopoly. Also assume a single buyer of the commodity, a monopsonist, faces the supply curve S and marginal resource cost curve MC. With this bilateral monopoly, the monopsonist processor wishes to buy Q_m and pay P_m; the monopolist farmer-group wishes to market Q_m and receive a price P_m'. While there is agreement on quantity, the model stipulates no unique price. Where the price will settle between P_m' and P_m depends on the bargaining power of the two antagonists. This indeterminateness could be eliminated either by farmers' integrating into the processing business or by the processor's integrating into the growing operation. If the integrated firm's marginal cost curve were S and it sold its product in an imperfect market with a marginal revenue MR, its output would be Q_m. However, if the integrated firm were confronted by a market situation where competition kept the price equal to marginal cost, then the price and output would be the competitive result, P_c and Q_c. The analysis shows that integration conceivably can improve market performance.

The most important conclusion relates to the thesis of this chapter: despite imperfection in the market in the form of only one buyer of farm output, represented, for example, by P_{mm} and Q_{mm} pricing and output, *farm resources will not receive low relative returns per unit if they are mobile*. Too few resources would be in farming and the nation's total resources will be earning lower returns than with a well-functioning market. But farm resource returns would be precisely equal to nonfarm resource returns at the margin.

Impact on Farm Sector if Input Supply Sector Behaves as Monopolist

We now examine the effect on farm wages and employment of monopoly selling of labor or other input used by farmers. The national labor supply is assumed to be S in Figure 8.2. The demand for labor in the farm sector VMP is assumed to be the same as the labor demand in the nonfarm sector. Total labor demand is therefore ΣVMP. If labor is sold competitively in both sectors, supply intersects demand at a wage rate W_{cc}, giving employment L_{cc} in each sector.

Suppose that a single labor union gains monopoly control over the sale of labor in the nonfarm sector.[1] The marginal revenue of labor is MR in the nonfarm sector. With labor

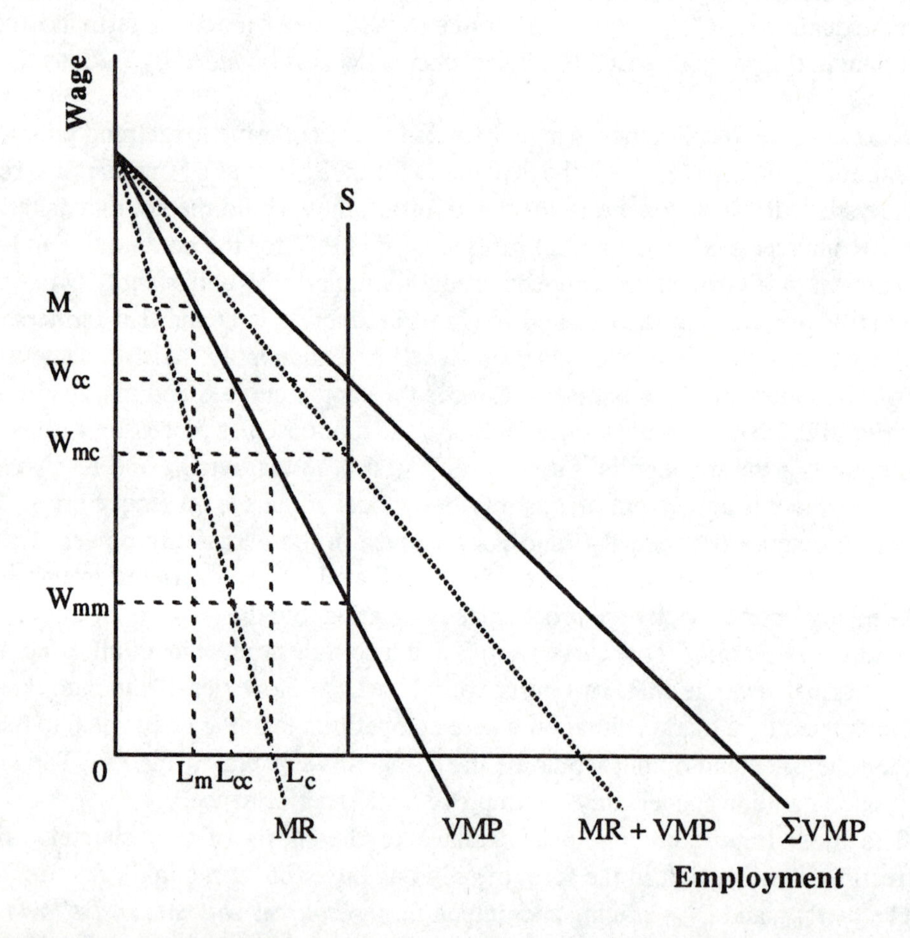

FIGURE 8.2. Hypothetical Demand Curves and the Supply Curve for Labor

---

[1]The assumption is that labor is allocated between sectors to maximize returns to all labor. It is as if one large labor allocator sold in a competitive structure in one market and as a single seller in the other market. If the nonfarm market were completely separated and operated independently of the total labor supply, the results would not necessarily be those above because the total "demand" would not be the sum of marginal revenue curves. This comment also applies to other parallel cases.

sold competitively in farming, the total labor marginal revenue from the two sectors is MR + VMP. This intersects supply at W_{mc}, the wage in the farm sector. However, M is the wage and L_m is the employment in the unionized sector. Based on the competitive norm, the wage is too low and the employment too large in agriculture. The opposite holds for the unionized nonfarm sector.

Again assume that the demand for labor is VMP in each sector but that labor is unionized into a monopoly seller in both sectors. The marginal revenue curve in each sector is MR. Optimum allocation for maximum wage income occurs where ΣMR = VMP intersects S at W_{mm}. While the marginal revenue is W_{mm}, the wage in each sector is W_{cc} and the employment is L_{cc}. The interesting result is that wages and employment are the same as under perfect competition. Having both sectors as monopoly unions is clearly preferred to having only one sector so organized. This result provides support for countervailing power. However, the lack of social cost with monopoly labor control in both sectors occurs only when supply is perfectly inelastic. In the real world, supply is not perfectly inelastic. This fact weakens the justification for countervailing power.

The outcome is similar if one perfectly discriminating monopolist allocates labor to the farm and nonfarm sector. Suppose the monopolist faces a perfectly elastic farm demand curve for labor, a horizontal line at W_{mc} in Figure 8.2. The nonfarm demand curve for labor is VMP and marginal revenue is MR. Defining S as the marginal cost curve of the discriminating monopolist, the profit maximizing position is where the sum of the marginal revenue curves (MR down to W_{mc}, then the horizontal line at W_{mc} to the right) equals marginal cost S. The wage in the nonagricultural sector is M and employment is L_m; the wage in the agricultural sector is W_{mc} and employment is $S - L_m$.

The conclusion is that returns to resources can remain low in farming relative to other sectors if an input such as labor is allocated to farming by a discriminating monopolist facing a more inelastic demand in the nonfarm sector than in the farm sector or by a union able to control employment and wages in the nonfarm sector but not in the farm sector. Whether imperfect competition in farm capital, labor, and other input markets is sufficiently high to cause persistent low returns is an empirical question to be addressed later.

Impact on Farm Labor of Monopoly Selling of Farm Products

We now complete the conceptual framework by examining the impact on farm resource returns of monopoly selling of farm products by the marketing sector.

Monopoly Selling Does Not Give Low Relative Returns to Farm Labor. Farm resources potentially can produce either farm products or nonfarm products. For example, in Figure 8.2 the national labor force is considered to be the resource. Its derived demand, or value of marginal product VMP, is assumed to be the same in the farm and nonfarm sector. Total labor supply S is assumed to be perfectly inelastic. Total labor demand is the sum of the demand in each sector, or ΣVMP in Figure 8.2. The equilibrium wage W_{cc} and employment L_{cc} (equal in each sector when both sectors sell products in a perfectly competitive market) are indicated by the intersection of supply curve S and the total demand curve ΣVMP.

Now assume that the nonfarm sector sells its product as a monopolist. Its marginal revenue product curve is MR. The farm sector continues to sell its products in a perfectly competitive market, and its demand for labor is the VMP curve. Labor wage in the economy is where MR + VMP intersects supply at a wage rate W_{mc}. While labor receives the same wage in each sector, the allocation of labor is only L_m to the nonfarm sector and L_c to the farm sector. Labor is exploited by nonfarm industry because its contribution to the value of output M considerably exceeds the wage W_{mc}. A reallocation of some labor from agriculture to nonfarm industry would increase the total value of output. Hence monopoly pricing and output represents a social cost.

Next assume that farm products also are sold monopolistically. The demand for farm labor then becomes the marginal revenue product MR in Figure 8.2. Total labor demand in the two sectors is MR + MR, which is the VMP curve. The curve intersects S at a wage rate W_{mm} received by labor in both sectors. Allocation of labor to each sector L_{cc} is the same as when both sectors sold their output competitively. Thus there is no malallocation of labor based on the competitive norm, but the wage rate W_{mm} is considerably below both W_{mc} and the value of marginal product W_{cc}. Labor in both sectors is exploited to the extent that it receives less than its marginal contribution to the value of output.

In summary, the example illustrates that monopoly selling of products in the nonfarm sector does not lead to relatively lower wages in agriculture but does lead to generally low wages and to malallocation of labor between sectors.

Marketing Margin Behavior. The economic behavior of agribusiness affects the behavior of marketing margins which in turn influences the level and variability of farm prices. The effect of marketing margins on farm and retail markets is depicted in Figure 8.3. The quantity is defined as the ingredients provided by farmers for items purchased by consumers. The supply of farm commodities at the farm level is S_F. The supply S_R of farm ingredients at the retail level is the farm supply plus marketing margins. The demand by consumers for farm output at the retail level is D_R. The derived demand D_F at the farm level is D_R less marketing margins. Marketing margins include transportation, processing, wholesaling, and retailing costs. With constant margins, the demand curves (and supply curves) at the farm and retail levels have the same slopes. Then the price elasticity at the farm level will bear the same ratio to the elasticity at the retail level as the farm price bears to the retail price at a particular quantity. This means that the price elasticity of demand for farm products is lower at the farm level than at the retail level. If margins rise per unit as quantity expands, this tends to make the demand for farm products even more inelastic. If margins decrease per unit as the quantity expands, this tends to make the demand for farm output less inelastic. In fact, marketing costs tend to remain somewhat stable as farm output expands and contracts.

The market initially is in equilibrium at quantity Q, farm price P_F, and retail price P_R. Now assume that elements of imperfect competition are introduced in the form of excess profits or advertising costs that neither educate consumers nor raise the demand for farm

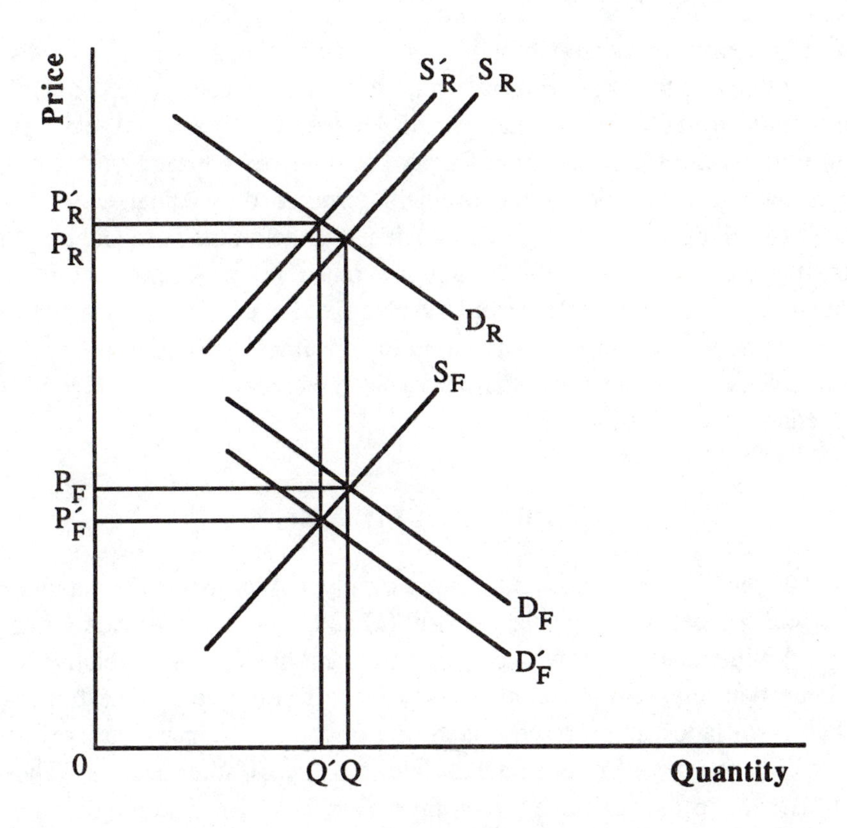

FIGURE 8.3. Marketing Margins; Farm and Retail Supply and Demand for Farm Commodity Q

commodities. These "costs" are assumed to raise the marketing margin by a fixed amount for all quantities and therefore to reduce demand at the farm level to D_F' and to reduce supply at the retail to S_R'. The result is a lower equilibrium farm quantity Q′, lower farm price P_F', and a higher consumer price P_R'. Conversely, viewing D_F' and S_R' as an initial position, part of the gains from removing wasteful advertising and excess profits from the marketing sector are passed on to farmers and part to consumers.

The distribution of benefits depends on the supply and demand elasticities. With marketing margins reduced by a given amount at all farm quantities as in Figure 8.3, the following are the guidelines for distribution of benefits: (1) if demand is perfectly inelastic, all benefits go to consumers and the quantity Q is unchanged; (2) if supply is perfectly inelastic, all benefits go to farmers and Q is unchanged; (3) if demand is perfectly elastic, all benefits go to farmers and the farm quantity Q is increased; and (4) if supply is perfectly elastic, all benefits go to consumers and the farm quantity Q is increased.

In fact, the actual distribution is somewhat indeterminate because both supply and demand for farm output are inelastic in the short run but not perfectly inelastic. Thus it is not possible to say exactly whether farmers or consumers gain the most benefits from increased marketing efficiency. However, aggregate supply and demand elasticities have been found to be somewhat comparable in magnitude (but opposite in sign, of course) in the short and intermediate run at the farm level. A crude inference therefore is that farmers and consumers share somewhat equally in the benefits of marketing efficiency — as depicted in Figure 8.3. However, farm supply becomes more elastic in the long run and benefits eventually pass in majority to consumers as specified by guideline (4) above.

Some measure of the impact of technological change on the farm and retail sectors is also apparent in the figure. As farm resources become more productive, supply S_F moves to the right and the farm price is depressed. With a fixed marketing margin, the benefits of the lower farm price are completely passed on to consumers. While benefits of improved farming technology go to consumers, farm prices and resource returns tend to be depressed in the short run before adjustments.

EMPIRICAL FINDINGS

Economic theory predicts that if (1) input supply, (2) farm, and (3) marketing sectors begin as perfect competition and subsequently (1) and (3) become monopolistic, then the farm sector (2) will experience low prices, incomes, and rates of return relative to the input supply and product marketing sectors. However, if resources of the farm sector are mobile, then farm labor, management, and equity capital after making adjustments to incentives will earn rates of return comparable to those of other sectors. Theory alone cannot estimate the impact of imperfect competition, however, and empirical analysis is essential to gauge the impact of the agribusiness sector on the farming economy.

Empirical evidence reveals that farm resources are not highly mobile in the short and intermediate run of up to five years, hence farmers experience annual and cyclical periods of low income and rates of return. Chapters 5 and 6 made the case that the annual and cyclical instability results from weather; monetary, fiscal, and trade policy at home and abroad; and from imperfect outlook expectations. This empirical section makes the case that imperfect competition in the private agribusiness sector has not been a significant element.

The classical analytical frameworks of pure competition, monopolistic competition, oligopoly, and monopoly here give way to the concept of *workable competition*, which is a more meaningful if less rigorous orientation. Market structure, conduct, and performance analysis operationalizes the concept of workable competition with analytical concepts providing a more adequate yardstick for the analysis of farm markets. Market *structure* refers to characteristics of the market, including the supply and demand parameters, number and size of buyers and sellers, differentiation of products, barriers to entry, and extent of market integration. Market *conduct* is the behavior of enterprises, methods of determining price, sales promotion, efforts to vary products, and incidence of predatory or exclusionary practices. Market *performance* is the economic efficiency of the market. It is measured by efficiency in procurement, plant utilization, and distribution, by the amount and type of sales promotion, innovative activity, and quality of product, and by the level of profits. A market need not have perfect information or perfect mobility to be efficient; it needs only to gather information or move resources if expected benefits exceed costs. A market that is characterized by a single large firm and high advertising costs may *not* be viewed as "bad" or inefficient if economies of size are so large relative to the market that only one firm can produce at an efficient level and supply the market, if it is innovative, if it does not charge unreasonable prices, and if its advertising is educational and its profits are in line with risk and capital costs.

It was noted earlier that input markets are as important as product markets to the economic well-being of farm people. Farm product markets are examined before turning to land, labor, and other input markets.

FARM PRODUCT MARKETS

This section is divided into three subsections. The first deals with marketing, especially food manufacturing and retailing where alleged excess profits and advertising are most prominent. The two remaining subsections specifically examine the grain export industry and the futures market, two subsectors frequently vilified by farm activists.

FOOD MARKETING

Consumers spent approximately $350 billion for food in 1985. Farmers received 31¢ of the dollar spent for food consumed at home and 14¢ of the dollar spent for food consumed away from home. As incomes rise and as more spouses work outside the home, demand for food processing and convenience grows. Consumers derive place, form, and time utility from processing, transportation, storing, wholesaling, and retailing functions performed by the marketing sector. Many farmers view the marketing sector contribution not as value added but as an unnecessary or at least overpriced cost. What appears to be diabolical (the falling share of the food dollar going to farmers) is normal and expected in an affluent economy where consumers prefer to spend additional dollars for food convenience rather than for more ingredients from the farm. The income elasticity of demand is higher for marketing services than for farm food ingredients. American consumers spend a lower percentage of their income for food than do consumers in any other nation while receiving a greater quantity, quality, and variety of food. At issue, however, is whether changes could be made in the marketing sector to increase efficiency and equity in meeting consumer demand while at the same time improving the farm economy. Such opportunities are not great for reasons given below.

1. Experts Note Some Inefficiency in the Marketing Sector But Disagree on the Magnitude. After-tax profit margins as a percent of sales were 3 percent for food manufacturers and 1 percent for retail food chains for each year from 1980 through 1985. Returns on equity averaging 13 percent for food manufacturing and retail food chains were in line with those of other industries and of adequate-size, well-managed family farms. That does not rule out opportunities to improve market performance.

A massive literature reports measures of the scope and importance of monopoly in the food marketing (including manufacturing, wholesaling, retailing, etc.). A landmark was the report of the National Commission on Food Marketing (1966) which reviewed the state of knowledge regarding the structure, conduct, and performance of the food marketing industry. Separate majority and minority reports each gave the industry high marks for efficiency — for moving a high volume of quality food to consumers at reasonable cost. However, the majority report called for greater government surveillance and reform of the marketing system while the minority report rejected the recommendations of the majority and instead called for no changes in the system before more careful analysis indicated such changes would be beneficial. The minority noted that strong legislation and regulatory powers already exist to avoid or control monopoly and predatory behavior, and to encourage conduct in the public interest. No new legislation was necessary, they contended.

Extensive subsequent studies (Parker and Connor, 1979; Marion, 1985; Connor et al., 1985) found considerable inefficiency and monopoly pricing in the food marketing industry but other analysts (O'Rourke and Greig, 1981; Bullock, 1981; Gisser, 1982) strongly objected to those findings.

Quail et al. (1986, p. 55) estimated that slaughter cattle prices would have been a modest 24 to 47 cents higher per hundredweight in four U.S. regions if the regions would have had lower beef-packer firm concentration ratios and hence more competition. Critics strenuously objected. Ward (1986) and Bullock (1986) rejected the conceptual and statistical models used by Quail et al. Ward accused the authors of underestimating economies of size — less concentration could have increased packer costs and reduced beef prices even more. Bullock noted that transportation costs and whether regions are surplus or deficit in beef production relative to consumption were not adequately accounted for by Quail et al. These factors according to Bullock might better explain price differences attributed to concentration and could support the conclusion that the beef packing industry is competitive.

In short, there is considerable agreement among experts that some inefficiency, extensive advertising, non-uniform pricing, and sizable profits are found in some components of the food marketing industry but the magnitudes are hotly disputed. Many experts have concluded that inefficiency is small relative to the size of the marketing sector and the standards of workable competition. Profit rates in the marketing sector comparable to those elsewhere and constituting only 5 percent of the food marketing bill are essential to attract and hold resources so that the industry can provide services demanded by consumers.

2. Even if the Highest Estimates of Inefficiency in the Marketing Sector Were Accepted, It Does Not Necessarily Follow That an Alternative System Would Entail Less Social Cost. Every industry, including the atomistic (many firms) farming industry and the more concentrated food marketing industry, contains inefficiency in that changes could be made to increase output per unit of input. But making such changes, especially through the political and public administrative process, entails costs. These costs in resources used to induce change are frequently greater than the cost of inefficiency being eliminated. Bruce Marion (1985, p. 422) and the NC 117 Committee state:

> Suppose the entire marketing system of a particular subsector were wiped out? Would we expect it to be replaced with all the same institutions and the same structure of firms at each level and coordinated vertically as before? Probably not. Yet most of us would be cautious in pressing forward possible improvements. In learning more about how market institutions work in the various subsectors, we have had to recognize an important diversity. It is no longer so tempting for a commodity specialist to assume that an institution that works well in one subsector will work equally well in others.

3. Even if Inefficiency in Marketing is Large and Could be Eliminated at Low Cost, Farm Problems Would Remain Essentially Unchanged. Whereas a case can be made that *some* parts of the food marketing industry such as food manufacturing are characterized by high concentration, pricing above socially optimal levels, and excessive product differentiation and advertising, it is not clear that an atomistically competitive agribusiness system would entail less social cost than the current system. Most of the social cost of imperfect competition in food marketing falls on consumers and not on farmers.

Suppose we take Parker and Connor's estimate of $12 billion loss from monopoly in food manufacturing in 1975, and arbitrarily double it to include costs of monopoly elsewhere and inflate the total to 1985 using the consumer food price index. The result is an estimated cost of monopoly of $36 billion for 1985, or 10 percent of the $350 billion consumers spent on food. A crude estimate based on supply and demand curve slopes is that consumers and food industry suppliers would split the savings 50:50 — for $18 billion apiece. Farmers supply just over one-fourth of inputs used by the marketing sector, hence could receive about $4.5 billion. Would this amount, outrageously high and unattainable according to most economists, solve farm price, income, and structure problems? It certainly would help in the short run. But benefits soon would be bid into land so comparative advantage between large and small farms would remain unchanged. The principal commercial farm income problem is annual and cyclical instability and the principal structure problem is technology, economies of size, and differences in comparative advantage among farms leading to the loss of family farms. Eliminating excess profits and waste from the marketing sector would have no lasting influence on these problems.

In the comprehensive reviews by Marion (1985) and by Connor et al. (1985), very little was said of the impact of inferior market performance on producers. An exception was the statement that "in markets for processing vegetables, hogs, and fed cattle, processors were able to significantly suppress farm-level commodity prices (Connor et al., 1985, p. 414). It is of interest that producers' profits have been favorable in the processed vegetable and hog subsectors on the average over the years. Even if some cattle feeders would have received an additional 24¢ to 47¢ per hundredweight from a less concentrated packer industry as calculated by Quail et al. (1986, p. 55) — a highly suspect conclusion as noted earlier — still, problems of beef cattle producers would have remained largely unchanged. Even if the findings of Quail et al. were accepted, an end to imperfect competition adding 47¢ per hundredweight to beef prices would be inconsequential and go almost unnoticed in a sector where beef prices swing $30 per hundredweight from the cattle cycle alone.

Monopoly power and efficiency losses therefrom have been found to be concentrated in a relatively few agribusiness industries; the breakfast cereal industry especially stands out. Using corn as an example, the breakfast cereal industry uses 25 million bushels or .3 percent of the approximately 8 billion bushels produced and consumed (Marion, 1987). Changes making the breakfast cereal industry more competitive might be desirable but would have no perceptible impact on corn sector demand and price. Similar statements could be made for wheat, rice, or oats.

4. Even if the Marketing Sector is Imperfectly Competitive, It Has Not Resulted in Chronic Low Rates of Return to Farmers. Theory cited earlier suggests that if farm resources are mobile, then adequate-size, well-managed farms will earn a return on the average comparable to returns on resources elsewhere whether the marketing sector is competitive or not.

Chapter 4 summarized profit rates by size of farm for aggregate output. Of interest is that adequate-size, well-managed farms covered all resource costs in each of several years analyzed from 1960 to 1984. These farms were within a family-size category although at the upper end. Smaller farms on the average lost money. On the average, small farms are less economically efficient than larger farms and their operators choose to subsidize farming with off-farm income and tax breaks. They too seem to be earning full social returns on the average because small-farm numbers have been growing based on census data for 1978 and 1982.

The implication is that adequate-size, well-managed farms adjust resource-use, output, and land values rather quickly to changing conditions. Agribusiness firms consistently paying below the resource costs of production will not be supplied. A higher commodity price might encourage commercial farms to expand their operations further, buying out and consolidating smaller, less efficient operations. There is no evidence that higher prices would result in more family farms over the long run.

It may be contended that farms grow in size to countervail the market pressure of large marketing firms with which they bargain. Incentives exist for farmers to gain market power whether the agribusiness sector is oligopolistic or atomistic and farmers frequently band together in marketing orders, cooperatives, or other collectives to bargain better. Collective bargaining to preserve the family farm does not necessarily serve the best interests of excluded family farmers, consumers, or society.

5. Less Advertising Would Reduce Demand for Farm Products. Although it is not possible to conclude a priori that oligopoly will be more or less efficient or pay more or less for farm output than would an atomized market structure, a good case can be made that oligopoly begets extensive product differentiation and advertising. Outlays are large for food advertising and may be one reason why the principal malnutrition problem in the U.S. today is chronically eating too much rather than too little. Although too much eating is socially undesirable, it benefits farmers as producers. A well functioning market providing optimal nutrition would reduce domestic demand for food by up to 15 percent (Tweeten, 1979, p. 382). An atomized food industry with less product differentiation, advertising, and controlled to serve the public interest likely would reduce the demand and price for farm output on the average.

The market studies cited earlier revealed higher costs in concentrated marketing industries and attributed some of these costs to higher wage rates. Perhaps firms also paid more for farm products. I am unaware of evidence that farmers are paid less for their produce by more concentrated than by less concentrated industries at the finished product level.

*6. **Cooperatives Provide Alternative Markets and a Yardstick of Performance.*** Presence of producer cooperatives reduces chances for exploitation of farmers. As apparent in Table 8.1, producer-owned cooperatives constitute a sizable share of the market (nearly one-third on the average) and are prominent in nearly all major categories of farm output and input. On the whole, cooperatives have increased market

TABLE 8.1. Number of Cooperatives and Cooperatives' Share of Farm Supplies Purchased and Products Marketed, 1973 and 1983

	Number of Co-ops		Share of Market	
	1973	1983	1973	1983
		(percent)		
Marketing				
All farm products	5257	4175	23	30
Milk and dairy products	657	418	76	77
Grain and soybeans	2567	2271	29	38
Cotton and products	507	488	21	31
Fruits, vegetables, and products	447	394	23	19
Dry beans and peas	47	23	23	18
Livestock and wool	748	402	9	11
Poultry and eggs	174	63	7	8
Supplies purchased				
All major farm supplies	5574	4243	23	27
Fertilizer and lime	3990	3226	36	38
Petroleum and products	2647	2562	38	38
Farm chemicals	3353	3202	19	35
Feed	3790	3097	18	19
Seed	3591	3121	17	14

SOURCE: Wissman (1985, pp. 18-19).

share but their share has decreased in fruits and vegetables. Cooperatives have a larger share of first-buyer markets than of retail markets, hence provide more market power in the former than the latter markets. A number of cooperatives have integrated vertically in recent years to operate in nearly all phases of marketing. Some cooperatives have consolidated or in other ways grown to a size providing countervailing power against large private firms. In fact, the predatory conduct of some large cooperatives has drawn successful litigation by federal antitrust agencies in recent decades.

7. The Marketing Sector Acts as a Shock Absorber for Prices. If the principal economic problem of commercial farmers is annual and cyclical price and income instability as contended, then attention is turned from the *level* to the *variability* of marketing margins (see Figure 8.3). Studies reveal that marketing firm behavior causes margins to behave like a shock absorber (Council on Wage and Price Stability, 1976, pp. 1, 2). For most commodities, margins widen when farm prices fall or retail prices rise. Margins narrow when farm commodity prices rise or retail prices fall. In less than one year, however, higher marketing costs tend to be fully passed to farmers and consumers.

Farm input prices seem to react differently than output prices. Tweeten (July 1983) found that prices paid by farmers overshoot in response to a change in the general price level (inflation). However, the overshoot is corrected in approximately three years. No overshoot tendency was apparent in prices received by farmers.

The market structure of the food industry causes per unit marketing margins to be quite stable from year to year despite variation in the volume of products handled. This places on farm and retail prices the main burden of adjustment to changing demand and supply of farm products. The result is substantial annual and long-term variation in prices received by farmers in the absence of buffer programs. An alternative structure of food marketing would not necessarily make a difference. The variability problem is better handled by buffer stocks and private risk bearing markets rather than by forced atomization of the food industry.

8. Even if the Marketing Sector Were to Reduce Farm Prices, It Would Not Necessarily Result in Fewer Family Farms. As indicated earlier, there is no evidence that reorganization of the marketing sector would raise farm commodity prices. But even if the current structure of the marketing sector causes farm price to be lower and more variable on average, it would not necessarily reduce the number of family farms. Size and number of farms are determined mostly by economies of size — which are dominated by production rather than market economies. In the absence of economies of size, family-size farms tend to replace failed family-size farms when economic conditions are unfavorable. Industrial-type corporations have little interest in entering a farming industry characterized by low and variable returns whether caused by agribusiness structure or other factors. The European Common Market with high rigid price supports has had a more rapid rate of decline in farm numbers than has the U.S. since 1959.

9. Some Marketing Firms are Vertically Integrated into Farming, But Such Integration is Not a Serious Threat to the Family Farm or to Food Production Efficiency. Vertical integration, in which firms involved in input supply or product marketing also control farm production, increased from 4 percent of farm output in 1960 to 7 percent in 1980 as noted in Chapter 1. It accounts for a small share of farm output but is increasing. Integration was highest in poultry and specialty crops and is becoming a larger though not a major component of fed cattle and hog production. Vertical integration compromises farmers' control of organizational management and marketing operations, but in the case of poultry has made it possible for limited-resource farm families to remain in farming. They have a chance for more stable and higher income along with productive use of their operational management and labor.

Concluding Comments. After reviewing recent literature my conclusions remain the same as written in the first edition of *Foundations of Farm Policy* published in 1970 and reproduced in the second edition (Tweeten, 1979, pp. 370-371):

> The proposition that the farm problem is due to a conspiracy in the market sector and can be eliminated by improving the performance of the marketing sector is a myth. There simply is no giant rathole in the marketing sector absorbing money that should go to farmers. Some feel that the solution to farm problems is to make the marketing industry atomistic. This is not possible. And, if it were possible, it would not be desirable. Economies in production and sales are large enough so that costs would rise if the industry were atomized, and consumers would end up paying more for a lower quality of food.
>
> Merger activity of a large firm in many instances increases workable competition, as, for example, when a large food chain absorbs a small firm about to go bankrupt in an area where the large chain has no stores. It also may be argued that profit rates, promotional activity, and other measures of market performance give unconvincing evidence of predatory behavior or other objectional practices of larger firms. Furthermore, it can be said that potential public regulations of operating practices can ensure high performance in an industry where concentration is very high. On the other hand, the pressure of effective competition is a much more reliable goad to efficiency than reliance on federal regulation of pricing or reliance on the "altruistic" desire of firms lacking competition to act in the public interest. Maintaining an environment of effective competition among firms is the most satisfactory public policy. In industries where economies of size are not large, antimerger and other public policies to preclude high concentration seem appropriate.

THE GRAIN EXPORT INDUSTRY

We now single out for scrutiny the grain marketing industry which has been labeled a villain by populist movements for over a century. The grain export industry has been blamed for high prices for consumers in the 1970s, low prices for producers in the 1980s, and for generally unstable prices. The popular perception of the grain export industry is illustrated by the following quote from Burbach (1976, p. 25):

> The five companies [Cargill, Continental, Bunge, Dreyfus, and Cook] maintain a strangle-hold over the world's grain supply and constitute a food cartel unprecedented in world history. The grain companies are not at the mercy of the free market.
>
> On the contrary, they use their enormous size to manipulate the free marketplace and to maximize profits at the expense of farmer and consumer alike.

The popular perception is that major grain companies constitute a cartel or shared monopoly in grain markets, that they manipulate price without restraint by government or any other institution, and that they cheat farmers and consumers through excessive profits and other costs creating unduly high marketing margins.

The grain export industry is indeed concentrated. The four largest firms in 1974-75 accounted for 58 percent of the U.S. food grain exports, 44 percent of the feed grain exports, and 42 percent of the oilseed and oilseed product exports (Wright and Krause 1976, p. E-13). An estimated 95 percent of world wheat sales have a state trading agency as either buyer or seller.

Performance Measures. For several reasons, the above concentration rates are not grounds for alarm.

1. The number of companies exporting agricultural products not only is large but is growing. The Foreign Agricultural Service of the USDA reported 395 firms actively exporting farm products in 1982-83, up from 263 in 1978-79. Of these, 108 were active in grain (excluding rice) and soybean exports in 1982-83. Firms enter and leave the industry — the rapid rise and demise of a once major exporter, Cook Industries, is an example.

2. Concentration ratios for export firms alone do not adequately reflect the degree of competition in the grain export industry. Grain exporters must compete with domestic merchandisers and processors for supplies of grain. The domestic grain industry is much less concentrated than the export industry.

3. Firms having Japanese ownership or affiliation and farmer-owned agricultural cooperatives increased their share of exports 5 percentage points between 1974-75 and 1980-81 at the expense of the five largest multinationals (GAO, 1982, p. 16). However, evidence (see Pakanati, Henneberry, and Warden 1986, p. 67) indicates that in more recent years ". . . both the number and the export market share of agricultural cooperatives have declined."

The best measure of the contribution of an industry to national goals such as equity and efficiency is not market structure but rather is performance. Economies of size characterize the export industry, especially in *information* — having a large enough staff to obtain extensive market intelligence and to seek out markets. Performance of the U.S. grain export industry is revealed by several studies:

1. Profit margins have been small, averaging less than 2¢ per bushel above costs on grain, for example. Grain exporting firms are aggressive in holding down costs by securing the most favorable rates possible on storage, transportation, insurance, and the other expenses of doing business.

2. The grain export industry practices efficient pricing. The industry relies heavily on the futures market, which as noted later is given high marks for pricing efficiency by expert analysts. GAO (1982, p. iii) found that the ". . . U.S. grain export system translates information about grain sales into price changes with reasonable efficiency."

3. The federal government maintains over 50 programs affecting the U.S. grain export trade. Some regulations such as requiring a specified percentage of PL 480 food aid shipments to move in U.S. bottoms and export embargoes reduce export performance. But others such as the USDA's Federal Inspection Service ensure grade and weight standards and the USDA's grain export reporting system collects and publishes grain sales so that a repetition of the 1972 experience is unlikely.[2]

4. Wilson and Anderson (1980) analyzed the performance of the Canadian grain export system and concluded that it lagged behind the U.S. system, although differences could not be attributed just to the state marketing board used in Canada. The Canadian system may be overrated however; its performance would be less satisfactory without use of efficient U.S. markets for price discovery. Other studies reviewed by Pakanati, Henneberry, and Warden (1986, p. 4) comparing performance of export systems in the U.S. and Canada were inconclusive.

Summary Comments. In-depth studies of the agricultural export industry produced conclusions such as

> The popular conception of the export industry as one controlled by a cartel of major multinational corporations is not only an oversimplified view, but a misconception [Conklin, 1982, p. 137].

> . . . [R]esearch has supported the [grain export] industry position that even though grain exports are dominated by a few large firms, the industry still remains competitive and efficient [Pakanati, Henneberry, and Warden, 1986, p.2].

---

[2] In 1972, the Soviet Union bought large quantities of grain from several exporters without the industry's awareness. The large early purchase should have triggered a sharp price increase in subsequent Soviet purchases. But because the market did not know the scope of the sale before large amounts had been purchased from several exporters, the U.S. grain industry and grain producers were denied the higher prices an efficient market would have provided.

It is ironic that the critics who accuse the current system of having too much market power wish to replace it with a single state marketing board or corporation (see the Weaver Bill, H.R. 4237, 96th Congress). Unlike the current system, a state board could indeed have great market power and could indeed exercise control over price or quantity (although not necessarily over both price and quantity at the same time). The U.S. could play a role in a grain cartel similar to the role played by Saudi Arabia in the OPEC oil cartel. A grain cartel would benefit U.S. producers for a few years at the expense of consumers at home and abroad. After approximately five years, however, U.S. producers as well as consumers worldwide might be worse off based on conventional estimates of export demand elasticity.

To be sure, the agricultural industry faces a large number of barriers to trade such as state trading, export subsidies, variable levies on imports, bilateral trade agreements, and high rigid price supports without production controls. The latter behavior characterizing the European Common Market features dumping of surpluses on world markets at subsidized prices. Such practices interfere with comparative advantage, reduce world trade, raise world price instability, and reduce general economic well-being. The biggest losers from such policies are the nations which practice them.

THE FUTURES MARKET

Futures markets historically have been viewed by populists as a conspiracy against farmers. That may help explain why in a 1986 survey of 29,000 active members of the National Cattlemen's Association 69 percent favored a ban on live and feeder cattle futures trading. In September 1984, farmers representing several national farm groups protested at the Chicago Board of Trade against the alleged contribution of futures market speculators to low commodity prices.

Critics charge that:

1. Futures markets are manipulated and hence are not efficient or fair markets. For example, commodity pooling into large quantity trades provide leverage so that a firm or individual can manipulate prices. Foreign traders or foreign affiliates of U.S. firms, unlike U.S. firms, need not report to the USDA sales of large amounts of grain, hence with insider information can profit in the futures market from large sales before the market is informed of the new supply-demand balance. It has been charged that "speculative short-selling of commodities should be a crime" (see GAO, 1985, p. 67).

2. No one should be allowed to trade in a commodity he/she never intends to physically deliver or take delivery on.

3. Futures markets lower average cash (spot) market prices for farm commodities.

4. Futures markets increase the variability of cash (spot) market prices.

These charges are evaluated in the following pages.

1. Do Large Grain Exporters and Other Speculators Manipulate Futures Markets to the Detriment of Producers? Grain exporters are concerned not only with profit but also with firm survival. Unfavorable outcomes from futures market speculation can threaten firm survival, hence grain exporters emphasize use of futures markets for hedging rather than for speculating. A study by the USDA of profits from grain exporters' futures market activities was inconclusive (GAO, 1985, p. 61). It would not be to the long-term advantage of grain exporters to engage in unfair practices which drive other participants, especially speculators who serve as the market for the hedge, from the futures market. To do so would reduce the ability of exporting firms to shift risk.

The Chicago Board of Trade (CBT) and the Commodity Futures Trading Commission (CFTC) share the objective of protecting against manipulation and abuse in futures markets. In general, CFTC has been more aggressive than CBT in carrying out market surveillance and regulatory action. CBT was criticized for lack of a large-trader reporting system to provide information on concentration in market positions and for lack of disciplinary action against traders when such action was warranted. By late 1984, CFTC reported that CBT had taken action to correct these shortcomings (GAO 1985, p. 51).

2. Should Participation in the Futures Markets be Restricted to Those Who Deliver or Take Delivery on the Physical Commodity? The number of futures contracts traded on the Chicago Board of Trade (CBT), by far the largest farm commodity futures market, increased from 9 million in 1968 to 140 million in 1983. The number of brokers, sellers, and other professionals registered with the Commodity Futures Trading Commission increased from 36,000 in 1979 to 56,000 in 1982. The value of contracts traded was in excess of $5 trillion in 1982, a number 31 times gross farm income of $162 billion. Does all this trading in nondelivered contracts by persons who have no interest in the physical commodity have any redeeming social value?

Yes. A thin market with few traders and few transactions is subject to wide price swings and to manipulation by the few. The advantages of a competitive, liquid market with many buyers and sellers are well documented. Frequent trading means that buyers or sellers can enter the market any time and complete a desired transaction with minimal delay and interference in that market.

Nearly all participants in farm production and marketing benefit from use of futures markets either directly or indirectly. However, the extensive use of futures markets by grain merchants to hedge risks stands in sharp contrast to producers' rare use of hedging. A USDA survey in 1978 found that 7 percent of farmers reported use of futures trading and about half of those who did trade were speculators rather than hedgers (GAO, 1985, p. 681). One reason for minimal direct use of futures markets is because many farmers forward contract sales to local elevators. Many producers do not realize that the local elevator hedges its forward purchases with sales of futures contracts. Large numbers of producers use futures markets as a guide to future profitability — futures prices are often the best predictors of future market prices available to producers. By providing a readily available alternative indicator of value futures markets eliminate the ability of large firms to dictate terms of trade.

3. Do Futures Markets Lower Average Farm Prices? The futures market performs two important functions which constitute the justification for such markets. One is *price discovery*. This is the process through which hundreds of buyers and sellers working with information from around the world pool their judgments through a competitive bidding process to arrive at a price that best represents the consensus of what commodity prices will be in the future based on information available today. Once established, the prices are widely disseminated and are used intensively at home and abroad to establish local market prices.

A second important function is *risk shifting*, also called hedging. This provides the opportunity for producers or others to shift the price risk associated with ownership of a commodity to others. Unlike hedgers who have an interest in the physical commodity itself and who wish to avoid price risk, speculators generally have no interest in the physical commodity. As might be expected, those who perform the useful function of accepting risk from others require compensation to do so. Compensation required is small, however. Most speculators lose money. Larger, professional speculators make money and that is a net cost to hedgers from shifting risk but the combined transaction cost and compensation to speculators is small in relation to commodity price.

A well-functioning market must provide efficient forward pricing.[3] Just and Rausser (1981) show that futures markets forecast as well as or better than commercial forecasters using sophisticated econometric models. Extensive analysis by GAO (1985, p. 84) provided ". . . no evidence suggesting that grain futures markets are not efficient." The USDA agreed with GAO's conclusion pertaining to pricing efficiency of the U.S. grain marketing system, noting that their own studies and those of others had shown this to be the case (GAO, 1985, p. viii).

Leuthold and Hartmann (1979, 1981) show that the futures markets for live hogs, live cattle, and frozen pork bellies do not reflect all available information. But futures perform as well as any other forecasting device on the average; if they did not, those who forecast more accurately would accumulate great wealth in the futures market.

If (as in-depth studies indicate) futures markets are efficient, they provide price signals which reduce inputs required to produce a given output (reduce cost) or increase output from a given amount of resources. Greater productivity made possible by pricing efficiency tends to increase innovative firms' profits in the short run but may reduce industry profits in the long run. Prices too may be lower in the long run because of productivity and output gains from efficient prices. In competitive industries such as agriculture and agribusiness, excess profits are passed to consumers as market forces adjust. Because of lower costs made possible by efficient pricing, lower commodity prices in the long run do not mean lower profit per unit.

---

[3]Fama (1970, p. 383) defines a market in which prices always reflect available information as *efficient*. It is more correct to define an efficient price discovery market as one which incorporates all information accessible at social costs less than the social benefits that information generates. That means knowledge will not be perfect in an efficient market; much potential information will be ignored because it isn't worth gathering.

Buyers and sellers able to hedge on the futures market can accept lower profit margins and make more sales than if they were unable to pass along risks to speculators. In this way the futures market enhances demand for farm products and spot market prices. Grain exporters on average receiving a profit of less than 2¢ per bushel would require a higher rate if they could not shift the risk of selling to speculators. The competitive structure of markets ensures that the earnings are passed on to farmers and consumers in the short run and solely to consumers in the long run because competition ensures that adequate-size, well-managed farms will make only a normal profit in the long run.

That long-term farm profit rates are unaffected by the futures market is supported by what meager empirical evidence is available. Based on a comprehensive review of studies analyzing the impact of futures trading on the level and volatility of spot prices, Britto (1985) found no conclusive theoretical or empirical evidence of positive or negative effects. The only empirical study (one by Emerson and Tomek) cited by Britto (p. 7) found that an increased volume of futures trading resulted in higher potato prices in Maine.

4. Does the Futures Market Increase Variability of Spot Markets? In theory, futures markets reduce variability of cash markets. By shifting risk of holding reserve stocks to speculators, more stocks are held to raise price when supplies are abundant and to reduce prices when supplies are short relative to demand. This ability to reduce cash market variability is mainly limited to seasonal (within-year) variability because most futures market contracts at maximum extend just over a year although some markets have contracts of up to two years.

Empirical results support the conclusion that futures markets reduce variation in cash prices for onions (Working, 1960; Gray, 1963), wheat (Tomek, 1971), cattle (Taylor and Leuthold, 1974), and pork bellies (Powers, 1970). Other studies show that futures markets are not only a low-cost source of information for use in cash markets but also improve efficiency of cash markets (Cox, 1976).

Concluding Comments. Ability to shift risk may slightly reduce farm prices and rates of return because of greater efficiency and less risk. But real returns adjusted for risk are no lower, and consumers and the nation as a whole are better off. Futures markets like other markets have flaws. Abuses have been found and many have been corrected. On the whole futures markets receive high marks for performance by those who have studied them in depth. They are an important part of the market for farm products. Futures markets are more competitive and responsive to information than are cash spot markets. Their demise would make not only the private but also the public sector less efficient.

Futures markets benefit not only those who use them but also those who do not use them. More family farmers could benefit from using the futures market to pass risk to others through hedging, but most family farmers are better off leaving speculation to professionals.

FARM INPUTS

As indicated earlier, monopoly in commodity markets will not result in relatively low farm income per capita or rates of return on resources if farm input markets function efficiently. Of concern is impediments to competition that would influence farm input prices and output and their impact on overall farm economic problems.

Farmers incurred $136 billion of production expenses in 1985. In this section, we examine the structure, conduct, and performance of several critical farm input markets that comprise the majority of farm expenses — fertilizers, pesticides, seeds, machinery, labor, credit, and land.

FERTILIZERS

The three main nutrients in commercial fertilizers are nitrogen, potash, and phosphate. Each of these nutrients may be viewed as separate subsectors of the industry when judging markets.

The nitrogen fertilizer industry consisted of over 60 producers in 1979 and the sector had modest four-firm and eight-firm concentration ratios of 35 percent and 45 percent respectively (Leibenluft, 1981, p. viii). Competition was judged to be relatively keen based on easy entry and exit to the industry, availability of supplies from foreign sources, presence of cooperatives, and overcapacity.

Several dozen companies were involved in the phosphate fertilizer industry with four firms accounting for 40 percent of sales and the largest eight firms accounting for 60 percent of sales. Competition in the phosphate subsector is adequate to ensure efficient pricing based on lack of substantial barriers to entry, relatively low concentration ratios, and presence of overcapacity.

Concentration in the potash subsector is greater than in other major subsectors of the fertilizer industry. The top four firms accounted for 75 percent of U.S. capacity in recent years. The firms are located in the southwest United States where accessible potash deposits are rapidly being depleted. An important portion of potash is being imported from Canada which now supplies over two-thirds of U.S. requirements. The Saskatchewan government is controlling an ever larger share of production and by the year 2000 will dominate the North American industry (Leibenluft, 1981, p. ix). Thus, potash prices in the future will be influenced strongly by policies of the Potash Corporation of Saskatchewan.

PESTICIDES

American farmers purchased $5 billion of pesticides in 1985. The industry is one of the most rapidly growing components of the agribusiness sector. Concentration at the manufacturing level is relatively high with four firms accounting for 57 percent of production and the eight largest accounting for 79 percent of production. Sales are even more concentrated in certain subsectors of the industry. For example, the four largest firms account for 87 percent of the corn herbicide market. Leinbenluft (1981, p. xi) concluded that "pesticide chemicals and submarkets may turn out to exhibit anti-competitive problems." The industry not only is concentrated, it exhibits substantial barriers to entry in the form of high research and development costs. Profit levels are high, although variable. Chemicals frequently come into favor and then go out of favor after a short time because of environmental problems, introduction of a superior competing product, or other factors.

On the whole, performance of the industry must be rated as satisfactory. Large sums are spent on research and development, and the rate of innovation of new pesticides has been rapid. The industry must compete not only within itself but also with mechanical and biological forms of pest control. Foreign firms are increasingly involved in development and production of pesticides, providing competition to American firms. The case for non-competitive behavior in the pesticide industry is inconclusive.

One of the major new barriers to entry is government regulations which now make it extremely expensive and time-consuming to introduce new products. Such restrictions create advantages for large, existing, well-capitalized firms able to exercise greater market power because of these regulations. To the extent that government regulations reduce environmental hazards at favorable benefit-cost ratios, society benefits. To the extent that regulation unduly inhibits product development and release, a significant source of increasing farm productivity, lower commodity prices, and excess capacity is restrained at a gain to producers as a whole but at a loss to early innovators and consumers.

Non-competitive behavior in the form of input price above competitive levels can benefit producers economically if the elasticity of demand for the input exceeds the elasticity of demand for farm output (Tweeten, 1979, p. 255). A higher price restrains quantity demanded of highly profitable and productive pesticide inputs, restricting input use and production. Given an inelastic demand for farm output, lower production raises receipts. Higher receipts coupled with reduced costs (given an elastic long-run input demand) make for a higher net farm income with higher input price. Too little is known of the elasticity of demand for chemicals to make definitive conclusions, but some data suggest that a higher price for fertilizer and pesticides brought about by a tax or monopoly could raise net farm income (Tweeten, 1979, pp. 256, 276).

SEED

Farmers spent $3.4 billion on seed purchases in 1985. The Plant Variety Protection Act of 1979, which granted patent protection to novel life forms, was designed to encourage innovation in the seed industry by allowing firms to appropriate compensation for research and development by charging monopoly prices for improved varieties. This along with the new biotechnology not only has raised expectations of major technological breakthroughs in seed production but also fear of exploitation of producers through pricing above competitive levels. This fear was intensified as many small seed companies were taken over by larger pesticide and pharmaceutical firms.

According to Leibenluft "such concerns [over exploitation] at the present time are not well-grounded" (1981, p. xi). This is because entry barriers are low, because market power is limited through competing seed varieties produced by agricultural experiment stations and by smaller seed companies, because many farmers have the option of growing their own seed, and because the pace of innovation is likely to be so rapid and unpredictable that no one firm can be assured for long of a dominant position.

The entry of large multinational firms into the seed industry injects a new dimension of market competition which as in the case of improved pesticides may be economically unfavorable to farmers. But the cause of fear is not poor performance but superior performance which will unleash productive new inputs. Resulting greater output could create excess capacity, low farm prices, and low farm income.

FARM MACHINERY

Despite financial stress, farmers purchased approximately $5 billion of farm machinery and equipment in 1985. One firm, John Deere, dominates the industry with approximately half the sales in major machinery categories. Formidable entry barriers exist in the industry, including establishing dealer networks, economies of size, recognition of a brand name, reputation for dependability and service, and large capital requirements. New entrants capable of obtaining a substantial market share, especially in full-line machinery manufacturing and sales, are not likely to appear. Concentration is increasing as a number of manufacturers have failed financially in recent years, including International Harvester which accounted for 25 percent of the two-wheel drive tractor market as recently as 1979. Economies of size are so substantial that only a few plants of optimal size could produce all of North America's requirements for tractors and other major machinery items such as combines.

Nonetheless, a number of factors inhibit opportunities for exploitation of farmers. Charging excessive prices reduces sales because demand is quite elastic for farm machinery — farmers can put off their purchases if prices are too high and make the old machinery do for another year. Farm machinery companies have been unprofitable in recent years. Even John Deere has had difficulty and the many machinery manufacturers and dealers who have lost money or gone bankrupt would be surprised to be accused of exploiting farmers.

The leading manufacturer, John Deere, has devoted a higher percentage of revenues to research and development than other manufacturers and has been highly innovative. Over the years, Deere has been the only major manufacturer with returns comparable to those of other industries. If Deere lowered its prices so that its return were in line with those of other full-line farm machinery firms, it eventually might be the only domestic firm left in the industry. The potential for expansion of Canadian, Japanese, Italian, German, and Soviet operations already active in farm machinery manufacture and U.S. sales provides some assurance that farmers will not be exploited by the farm machinery industry despite high and growing concentration ratios in the domestic industry alone.

CREDIT

Interest ($20 billion in 1984) is the largest single farm expense. Thus the credit market is of more than passing interest. The high real interest rate (market rate less inflation rate) was the most important factor behind rising farm costs and declining net income, assets, and equity in the first half of the 1980s, hence played the central role in financial stress.

Credit market shares in 1985 were as follows:

	Credit Market, 1985	
Lending Institution	**Non-real Estate Debt[a]**	**Real Estate Debt**
	(Percent)	
Cooperatives	26	60
Federal Government	20	12
Other	54	28
Total	100	100

[a]Excludes farm household loans, and excludes CCC and individuals as sources.

The cooperative Farm Credit Administration lenders (Production Credit Associations for non-real estate and Federal Land Banks for real estate debt) and the federal Farmers Home Administration account for such a large share of the credit market that a case for exploitation by the private sector cannot be made. In some rural areas, local banks may wish to charge excessive interest rates. But availability of cooperative credit sources and of private banks at other locations (easily accessible with improved roads and motor vehicles) precludes exploitation of borrowers.

It may be argued that easy access in the past to non-real estate farm credit through cooperative and federal sources encouraged excessive inputs, output, and low farm prices. And excessive use of real estate credit encouraged excessive prices for land and overextension of debt in the 1970s, helping to establish preconditions for financial stress in the 1980s. The main cause of financial stress in the 1980s, high real interest rates, was not the commercial banking system or Federal Reserve System but instead mainly was inflation and overexpansion in the 1970s and large full-employment federal deficits in the 1980s. Congress and the executive branch contributed to financial stress in the 1980s by encouraging lax lending standards and overly generous lending by cooperative and federal lending institutions in the 1970s.

It is difficult to justify subsidized credit to an industry troubled by excessive capital and overcapacity to produce. The Farmers Home Administration was all but paralyzed in collecting interest and pricipal even from well-to-do borrowers in the 1980s. The principal need of the farm credit agency is to diversify among industries and secondary mortgage markets so that when farming slumps suppliers do not have to be bailed out by taxpayers.

LABOR

On the whole, farmers do not pay higher prices for inputs than do other sectors so there is no reason to conclude that farmers are victims of a discriminating monopolist selling inputs to farmers. Rational farmers will not buy inputs unless the input price is below the expected marginal value of the product they produce. And if a mistake is made in purchasing too much input, farmers can avoid buying more and work off excess capacity. Even machinery which is quite fixed to agriculture can be reduced in service given off by about 10 percent in 3 years as explained in Chapter 4. The farming industry will not incur chronic low returns if farm operator and family labor and equity capital are mobile. Farm financial capital is highly mobile.

In theory, imperfect competition in just one market, labor, could explain farm economic returns chronically below returns elsewhere in the economy (see earlier sections of this chapter and Tweeten, 1979, pp. 189-194). If farm operator, family, and hired labor does not respond to economic signals because of lack of knowledge of alternatives or emotional ties to the farm, then returns can be low in farming for extended periods. The earlier conceptual framework in Figure 8.2 indicated that labor bargaining power discrimination in nonfarm industry could keep farm labor returns in equilibrium permanently below returns in farming. Labor unions account for only less than 20 percent of the U.S. labor force and are incapable of preventing farmers from entering nonfarm jobs, however.

Studies show that farm labor is not very responsive to change in economic conditions in the short run of up to five years, but labor is responsive in the long run to ten years or more (Heady and Tweeten, 1963). Human resource development programs such as general and vocational education; job information, counseling, and relocation assistance; as well as removal of barriers to mobility such as the union shop and race, sex, and age discrimination can accomplish more than any other public effort to improve economic conditions for farm people in the long run. These are discussed in Chapter 10.

LAND

Although only 2-3 percent of farmland changes hands each year, it is a major market. The land market potentially has thousands of participants although many states choose to limit who can own land.

Approximately one-fifth of farmland is owned by nonfarmers (excluding retired farmers and their spouses) as noted in Chapter 1. Nonfamily corporations own a minor portion of U.S. farmland, approximately 3 percent of all farmland in 1978.

Land is so fixed to farming that its price is determined by economic conditions in agriculture. Because family farmers account for most land purchases and sales and hence play the decisive role in setting land values, it is not possible to blame high land values on nonfarmers or large corporations. Studies indicate that the land market is efficient, pricing according to reasonable expectations of future interest rates and returns from farming (Tweeten, 1986). Operators of adequate-size, well-managed farms bid for land until they earn as much from land as from comparable investments elsewhere. Other operators and nonfarm investors must pay this same land price. Of course, land values are too high for inadequately-sized and poorly-managed farms.

Farmland ownership is concentrated about like commodity sales, with 5 percent of farm owners holding 50 percent of farmland. This has led to fears noted in Chapter 3 that the U.S. is becoming another Central America with attendant social unrest, revolution, and urgent need for land reform. Such fears are poorly founded. Unlike Central America, the vast majority of U.S. wealth is human and other nonland capital which is distributed much more evenly than farmland. The key to avoiding exploitation is to have alternatives, and most farm operators do — more than half already work off the farm. Economies of size extend beyond traditional family-size farmers as noted in Chapter 4. Limiting farm size would reduce efficiency and raise food costs. Numerous options to control extensive concentration of farm wealth have been listed (Tweeten, 1984, pp. 54-58). There is no reason to believe the government would do a better job than the market of establishing farm size. What is the right size for one farm operator is the wrong size for another. It is notable that one of the potentially most effective options, higher taxes on estate and other transfers from the current to the next generation, have been opposed by the very farm populists who seem most concerned by loss of the family farm.

CONCLUSIONS

Farming has been far more influenced by favorable performance of agribusiness bringing increased productivity than by unfavorable performance bringing high input prices or low commodity prices. Productivity gains have brought massive benefits to society as a whole and to farmers in the long run because farm income per capita has trended toward national income per capita.

To be sure, farmers experience annual and cyclical economic setbacks, the latter especially apparent in the 1980s. The economic instability which is the heart of the farm problem is not the product of a concentrated agribusiness sector or of productivity gains. In recent years cyclical changes have been mostly the result of decisions of governments. Examples include decisions by the Soviet Union to enter world grain markets to maintain consumption in the early 1970s, of OPEC to form a cartel to raise oil prices, of countries to expand money supply and credit at an unsustainable rate to accommodate higher oil prices in the 1970s, and of the U.S. government to run very large full-employment budget deficits in the 1980s.

The measures have indeed caused farmers severe economic pain. However, the policies are not the result of a conspiracy of agribusiness or anyone else against farmers. Universal concern is apparent for improving the well-being of farm people among people everywhere, including politicians, bureaucrats, and agribusiness persons.

Then how does one explain government policies such as those above which have caused severe hardship even to families on well-managed, adequate-size farms? Numerous reasons can be listed but most fall under the category of greed, bungling, or unintended side effects. Examples of unintended side effects include less U.S. farm exports because of (1) embargoes and other measures taken by President Carter to show displeasure with the Soviet invasion of Afghanistan and (2) huge export subsidies paid by the European Common Market to dump surpluses generated by high rigid price supports without production controls. An example of greed is the push for lower tax rates and higher government spending which led to large full-employment deficits, high real interest and exchange rates, and to lower exports in the 1980s. An example of bungling is the "Laffer Curve" notion that lower tax rates in the 1980s would raise the tax take to pay for greater government outlays. The list could be extended.

An informal social contract always exists between any sector, such as agriculture, and society. Under a high-quality contract the sector agrees to work within the system and contribute to society whereas society agrees to avoid arbitrary policies unnecessarily disadvantaging a sector without compensation. A breakdown of the social contract leads to alienation and, in the extreme, to anarchy and revolution. The quality of the social contract with farmers will be improved if government avoids policies such as the erratic and overly expansionary monetary and credit policies of the 1970s and full-employment deficit spending of the 1980s and pursues policies such as free trade, antitrust action, and a safety net for the disadvantaged. The public could improve its side of the contract by discouraging use of "junk bonds," leverage buyouts, and other high-risk devices encouraging conglomerate corporate mergers and concentration to gain quick financial profits rather than to improve economic efficiency. Farmers will improve their end of the contract by relying on the market (without supply control) rather than transfers from taxpayers (especially from lower-income-wealth taxpayers to higher-income-wealth farmers) and making the best possible use of private risk management institutions such as the futures and option markets, storage, and insurance.

The major source of decline in number and increase in size of commercial family farms (small farms increased in numbers from 1978 to 1982 according to census data) has been technology, primarily the tractor and its complements (Tweeten, 1984, pp. 21-44). Such technology is not the result of monopoly structure or subpar performance of agribusiness. Scale-influencing technologies would have caused losses in commercial farm numbers even if farm prices would have been much higher.

One cannot help but be struck by the stark contrast between vilification of agribusiness industries by farm activists and the almost complete absence of evidence justifying such vilification through numerous in-depth economic studies of agribusiness.[4] There is no evidence that farm problems of annual and cyclical income instability and squeezing out of commercial family farms would be any different today if the agribusiness sector were perfectly competitive. Finding scapegoats for farm problems in bankers, the multinational grain trade, the futures market, the Trilateral Commission, or in some ethnic

---

[4]It is easy for activists to dismiss such studies as performed by economists who are part of the coverup. The economists performed such studies did not begin with the presumption that agribusiness firms were innocent until proven guilty although the analysts commendably attempted to be professional and objective.

group at best detracts from finding the real causes of farm problems and at worst establishes a climate of hate, violence, and even murder.

REFERENCES

Britto, Ronald. 1985. Futures trading and the level and volatility of spot prices: A survey. Working Paper CSFM-112. New York, N.Y.: Center for the Study of Futures Markets, Columbia University.

Bullock, J. Bruce. 1986. Evaluation of NC 117 Working Paper WP-89. (Mimeo) Columbia: Department of Agricultural Economics, University of Missouri.

Bullock, J. Bruce. 1981. Estimate of consumer loss due to monopoly in the U.S. food-manufacturing industries: Comment. *American Journal of Agricultural Economics* 63:290-292.

Burbach, Roger. July 1976. The great grain robbery. *The Progressive Farmer*, pp. 24-27.

Conklin, Neilson. June 1982. An economic analysis of the pricing efficiency and market organization of the U.S. grain export system. GAO/CED-82-61S. Washington, D.C.: General Accounting Office.

Connor, John, Richard Rogers, Bruce Marion, and Willard Mueller. 1985. *The Food Manufacturing Industries: Structure, Strategies, Performance, and Policies.* Lexington, MA: Lexington Books.

Council on Wage and Price Stability. November 1976. The responsiveness of wholesale and retail food prices to changes in the cost of food production and distribution. Staff Report. Washington, D.C.: Executive Office of the President.

Cox, C.C. 1976. Futures trading and market information. *Journal of Political Economy* 84:1215-1237.

Fama, E. 1970. Efficient capital markets: A review of theory and the empirical work. *Journal of Finance* 25:383-417.

General Accounting Office (GAO), U.S. Congress. April 1985. Controls over export sales reporting and futures trading help ensure fairness, integrity, and pricing efficiency in the U.S. grain marketing system. GAO/RCED-85-20. Washington, D.C.: GAO.

General Accounting Office (GAO), U.S. Congress. June 1982. Market structure and pricing efficiency of U.S. grain export system. GAO/CED-815-61. Washington, D.C.: GAO.

Gisser, Micah. 1982. Welfare implications of oligopoly in U.S. food manufacturing. *American Journal of Agricultural Economics* 64:616-24.

Gray, Roger. 1963. Onions revisited. *Journal of Farm Economics* 45:273-276.

Heady, Earl O. and Luther Tweeten. 1963. *Resource Demand and Structure of the Agricultural Industry.* Ames: Iowa State University Press.

Just, Richard and Gordon Rausser. 1981. Commodity price forecasting with large-scale econometric models and the futures market. *American Journal of Agricultural Economics* 63:197-208.

Leibenluft, Robert. February 1981. Competition in farm inputs: An examination of four industries. Washington, D.C.: Office of Policy Planning, Federal Trade Commission.

Leuthold, R.M. and P.A. Hartmann. 1979. A semi-strong form evaluation of the efficiency of the hog futures market. *American Journal of Agricultural Economics* 61:482-89.

Leuthold, R.M. and P.A. Hartmann. 1981. An evaluation of the forward-pricing efficiency of livestock futures markets. *North Central Journal of Agricultural Economics* 3:71-80.

Marion, Bruce. 1987. Is the family farm being squeezed out of business by monopolies? Proceedings of conference *Is There a Conspiracy Against Family Farmers?* Ames: Religious Studies Program, Iowa State University.

Marion, Bruce. 1985. *The Organization and Performance of the U.S. Food System.* Lexington, MA: Lexington Books.

National Commission on Food Marketing. 1966. *Food from Farmer to Consumer.* Washington, D.C.: U.S. Government Printing Office.

O'Rourke, A. Desmond and W. Smith Greig. 1981. Estimates of consumer loss due to monopoly in the U.S. food-manufacturing industries: Comment. *American Journal of Agricultural Economics* 63:285-289.

Pakanati, Venugopal, David Henneberry, and Thomas Warden. May 1986. The U.S. grain export industry: Survey responses and analysis. Stillwater: Department of Agricultural Economics.

Parker, Russell and John Connor. 1979. Estimates of consumer loss due to monopoly in the U.S. food-manufacturing industries. *American Journal of Agricultural Economics* 61:626-39.

Powers, M.J. 1970. Does futures trading reduce price fluctuations in cash markets? *American Economic Review* 60:460-64.

Quail, Gwen, Bruce Marion, Frederick Geithman, and Jeffrey Marquardt. May 1986. The impact of packer-buyer concentration on live cattle prices. NC 117 Working Paper WP-89. Madison: Department of Agricultural Economics, University of Wisconsin.

Smith, Tony. 1986. Social scientists are not neutral onlookers to agricultural policy. Proceedings of conference *Is There a Moral Obligation to Save the Family Farm?* Ames: Religious Studies Program, Iowa State University.

Taylor, G.S. and R.M. Leuthold. 1974. The influence of futures trading on cash cattle price variations. *Food Research Institute Studies* 3:29-35.

Tomek, William. 1971. A note on historical wheat prices and futures trading. *Food Research Institute Studies* 10:109-113.

Tweeten, Luther. 1987. Is the family farm being squeezed out of business by monopolies? Another view. Proceedings of conference *Is There a Conspiracy Against Family Farmers?* Ames: Religious Studies Program, Iowa State University.

Tweeten, Luther. Fall 1986. A note on explaining farmland price changes in the seventies and eighties. *Agricultural Economics Research* 38:25-30.

Tweeten, Luther. 1984. Causes and consequences of structural change in the farming industry. NPA Report No. 207. Washington, D.C.: National Planning Association.

Tweeten Luther. July 1983. Impact of federal fiscal-monetary policy on farm structure. *Southern Journal of Agricultural Economics* 15:61-71.

Tweeten Luther. 1979. *Foundations of Farm Policy*. Lincoln: University of Nebraska Press.

U.S. Department of Agriculture. July 1986. Food cost review, 1985. Agri. Econ. Rept. No. 559. Washington, D.C.: Economic Research Service, USDA.

Ward, Clement. 1986. The impact of packer-buyer concentration on live cattle prices: A review and comments. (Mimeo) Stillwater: Department of Agricultural Economics, Oklahoma State University.

Wilson, W.W. and D.E. Anderson. 1980. The Canadian grain marketing system. Paper presented at American Agricultural Economics Association annual meeting, Urbana, Illinois.

Wissman, Roger. 1985. Coop share of marketing, purchasing stabilizes after period of growth. In *Farmer Cooperative*. Washington, D.C.: USDA, Agricultural Cooperative Service.

Working, Holbrook. 1960. Price effects of futures trading. *Food Research Institute Studies* 1:3-31.

Wright, Bruce and Kenneth Krause. April 1976. Foreign direct investment in the U.S. grain trade. *Report to the Congress: Foreign Direct Investment in the United States*. Volume 4, Appendix E. Washington, D.C.: U.S. Department of Commerce.

CHAPTER NINE

Environmental and Natural Resources

Publication of Rachael Carlson's *Silent Spring* in 1962 initiated an educational process eventually making virtually all American's environmentalists. Remaining differences are mostly of degree, but they are spirited.

Flora (1987, p. 5) for example is the skeptic:

> Researchers, university professors, and government regulators place great emphasis on what they perceive as factual information. On the other hand, they, almost without exception, use terminology like "may," "potential," "possible," "conceivable," " latent," "hidden," or "conceivable" dangers in the use of pesticides. Somewhere, sometime, somebody should demand, "show us the facts."

He went on to note that we are needlessly worried that pesticides will *unintentionally* wipe out a beneficial species when those same pesticides have been unable to make a dent in the targeted harmful species we *intended* to wipe out.

At the other extreme are a host of writers who call for conservation of environmental and natural resources before exploitation and excessive use of agricultural chemicals fouls our nest and destroys our planet. The Cornucopia Project (1981, p. 34) writes:

> Together erosion and development waste about 34 square miles of [cropland] every day. ...continual, heavy use of herbicides and pesticides can destroy the living organisms which make up 1 to 5 percent of the normal soil. Repeated plowing does the same thing by increasing oxidation losses. The destruction of this organic matter reduces tilth, the ability of the soil to hold water and bind nutrients in forms that will resist leaching but are available to plants. Such "dead" earth is useful only for holding up plants, and any necessary nutrients must be added to the soil.

Given that error inevitably attends experimentation and the impossibility of *proving* any conclusion regarding natural and environmental resources, two extreme interpretations are possible. One is that inability to prove conclusively that any chemical or practice causes cancer in humans best justifies erring on the side of freedom of decisions by making no environmental regulations. The other extreme is that lack of precise knowledge best justifies erring on the side of caution and making no new or modified agricultural chemicals available. Implementing the first interpretation could entail massive social costs of environmental degradation. Implementation of the second interpretation could entail massive social costs from lost opportunity to reduce hunger and death through careful use of chemicals and natural resources. The market alone will not resolve these issues. Environmental and natural resource issues are serious economic problems of agriculture which must be addressed using the best information available.

This chapter first provides a brief overview and history of environmental and natural resource problems and policy responses. Conceptual issues are then reviewed before examining in greater depth the natural resource and environmental concerns of agriculture.

OVERVIEW AND HISTORY

Passage of the National Environmental Policy Act of 1969 (NEPA) embarked the United States upon a wide-ranging national program of environmental protection and management. Prior to the enactment of NEPA, the federal role in environmental regulation was generally limited to the management of publicly owned lands and encouraging soil conservation. Prior to the 1970s, many separate pieces of specific environmental legislation were enacted. Earlier legislation tended to be massive responses to specific issues. Ecologists, other scientists, and the public at large by the 1950s and 1960s began to view the various aspects of the environment in a nationwide perspective. The result was the enactment of NEPA and a major reexamination of the environmental legislation existing at that time.

The federal government by 1979 was writing and enforcing regulations in numerous areas including occupational health and safety, resource recovery, water quality, air quality, pesticides, toxic chemicals, hazardous wastes, mine safety, coastal zones, ocean pollution and the outer continental shelf, and the upper atmosphere. Implementation of the recent major environmental legislation or amendments falls within the purview of the U.S. Environmental Protection Agency (EPA). Specific legislation includes the Clean Air Act; the Clean Water Act; the Safe Drinking Water Act; the Marine Protection, Research, and Sanctuaries Act; the Federal Insecticide, Fungicide, and Rodenticide Act; the Toxic Substances Control Act; the Resource Conservation and Recovery Act; and the Comprehensive Environmental Response, Compensation, and Liability Act (Superfund). These Acts and their implications are discussed more fully in various issues of *Environmental Quality* (see CEQ, 1984 and other issues).

The United States devotes massive resources to reducing pollution. In 1984, public and private expenditures for pollution abatement and control totalled $69 billion or nearly 2 percent of GNP (Council of Economic Advisors, 1987, p. 202). Of the total, 45 percent was for control of air emissions and 38 percent of water pollution. National Ambient Air Quality Standards for concentrations have been reached for sulfur dioxide and nitrogen dioxide in nearly all parts of the country and for carbon monoxide and total suspended particles have been met in most parts (Council of Economic Advisors, 1987, p. 202). It is difficult to determine whether benefits of these gains exceed costs.

Surface-level ozone is the largest remaining problem and much concern remains concerning release of chlorofluorocarbons (such as found in freon gas) into the atmosphere where it eventually rises to the stratosphere to react with and break down the ozone layer. The ozone layer in turn protects the earth from ultraviolet radiation which can cause skin cancer and reduce yields of some crops. Like carbon dioxide, whose level may be increasing from deforestation and from burning of fossil fuels, chlorofluorocarbons may contribute to global warming through a greenhouse effect. This could permit crop production to move further north in the northern hemisphere and south in the southern hemisphere. It also could cause drouth, desertification, and flooding, the latter as the water level is raised by melting the polar ice cap.

The Occupational Safety and Health Administration's (OSHA) sweeping mandate is to ensure that "so far as possible every working man and woman in the nation [has] safe and healthful working conditions." OSHA has issued several thousand workplace standards, the large majority of which were adopted soon after OSHA's formation. Some of the most obviously ineffective of these regulations have since been revoked. Most of OSHA's rules deal with safety rather than health hazards. Employers continue to complain that OSHA's regulations are costly, but no comprehensive estimates of compliance costs have been made. Many of OSHA's rules for farmers were found to be unworkable and were withdrawn.

Some of studies have found that OSHA's activities have not been effective in promoting workplace safety (Council of Economic Advisors, 1987, p. 199). Before the establishment of OSHA, evidence was lacking that working conditions were safer or healthier in states with stricter regulatory standards. Over recent decades, the job fatality rate has declined fairly steadily by more than 2 percent per year. The advent of OSHA did not change this rate of decline. One recent study, however, found that OSHA's activities have modestly reduced work injuries. It is more difficult to assess OSHA's effects on health because of the time lag between a worker's exposure to a toxic chemical or environmental hazard and the manifestation of disease.

The Supreme Court has interpreted OSHA's legislative mandate as prohibiting the balancing of costs and benefits in formulating health regulations. OSHA has adopted a restricted cost-effectiveness approach in accordance with this decision, allowing lowest cost methods of compliance in achieving a given technical standard. Studies of past OSHA rules show that costs typically exceed expected benefits (Council of economic Advisors, 1987, p. 199). The Council of Economic Advisors noted one exception, the OSHA hazard communication standard, requiring that workers be informed of workplace hazards, but analysts hotly dispute such results. Controversy is intense over benefits and costs of measures to influence the environment.

CONCEPTUAL ISSUES

Economic analysis is most straightforward where markets work best. It is unfortunate that economics is needed most and is especially difficult to apply in a field where markets work least — in allocating natural and environmental resources. Market failure affecting these resources arises from interrelated characteristics referred to as common property, public goods, externalities, and institutional property rights.

An example of *common property,* the commons, is an underground reservoir of water which is the property of all but the responsibility of none. The rational strategy for any farmer is to utilize a fixed pool or reservoir of water open to the public for irrigation "before the hoarders get it." Of course, the water has opportunity cost in terms of value to a society with limited water supplies, especially in the future. It must be rationed over time to yield fullest benefits. Unless that opportunity cost is charged through property rights and collective action, water will be exploited at an inefficient and wasteful rate in excess of the socially optimal level. If access to the pool is limited to one or a few users to avert excessive use, then the opposite danger — cartelization — may impose equally onerous costs on society.

The "tragedy of the commons" also occurs with another common resource, air, with its ability to absorb pollutants. In the absence of constraints polluters excessively use ambient air to absorb emissions.

The second case of market failure, *externalities*, was discussed in Chapter 2 and is related to the common property problem. Unrestricted use of waste absorption services of ambient air and public waters reduces recreational, life support, and amenity services of these resources. Because users of the waste absorption service do not bear the cost, use for that purpose is excessive. However, use for other purposes is below the socially optimal level. A farmer who allows nitrogen fertilizer and pesticides to seep into groundwater and poison his neighbors' groundwater supplies does not pay the cost of such action. He may not even be aware of the damage, hence will use excessively agricultural chemicals. If his and only his groundwater were affected, he would internalize costs and use less chemicals.

A third market failure arises because many natural resources are *public goods*. For example, the flood control benefits of a dam are available to all downstream parties whether or not they pay for the dam. Benefits of the safe agricultural chemical not leaving a residue in food replacing a dangerous chemical may be unrecognizable initially by consumers. A requirement that all producers use a safe chemical benefits all, whether a specific consumer pays for the regulation or not. In such case, consumers prefer to be free riders, letting someone else pay. Consequently, no one pays and the market alone will not provide the socially desired outcome. Public intervention is necessary.

A fourth case of market failure is *institutional property rights*. An example is the over 700 million acres of public lands in federal hands because no private claimants initially thought the land worthwhile. Whether waste and inefficiency result depends on how well the lands are administered by the government. Some of the lands are excessively grazed, some are underutilized, and others have been leased to the private sector at windfall prices with little accountability. The Council of Environmental Quality (1981, p. 13) contends the history of public lands in the West demonstrates that the government undervalues public resources because of pressures imposed by self-serving users.

It should be noted that all of the above instances of market failure for natural and environmental resources are variants of the same problem — resources are not used to the point where marginal social cost equals marginal social revenue as noted earlier in Figure 2.5A. Another difficulty from failure of markets for environmental goods is that market signals — higher prices — do not warn of shortages and need for conservation and substitution. Lack of substitutes for productive soil, clean air and water, and biological diversity also may constrain ability to respond even if price signals worked.

Two basic approaches to allocation of environmental resources are widely used. The *regulatory approach* is for the government to control directly the use of water, air, and other resources. For example, a farmer might be allowed to use no more than a fixed poundage of a pesticide, so many gallons of irrigation water, and lose no more than 5 tons of soil per acre. An advantage of the approach is that it gives an impression of certainty and equitable sharing of burden.

In contrast the *economic approach* views problems of the environment as arising from lack of property rights. By attaching property right, costs and benefits can be exchanged in the market for environmental and natural resources. The economic approach recognizes that a goal of reducing pollution by a given amount is most efficiently achieved by emphasizing cutbacks where abatement costs are least. And the disposal of pollutants is emphasized among those uses where absorption costs are least. This maximizes net social gain or minimizes net social cost of natural and environmental resources. Owners can be expected to bargain and exchange property rights in the market until expected social benefits equal expected social cost. The difficulty is how to establish and maintain property rights.

In 1981, the Council of Environmental Quality noted the inflexibility of the "command-and-control" strategy and the minor role of economic incentives in past environmental policy. In addition to calling for more global cooperation, it listed three major goals for U.S. environmental policy (Randall, 1987, p. 380):

1. Balancing the costs and benefits of environmental controls;
2. Allowing market incentives to work in environmental policy; and,
3. Decentralizing responsibilities for environmental improvement.

Limited success by 1987 had been achieved in accomplishing these goals with emissions trading, a strategy begun in the 1970s and notable for three innovations:

1. "Bubbles," allowing emissions from several neighboring outlets of one or more firms to be combined and treated for purposes of regulation as one source. This allowed flexibility within the outlets to achieve a prescribed reduction in overall emissions as the lowest cost. The strategy of the "bubble" allowed restraining of sources with the greatest emission cutback per dollar of cost at the margin.

2. "Banks" of emission credits allowed a firm to stockpile emission reductions which in turn could be traded.

3. "Offsets" in the form of emission credits could be purchased from others by new or expanding firms to allow emissions without raising overall pollution in the region.

Such measures moved toward a market solution. In theory, they behaved like a certificate scheme requiring each firm to purchase a certificate covering all emissions. The government organization providing the certificates would adjust the price until marginal benefits equal marginal costs. Competition among bidders would ensure that pollution is controlled at least cost. This could minimize costs of pollution abatement but would not ensure optimality on the benefit side. Costs and bearers of cost of pollution control are rather easily identified; identification of beneficiaries and how much each benefits is an elusive if not unreachable objective. Hence the price system will continue to allocate environmental goods imperfectly at best and many of the issues will be resolved only through public dialogue and decisions.

Other candidates for injection of economics into environmental and natural resource policy include:

1. An increase in the price of publicly produced irrigation water to its market level. There would be reason to subsidize overuse of water and other resources if the beneficiaries were low income families. But the current system targets high income families.

2. An effluence charge on waste discharges into public water courses and the air mantle.

3. A national system of mandatory flood control insurance with premiums set equal to expected cost.

4. An increase in the federally administered price for timber rights, grazing rights, and mineral leases to the full market value. If studies indicate that private owners take as good care of land as the public which leases grazing rights on public lands, then public grazing lands might be privatized.

Despite scattered, valiant efforts by a few economists for objective analysis, on the whole society has not been well served by those who evaluate the feasibility of environmental and natural resource projects.[1] A bipartisan *Office of Policy Analysis and Evaluation* established by Congress, patterned after other agencies such as the Congressional Budget Office, and somewhat insulated from narrow political pressures could provide more objective and reliable sources of information meeting high standards set by appropriate professional organizations. This could reduce the undesirable practice of agencies hiring those who evaluate or determine feasibility of the agency's own projects.

---

[1]When economic analysis is politicized, frequent casualties are the quality of analysis and sometimes the analysts themselves. For example agricultural economists at the University of Arizona nearly lost their jobs when they found that it would be economically more feasible to pay farmers to give up irrigation water for the growing urban needs of Arizona than to divert water from the Colorado River. A professor at Oklahoma State University almost lost his job when he reexamined benefit-cost calculations done by others for the Arkansas Waterway Development and found the benefit-cost ratio to be considerably lower than the original estimate used to justify the project.

NATURAL RESOURCES FOR FUTURE AGRICULTURAL PRODUCTION[2]

U.S. agriculture historically has benefitted from an abundant and productive natural resource base. However, the adequacy of the resource base for sustained future agricultural production has been a recurring concern. The agricultural resource base appears to be more than adequate to meet foreseeable future needs if public policy is supportive. Resource issues needing attention include soil erosion, urban encroachment on agricultural lands, depletion or pollution of groundwater supplies, and quality of air.

SOIL EROSION

Loss of topsoil to wind and water erosion can seriously reduce soil productivity and crop yields. Erosion reduces productivity by carrying away soil nutrients, reducing available water-holding capacity of the soil, and restricting the crop rooting zone. Technology, such as the use of fertilizer and other nutrients in combination with farm management practices, can compensate for most of the productivity losses, at least in the short run, but at a cost. According to the 1982 National Resources Inventory (NRI) conducted by the Soil Conservation Service, the national average of sheet and rill erosion on cropland was 4.4 tons per acre per year compared to 5.1 tons per acre in 1977. Soil loss tolerance (T), defined by USDA as the maximum average annual soil loss that will permit a high level of production on a specific soil, normally is about 5 tons per acre. Data on *average* erosion rates across the nation conceal the fact that some soils and regions have serious problems. About 44 percent of all cropland in the U.S. is eroding at levels greater than the soil loss tolerance. The most serious soil erosion problems occur on relatively few acres: The Missouri River basin in Iowa and Missouri, the Southern High Plains (mainly wind erosion) and the lower Colorado River basin (Figure 9.1). In 14 intensely cropped areas in the U.S., average erosion rates on cultivated cropland exceed 10 tons per acre (Lee, 1984). Erosion on those lands without excess erosion totalled only 18 percent of all soil erosion on the nation's cropland. Nationwide this is a general improvement compared to 1977 conditions.

Erosion on pastureland and rangeland averaged 1.4 tons per acre; grazed forestland erosion rates averaged 2.3 tons per acre; and soil loss on ungrazed forestland averaged 0.7 ton per acre. Almost all pastureland and forestland is eroding at less than T, and roughly 12 percent of rangeland is eroding at levels exceeding 2T. Thus the principal problem is cropland.

Wind erosion data were collected for all states (except Alaska) for the first time in the 1982 NRI. Wind erosion on cultivated cropland averaged 3.3 tons per acre and on rangeland averaged 1.5 tons per acre (CEQ, 1984, p. 287).

Soil erosion increases both short-term and long-term farm production cost. Erosion removes fertilizer, water holding capacity, soil fertility, and pesticides. Newly planted

---

[2]Parts of this section are from the Council for Agricultural Science and Technology (forthcoming).

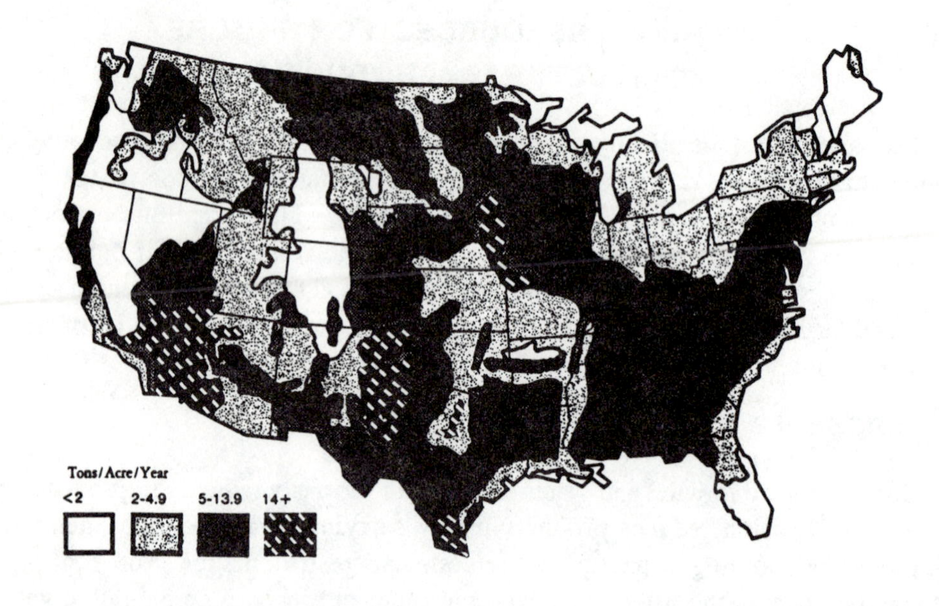

SOURCE: CEQ (1984, p. 287).

FIGURE 9.1. Average Annual Sheet, Rill, and Wind Erosion on Cropland

crops are damaged. Topsoils developed over geological time are stripped away to expose clay, bedrock, or other root-impermeable, non-productive materials.

The American Agricultural Economics Association Soil Conservation Policy Task Force (1986) estimated the 1986 erosion-induced productivity losses for soil used for corn and soybeans to be only $40 million per year, with accumulated present values of erosion over 100 years of $4.3 billion to $17 billion, depending on the discount rate . These estimates do not include the costs of offsetting management practices, the costs of erosion reduction measures such as terraces, and the costs of damage to growing-crops from deposition of wind-blown soil.

Another study of soil productivity losses concluded that if present levels of wind and sheet and rill erosion continue for another 100 years, other things equal, productivity on soils with the greatest erosion problems nationwide might decline about 4 percent (Alt and Putman, 1987). Of course, productivity losses in some regions will be much larger than the national average. Findings of Alt and Putman are reinforced by other studies. A U.S. Department of Agriculture study (1986) used the Erosion Productivity Impact Calculator (EPIC) to simulate the effects of weather, cropping rotations, plant growth, and related processes on soil erosion over periods of 50 to 100 years. The analysis indicated that if 1982 rates of erosion continued for 100 years on the 217 million acres of land where erosion is the principal soil problem, crop yields on those acres would be 3 percent less at the end of the period than they otherwise would be. Improvements in technology and other factors more than offset losses due to soil erosion so that corn yields were expected to increase 76-102 percent and soybean yields 119-122 percent from 1982 to 2030 — sufficient to allow U.S. land in crops to decline 30 percent during the period under the "most likely" scenario.

Soil scientists at the University of Minnesota developed a Productivity Impact (PI) model to estimate the long-term effects of erosion on soil productivity. Pierce et al. (1984) applied the model to 97 million acres of cropland in the Cornbelt, concluding that 100 years of 1977 soil erosion rates would reduce corn yields 4 percent, other things equal.

Crosson (1986) used regression analysis to estimate the impact of erosion on corn, wheat, and soybean yields using data from 91 million acres in the Cornbelt and Northern Plains. He estimated that yields of corn would fall 5-6 percent from erosion continued for 50 years at 1982 rates. Later, Crosson (1987, p. 14) concluded that the EPIC, PI, and Crosson (1986) studies ". . . give similar estimates of yield losses in the American Midwest after 50 years of erosion. Considering the very different analytical approaches underlying the three sets of estimates, the similarity of results is impressive."

Competition Among Sectors for Land

Rural land lost to urban and built-up uses nationally has ranged from .9 to 1.1 million acres per year over the last several decades (CAST, 1988). Recent data from the 1982 National Resources Inventory suggest that between 1967 and 1982 the urban, built-up, and rural transportation uses increased about .9 million acres a year (Lee, 1984). These estimates reflect the loss of all rural land, not just cropland or agricultural land. The 1975 Potential Cropland Study estimated that less than 40 percent of land converted to urban and built-up uses between 1967 and 1975 was from prime farmland, considered by the Soil Conservation Service to be the best farmland. Not all of this land was in crops. Thus loss of prime farmland to urban and built-up uses probably is less than 360,000 acres annually.

Of concern is how much agricultural land is available for cropping. The 1977 and 1982 National Resources Inventories indicated that for price-cost relationships of the respective study years, 127 million acres had high and 153 million acres of land had medium potential for conversion to cropland. This land historically has been very unresponsive to economic incentives to use in agriculture. Some potential cropland is available for development if economic conditions are favorable. But even considering the loss of some rural land to urban development, the future supply of cropland is adequate. Farm output will depend far more on technology and on human and material capital inputs than on shifts in land use. The loss of amenity values and open space in developing areas is of continuing concern, however.

Water as a Limiting Factor

Water is essential for U.S. agriculture. Through crop irrigation, agriculture consumes 80 to 85 percent of the fresh water resources of the United States (Gibbs and Carlson, 1985, p. 34). The western United States is particularly dependent on irrigation;

more than one-half of the West's agricultural production comes from irrigated lands (Frederick, 1982). Past expansion of western irrigation was stimulated by inexpensive federally subsidized water but new federal water projects are limited and new irrigation water supplies will be more costly. Nonagricultural demands for water in the West increasingly will compete with agricultural uses.

Aquifer mining, such as the Ogallala in the Great Plains, has lengthened pumping distances, making groundwater expensive for irrigation (Sloggett, 1985). Irrigation becoming uneconomic on many areas in the Southern High Plains has been offset by irrigation expansion in the Northern Plains where water supplies are more abundant. The net impact on farm production has not been large in the past because dryland cropping continues but adjustment costs of dislocation for individual farmers and communities are sometimes severe.

Current irrigation levels with average precipitation result in "mining of over 22 million acre feet of water from aquifers west of the Missouri-Mississippi Rivers" (Council of Economic Advisors, 1987, p. 151). Nationally, almost one-fourth of groundwater used by agriculture is not replenished. About one-quarter of the irrigated land in the West depends heavily on nonrenewable water supplies, and the productivity of the several million additional acres is threatened by rising salt levels. Management technologies to conserve water while maintaining or increasing crop production become profitable as the price of water for irrigation increases. Humid areas in the East and Midwest may increase irrigation to reduce yield variability and raise production.

The Council of Economic Advisors (1987, p. 162) noted that the emerging revolution in biotechnology along with "more efficient use of resources, more effective management, and regional shifts in production patterns could, under the right circumstances, expand the production of agricultural products within the United States [and lead] to a 2.4 percent productivity growth rate forecast for U.S. agriculture as a whole." Tutwiler and Rossmiller (1987, p. 22) propose that "the probability is high that the rate of productivity growth in the United States will be at 2.4 percent annually." These predicted future productivity advances are only guesses but imply that supply is likely to shift faster than demand to the right, tending to lower real farm prices and erase fears that declining groundwater supplies or soil erosion threaten future food and fiber supplies.

Most analysts do not consider water scarcity to be a long-term threat to the viability of U.S. agriculture (see Frederick, 1982). However, changes in the institutional rules governing water allocation may be needed for the wise use of limited water supplies.

Water Quality

Water salinity will limit irrigation in some areas. An estimated 20-25 percent (about 10 million acres) of all irrigated land in the U.S. suffers from salt-caused yield reduction (El-Ashry, Schilfgaarde, and Shiffman, 1985). Irrigation return flows are a major cause of salinity problems in the semiarid western states where significant quantities of salts occur naturally in rock and soils. In affected river basins, salinity has progressively increased as

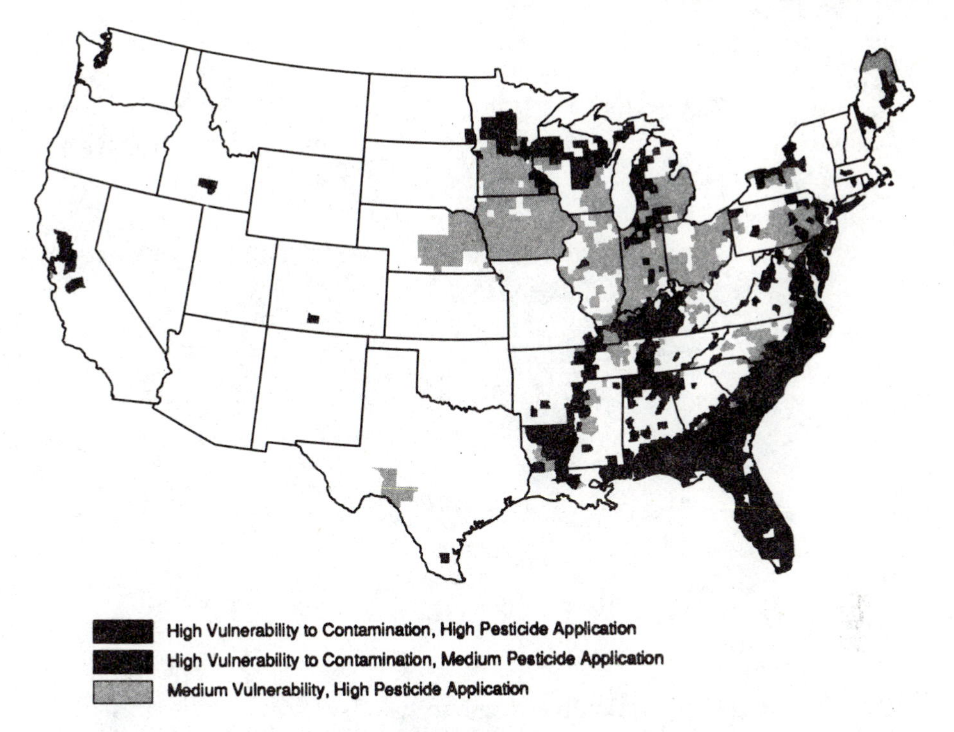

SOURCE: U.S. Department of Agriculture (January 1987, p. 26).

FIGURE 9.2. Potential Groundwater Contamination from Pesticide Use

water resources have been developed. This trend is expected to continue unless comprehensive water quality management schemes are implemented.

Nitrate and pesticide contamination of groundwater used for crop irrigation is being detected in some areas and will increase. The impact of these contaminants on crop yields is not yet known. Potential harmful effects on yields must be ascertained. Drinking water contamination is a serious problem.

The southern Coastal Plain (including Florida), the central Atlantic region, the Mississippi Delta, the midwestern Cornbelt, western Kentucky, and the Central Valley of California are the major regions with predicted pesticide contamination potential (Figure 9.2). Smaller areas in the Northeast, Texas, and Idaho also have potential contamination (Nielsen and Lee, 1987, p. 25).

The regions shown in Figure 9.2 to have potential groundwater contamination from pesticides correspond with production of pesticide-intensive crops such as corn and soybeans. Tobacco, cotton, rice, and peanut production in the Southeast and Delta also show high pesticide use. Fruit and vegetable production is associated with high pesticide use in Florida, California, and portions of the Northeast and Lake States. It is cautioned that Figure 9.2 is based on estimated rather than actual levels of contaminants in groundwater.

Nitrate-nitrogen contamination of groundwater from agricultural fertilizer use is concentrated in the central Great Plains; the Palouse and western Washington state; portions

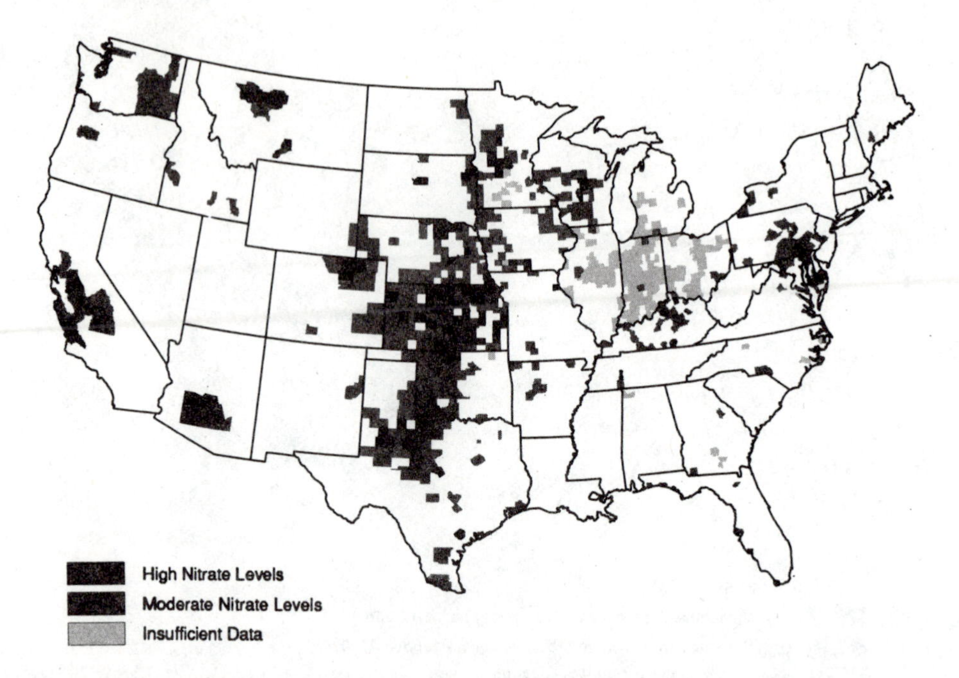

SOURCE: U.S. Department of Agriculture (January 1987, p. 26).

FIGURE 9.3. Nitrate-Nitrogen Distribution in Groundwater in Agricultural Areas

of Montana; southwest Arizona; the irrigated fruit, vegetable, and cotton-growing areas of California; portions of the upper Cornbelt; southeast Pennsylvania; Maryland; and Delaware (Figure 9.3). In many cases, the areas highlighted in Figure 9.3 represent a combination of fertilizer applications and irrigation, particularly in California, the Palouse area in Washington, northern Texas, and portions of Kansas and Oklahoma.

Combining the above results, areas of potential contamination from pesticide and fertilizer use account for 1,437 counties, or about 46 percent of the counties in the conterminous States. Less than one-fourth of counties identified to have potential contamination from agricultural chemicals exhibit both high pesticide and nitrate contamination potential. These 314 counties are located chiefly within the Cornbelt, the Lake States, and the Northeast.

These 1,437 counties with pesticide or nitrate contamination are cropped intensively, with 33 percent of all land area in crops compared with 16 percent nationwide. Over 70 percent of the crop acreage in the sample is devoted to corn, wheat, and soybeans. Though strongly agricultural, these counties are heavily populated, with 27 percent of the land but 47 percent of the population.

People who live where the contamination potential from agriculture is high and consume mostly groundwater most likely will incur the highest costs. These costs include monitoring and detection activities, adverse health effects, the installation of water filters, or the use of bottled water.

The areas with the most people relying on groundwater are scattered throughout the South, Northeast, Midwest, and portions of the West. Over 19 million people in these counties obtain their drinking water from private wells. Over 65 percent of these people live in areas where only potential pesticide contamination is indicated, while less than 10 percent live in areas with only potential nitrates. The remainder reside in areas with a potential for both pesticides and nitrates. As with private wells, the majority of people using public groundwater supplies (68 percent) reside in areas with only potential pesticide contamination. The remainder are divided nearly equally into areas of only potential nitrate contamination and both pesticide and nitrate contamination.

In summary, agricultural chemical contamination of groundwater could affect 53.8 million people. This occurs because of the density of population in the areas affected and their heavy reliance on groundwater and private wells. Despite the large population potentially affected, data indicate that groundwater contamination from agricultural chemicals is not national in scope. Rather, areas of potential contamination appear to be regional, which suggests that targeting is needed for any prevention strategy.

Farmers' private wells are affected when agricultural chemicals reach groundwater. Informed farmers have strong incentives to reduce or minimize activities that pollute. Unfortunately, little advice is available to farmers about the impact of agricultural practices, such as conservation tillage, on groundwater quality. While farmer education programs are needed, their success will in part depend on well-documented research, much of which is just being initiated. While the findings thus far indicate a significant potential for groundwater contamination from agricultural chemicals, the data do not determine the magnitude of the costs from such contamination.

AIR QUALITY

Atmospheric pollutants can negatively effect agricultural production and yields. Ozone is the air pollutant affecting vegetation to the greatest extent in the United States (Heck et al., 1982). Ozone alone and in combination with sulfur and nitrogen oxides may account for 90 percent of crop losses in the United States caused by air pollution (Gibbs and Carlson, 1985, p. 34).

Highest ozone concentration in the U.S. is during the growing season in the Southeast, but ambient levels of ozone across most agricultural regions of the country are high enough to have measurable impacts on crop yield. The annual effects on crop production from ambient ozone are estimated to be comparable to losses from pests and diseases. The economic benefits to producers and consumers of agricultural commodities from a 25 percent decrease in ambient ozone levels have been estimated to range from $1 billion to $2 billion (Adams, Hamilton, and McCarl, 1984).

Carbon dioxide levels rise from burning of fossil fuels. Levels also rise with deforestation and desertification which destroy vegetation converting carbon dioxide to oxygen. Rising carbon dioxide levels could increase temperatures, raise water levels by melting ice, and shift cropping in the direction of the poles. The process is slow, not easily predicted, and in need of further monitoring. Increasing levels of carbon dioxide may enhance plant growth (Gibbs and Carlson, 1985, p. 34).

No one yet can predict accurately the adverse effects of acid rain on crop productivity. Acid rain and other gaseous pollutants may cause extensive damage to forests and lakes but are not estimated to cause significant crop damage in the U.S. However, accurate estimates of pollutant damage are not available. Recently, gaseous concentrations of volatilized pesticides have been found over treated cropland in California (Glotfelty, Seiber, and Liljedahl, 1987). Chemical reactions in the atmosphere can both reconcentrate these pesticides and transform them into more toxic substances than the original chemicals. Information is not yet available to appraise the biological impacts on crops.

OFF-SITE IMPACTS ON THE ENVIRONMENT AND HUMAN HEALTH

Concern is growing about adverse impacts of agricultural management practices and chemicals on the environment and human health. Agencies previously concerned only with on-farm productivity impacts, such as the Soil Conservation Service, are now considering off-site environmental impacts of soil erosion. The impacts of agricultural chemicals on water and air quality and food safety are attracting increasing attention from regulatory agencies, both state and federal. The costs to farmers of regulating agricultural activities or agricultural inputs are measurable, but the benefits to society are difficult to quantify. Public concern over unavoidable risks is often greater than over clearly identified, avoidable risks. Agriculture will need to adapt to public demands for lessening environmental and human health risks.

OFF-SITE IMPACTS OF SOIL EROSION

Concern over the productivity impacts of soil loss has existed since the Dust Bowl, but study of the off-site impacts of soil erosion is relatively recent. Nonpoint source water pollution problems caused by sediment, animal wastes, fertilizers, pesticides, and other contaminants carried off by storm water from fields are estimated to cause a wide variety of in-stream and off-stream damages. The Conservation Foundation has estimated that the annual off-site damage costs from soil erosion range from $3.2 to $13 billion dollars. Of this, cropland's share is estimated to be $2.2 billion dollars annually (Clark, Haverkamp, and Chapman, 1985). Wind erosion has been estimated to impose similar off-site costs (Huszar and Piper, 1986). The off-site benefits of erosion control expenditures of three major USDA programs were found to exceed the on-site productivity benefits in a recent study (Strohbehn, 1986).

These findings can influence public policy regarding soil erosion control. If productivity losses for farmers are the major concern, a voluntary program to encourage adoption and implementation of soil conservation plans might be acceptable. Any farmer who does not adopt appropriate soil conservation measures is impairing his own future productivity, profits, and real estate value. And if soil erosion were deemed by the public to have only a minor impact on on-site farming productivity, conservation might be viewed as a small public issue.

However, focus on large off-site damages makes a solely voluntary program more difficult for environmentalists to accept. Off-site damages impose costs on others. If voluntary programs are not effective, mandatory programs may be proposed. Any successful soil conservation program will have to meet off-site as well as on-site erosion reduction goals.

Agricultural Chemicals

Of growing concern is the impact on the environment and human health of agricultural chemicals used to enhance production. Highly publicized events associated with the Kesterson Reservoir, where concentrated salts and contaminants from irrigated agriculture in California have damaged wildlife and the aquatic environment, emphasize the sensitive interactions between agriculture and the environment. Pesticides, commercial fertilizers, and animal wastes are major contributors to nonpoint-(widely dispersed) source pollution. Contaminants can reach surface waters with sediment from soil erosion. Groundwater can be contaminated in some regions by leaching of pesticides and nitrates. Irrigation and some tillage practices can further increase leaching. Residues in foods from pesticides, antibiotics, and additives such as hormones are a public health problem.

Public concern about environmental and human health risks is real, but the extent of damages is very difficult to assess. Studies on humans are limited, in some cases produce contradictory results, and are differently interpreted. It is difficult to prove conclusively that exposure to low dosages of any chemical contributes to cancer or other forms of human illness. Based on risk assessments, the Environment Protection Agency (EPA) has cancelled the use of pesticides such as DDT and chlordane. More are currently under study. The *Journal of the American Medical Association* recently reported a link between the use of 2,4-D by farmers in Kansas and certain forms of cancer (Hoar et al., 1986).

Risks from use of agricultural chemicals are difficult to assess within a *cost-benefit* or *risk-benefit* framework. Cost-benefit analysis is troubled by inability or unwillingness to place a dollar value on morbidity or mortality. Risk-benefit analysis attempts to measure number of lives lost relative to benefits generated by a residue but is troubled by inadequate data and conceptual problems. For example, a study of groundwater contamination from agricultural chemicals estimated that it would cost private well owners in potentially contaminated counties $.9 to $2.2 billion in initial monitoring costs to detect potential problems (Nielsen and Lee, 1987). However, the costs to agriculture of alternative management, cropping patterns, pest control strategies, or regulatory actions are unknown. Substantial data are required to compare the benefits of environmental protection with the costs of programs and policies on the agricultural sector. The public and policymakers may choose to err on the side of caution, possibly requiring the agricultural sector to choose between voluntarily implementing measures to reduce chemical impacts on the environment or face regulatory measures that will require them to do so.

Food Safety and Quality

The Food and Drug Administration attempts to ensure the purity of foods and drugs. The U.S. Department of Agriculture inspects food to ensure that it is free of pesticide residues, bacteria, and other pathogens. Much of this inspection is random. Despite regulations such as banning of DES to ensure that livestock and produce coming to market will be free of contaminants, consumers are fearful that antibiotics routinely fed to broilers or hogs can escape the inspection. Cases of salmonella poisoning in meat have been traced back to subtherapeutic antibiotic feeding to animals for faster weight gain.

Laws and regulations against use of chemicals in farm commodities include the Delaney clause of the federal Food, Drug, and Cosmetic Act of 1938. It specifies that any agent found to cause cancer in either human beings or animals must be removed from the food supply. Much controversy surrounds the Delaney clause. Sceptics ask troublesome questions:

1. The test is usually not on humans but rats or bacteria. To what extent do results from rats extend to humans?

2. Instruments have become more refined to detect ever smaller traces of carcinogenic substances. Should a substance which requires massive doses to cause cancer in rats be banned although it is only found in one part per billion in food or water? Does zero tolerance make sense? A substance such as cyclamate or saccharine sweeteners may cause cancer in rats at very heavy dosages equivalent to a human consuming say the equivalent of 100 gallons or pounds per day. But if it does not cause cancer in small doses and if humans consume at most only an ounce per day, should it be banned entirely? As instruments become ever more sensitive, the zero tolerance requirement bans more products with less justification.

3. Benefits and risks need to be balanced with costs. Suppose that a food additive with no substitute saves 2,000 lives per year by reducing cholesterol in the blood stream but causes 2 deaths per year from cancer. Should it be banned by the Delaney clause?

TECHNOLOGY AND THE ENVIRONMENT

The Office of Technology Assessment (OTA, 1986) identified about 150 emerging technologies in 28 technological areas as part of a study, *Technology, Public Policy, and the Changing Structure of American Agriculture*. These technologies range from biotechnologies and regenerative and organic farming to informational systems. Most of these emerging technologies are expected to reduce the land and water requirements for meeting future agricultural output demand. The technologies are thought to have beneficial effects relative to soil erosion, wildlife habitat, and the risks associated with the use of agricultural chemicals. Yield-increasing technologies can reduce requirements for land and can permit conversion of cropland to pasture, forest, and recreational uses more consistent with soil conservation. Development of biological resistance to pests reduces dependence on chemical pesticides. As noted by OTA, however, not all technologies will reduce dependence on chemicals. New conservation tillage technologies may reduce erosion and threats to wildlife while increasing the use of agricultural chemicals. Development of new technologies will occur in a social and political atmosphere requiring environmental and public health consequences to be explicitly considered.

Breeders of crops and livestock have traditionally shifted genes within species to change plants and animals for beneficial use. Recombinant DNA (rDNA) and other techniques of the new biotechnology permit genes to be selectively transferred among species with precision and speed. These powerful tools of biotechnology have the potential to transform agriculture for better or for worse. Plants may be developed to resist drouth, pests, frost, or salinity while providing improved nutrition. Bovine and porcine growth hormones have been produced in volume at low cost by simple bacteria. These are not new chemicals, merely natural ones produced by introducing the genes for hormone production into simple organisms. Supplementing cattle or hog hormones with these hormones produced by simple organisms markedly improve volume and efficiency of milk and pork production.

A fear is that introduction of genes into plants or animals by rDNA techniques will create and unleash an altered organism which cannot be controlled and will stalk the earth like the black death or a Frankenstein monster. To avoid problems and address concerns, the National Institutes of Health (NIH) in 1974 established a special Recombinant DNA Advisory Committee (called the RAC) which provided advice and in 1976 developed the NIH "Guidelines for Research Involving Recombinant DNA Molecules" for use in federally funded or conducted research. Subsequent experience gained from rDNA laboratory research and risk assessment experiments have dampened many earlier concerns about dangers of rDNA research and has led to less stringent guidelines.

As basic and applied research and commercial applications in the new biotechnology mount, so too do the number of regulatory agencies involved. These include but are by no means limited to the U.S. Department of Agriculture, Food and Drug Administration, Environmental Protection Agency, and National Science Foundation (see CEQ, 1984, pp. 462-67). Consequently, a Cabinet Council Working Group on Biotechnology comprised of these and other federal agencies and departments was formed in 1984 to review regulatory requirements (especially those of NIH) and recommend appropriate administrative or legislative actions. One result was a framework coordinating federal regulation of biotechnology and for review of applications for new microorganisms (Council of Economic Advisors, 1987, p. 206).

The timeliness of the Cabinet Council Working Group was highlighted a few weeks after its establishment by an opinion by District Judge John Sirica in the case of *Foundation for Economic Trends v. Heckler*. The case awarded a preliminary injunction preventing the conduct of a particular field test of an altered microorganism and further prohibiting other environmental applications by federal grants until such time as the government prepared an *environmental impact statement* as required by the National Environmental Policy Act. On appeal, the appellate count required the preparation of the less detailed *environmental assessment* and lifted the injunction on other experiments.

Many social scientists and social activists call for mandatory filing of a *socio-economic impact statement* before a firm or agency undertakes research on technology. While a strong case can be made for more research to anticipate socio-economic impacts, the case for a binding legal structure to impede or halt research on technology is weak at best.

The device undoubtedly would be used to stop or delay resource development projects on the basis of procedure rather than merit as measured by costs and benefits. Social scientists have not been skillful at anticipating benefit-cost ratios from new technology. One example is from a seminar in 1966 at the Delhi School of Economics in India to anticipate the consequences from massive introduction of high-yielding green revolution seeds of dwarf wheat and rice varieties. David Hopper (1978, p. 69) reports:

> The seminar participants were government bureaucrats, scholars from agricultural and general universities, a sprinkling of foreign advisors and expatriate technical assistants, and a few political leaders, including, when time permitted, India's Minister of Agriculture.
>
> Within the first few hours of a three-day meeting, the discussion focused on a call by many participants for government prohibition of further imports of high-yielding seeds and for government efforts to ban the spread to farmers of the genetic stocks of dwarf materials then available on the research stations of the nation. Despite the protests of the few, the meeting carried a clear consensus for prohibiting the entry and use of the new varieties. Fortunately for the nation's hungry masses, the politicians ignored the consensus.

The second example began several years ago with a lawsuit by California Rural Legal Assistance (CRLA) attorneys. With advice and encouragement from some social scientists, attorneys filed suit against the University of California on behalf of farm workers whose jobs might be eliminated by labor-saving machinery under development at the University of California. Although the lawsuit applied only to labor-saving machinery for agriculture, many in the academic community feared that it would set a dangerous precedent for all applied research. CRLA charged that mechanization research displaced farm workers, eliminated small family farms, diminished the quality of rural life, and harmed consumers (Martin and Olmstead, 1984, p. 25). Proponents of publicly supported mechanization research claimed that such research reduced costs of food to consumers, kept the United States competitive with less developed countries in production of commodities, and eliminated low-skill jobs often characterized by substantial drudgery and backbreaking labor while retaining food production and processing industries in the United States. Proponents contended that millions of Americans were freed by farm mechanization to work off farms to supply goods and services such as recreation, health, and education highly sought by Americans. They reasoned that a diversified public institution such as the University of California-Davis could perform a unique role in development of technology such as the tomato harvester. A large diversified research institution can employ an integrated systems approach to technology by simultaneously working on mechanization, variety breeding, and socio-economic research which a private firm cannot do at tolerable cost and risk.

The third example was an effort of the Foundation on Economic Trends (see earlier comment) and the Humane Society to stop the National Institutes of Health (NIH) from financing research involving the transfer of genes from one mammalian species into another (McDonald, 1984, p. 7ff). The two groups also wanted NIH to withhold financial support from any institution conducting such experiments. The two groups filed a lawsuit against the U.S. Department of Agriculture to halt a study to transfer human growth-hormone genes into pigs and sheep, a study conducted jointly by the U.S. Department of Agriculture and the University of Pennsylvania. The Foundation for Economic Trends persuaded a federal judge to stop some types of genetic-engineering research approved by NIH. The University of California appealed an order by a federal judge stopping all experiments involving releases of genetically engineered organisms.

At issue is the role of society in imposing restraints on the new biotechnology research. Restraint is unavoidable and essential: a scientist cannot be free to unleash intentionally or inadvertently a genetically engineered microorganism that would do great harm. On the other hand, excessive restraint could forego benefits of eliminating genetic disorders, slowing the aging processes, curing cancer and other diseases, and ending hunger for millions of people. Other nations will advance faster technologically and economically if our research and experimentation is forbidden or too tightly constrained by rules and regulations. To release, prematurely or otherwise, technology incapable of generating more social benefits than costs is a Type I error. To forego benefits by withholding release of a technology with a favorable social benefit-cost ratio is a Type II error. Clearly, Type I and II errors must be balanced against each other. To contend as many social activists do that bioengineering research should be stopped until the social, moral, and political issues are resolved is to contend that the moratorium on research be as interminable as the debate itself. Research will go on but tough questions must be faced. One molecular biologist, noting that research is necessary to preserve human life, asked, "Whose rights are to be defended, the right of a mouse to its genetic heritage or the rights of human beings to health and happiness?" (see McDonald, 1984, p. 8).

Modern biotechnology, computers, and telecommunications will change the environment and structure of agriculture in profound ways in the future. The pace of technological and structural change along with measures to cushion impacts on losers in the process will be influenced by public policy. Social scientists will provide a major input into that public policy. Simply calling for a halt to research on technology until the social, moral, and political issues are resolved will not do. Social scientists will need to do a better job than they have in the past in appraising the prospective benefits and costs of technology so they can be a constructive part of the inevitable debate over the appropriate public role in guiding technological research and consequent induced farm structural change.

FERTILIZER AND ENERGY

Agriculture relies heavily on natural resources. Some such as land, water, and air were discussed earlier. Also of concern is future availability of fertilizers and energy.

Analysis (Yeh et al., 1977, pp. 46, 47) indicated that world reserves of potash are adequate for the indefinite future but U.S. supplies are rapidly being depleted. Of greater concern are world phosphate reserves estimated in 1977 to last 150 years at 1976 costs per unit. Landsberg, Tilton, and Haas (1982, p. 81) reported estimates from the U.S. Bureau of Mines of potash (K_2O) reserves sufficient to last for 3,638 years at 1974 consumption rates and for 107 years at a consumption growth rate of 5 percent per year — well under the 1947-74 annual growth rate of 9.0 percent. The same authors reported phosphate rock reserves adequate for 128 years at a 1974 consumption rate and for 41 years at a 5 percent annual growth rate — well below the 7.3 percent growth rate of 1947-74.

A more recent analysis (Fantel et al., 1985) estimated that world known phosphate rock reserves will last for 245 years at 1981 annual usage. Adding to known reserves the reserves anticipated by experts to exist, phosphate rock reserves were estimated to last 653 years at 1981 annual usage. However, fertilizer usage is rising, and the U.S. Bureau of Mines and Geological Survey projects a 3.6 percent annual increase in phosphate use. Applying this figure to future use, the onerous portents of exponential growth become clear: Demonstrated phosphate rock reserves are adequate for only 61 years and known plus anticipated reserves are adequate for only 88 years.

Reserves of fossil fuels, fertilizers, and other minerals do not run out; they become uneconomic to utilize. Prices rise to ration supplies as supplies diminish. Worldwide production of phosphate rock totaled 145 million metric tonnes in 1981. Fantel et al. (1985, pp. 20-22) estimated that 1.6 billion tonnes of phosphate rock are recoverable in market economies at a cost of less than $30 per tonne (55 percent of world total reserve in the U.S.), 11 billion tonnes are recoverable at less than $40 per tonne (13 percent in the U.S.), and 16 billion tonnes recoverable at production costs of less than $50 per tonne (21 percent in the U.S.).

Yeh et al. (1977, p. 47) noted that, "In the case of phosphate and potash, troublesome issues could emerge regarding availability of supplies in the face of possible international political exigencies." Massive potash reserves are in Saskatchewan; more problematic is accessibility of large phosphate reserves in northwest Africa, a region characterized by political instability from time to time.

The third principal fertilizer ingredient, nitrogen, is abundant in the air but processing requires hydrocarbons, primarily natural gas. Petroleum is also critical to power machinery and equipment and is used to produce pesticides. The productivity of U.S. agriculture depends heavily on petroleum.

The food and agriculture industry is not a disproportionately heavy user of fossil fuels — use is nearly equal to its share of gross national product. Analysts are not in full agreement but one quite recent study reported, "All our analyses indicate that by the year 2020 domestic U.S. oil supplies will, effectively, be depleted" and went on to add that "economic domestic [natural] gas supplies will also be depleted by 2020" (Carrying

Capacity, 1986, pp. 18, 19). A report by the Rockefeller Foundation (Sivard, 1980, p. 14) indicated that world natural gas reserves are adequate to last 51 years and oil reserves 28 years at 1979 rates of use.

Not all analyses are that pessimistic but it is well to contemplate the consequences of worst-case domestic scenarios coupled with inaccessible or very expensive foreign oil and gas supplies. Several options are available. The U.S. has coal reserves to last hundreds of years at current consumption rates (Yeh et al., 1977, p. 47). The solid, liquid, and gaseous energy from coal could provide adequate energy supplies but at large environmental costs based on current technology. Many scientists are optimistic that safe, low-cost, abundant energy will become available from nuclear fusion (the same process that powers the sun) within 100 years and well before fossil fuels are depleted.

The potential for additional life from existing sources of energy is great if account is taken of opportunities for substitution, technological innovation, and conservation. With sufficient incentives, wind, solar, biomass, gasahol, and nuclear fission and breeder reactor technology also can provide energy supplies. The need for hydrocarbons can be reduced by using collectable manure, growing leguminous green manure, and using legumes within the farming system, and by developing nitrogen-fixing capabilities in grasses (including grains) through biotechnology. There is basis for optimism but not complacency. The nation is highly dependent on imported energy. A strategy to maintain viability of agriculture and the nation is to build energy reserves for meeting short-term energy crises, to provide incentives for conservation of energy for the intermediate-run, and to pursue a strong program of science and technology to develop low-cost, safe, and abundant domestic energy sources not dependent on fossil fuel for the long run. Each of these efforts can be pursued simultaneously.

CONCLUSIONS AND POLICY OPTIONS

ENVIRONMENT

The public is concerned over environmental impacts of modern farming methods and of pesticides, fertilizers, and other chemicals on food quality. The appropriate response by the agricultural community is to search for the truth, report the truth fully insofar as it can be known, and support appropriate private and public actions to promote the general welfare. Restrictions on use of pesticides and other chemicals curtailed to protect food and water will not threaten the adequacy of food supplies or the economic vitality of agriculture. In fact, less agricultural output could raise farm income in the short and intermediate run.

Farmers have a stake in a safe food and water supply not only for consumers at home and abroad, but also for themselves as consumers of food and water. Although the market can encourage safe food by paying a premium for high quality products, the market alone and producers' good intentions alone are inadequate. Standards, inspection, and testing under public authority is required in production and marketing to ensure that foods labeled as organically grown or meeting other standards are in fact grown and marketed as labeled.

The U.S. has the wherewithal to provide a food supply which not only is dependable in the long run but avoids health risks to consumers from additives or other sources. All Americans deserve the opportunity to obtain abundant, nutritious, and safe food supplies. Full disclosure, access to information, and increased research are essential in resolving issues of food safety.

Acceptable tradeoffs between risk and benefit must be determined by consumers and the political process. Risk-benefit analysis, though in a formative stage and still frequently flawed by inadequate data, potentially has much to offer. Careful information gathering, analysis, and dissemination are critical to sound decisions on such matters. If the decision is made to release a substance that entails risks but expected benefits exceed costs, then attention can be paid to focusing use on groups with less risk, and informing potential users so each can make a personal decision whether benefits exceed risks of use.

NATURAL RESOURCES

Most analysts conclude that the soil and water resources in the United States are adequate to provide for future agricultural viability. If increased demands for food and fiber emerge, the cropland base can be expanded and utilized more intensively. If cost/price relationships are favorable, new technologies will be developed to increase yields and conserve land and water. Air quality changes, on the other hand, will be affected by technological changes that are predominantly beyond the control of agriculture.

Concern about the resource base extends beyond agricultural productivity. Yield-enhancing technology does not replace the need for careful stewardship of the resource base. Some resource decisions are economically irreversible. Once land conversion occurs, for example, it is difficult to return it to agricultural use. The considerable uncertainty about future technological growth and demand for resources was noted in the previous section. Use of resources now may preclude options for future generations whose tastes, preferences, and needs for resources may differ from ours. This highlights the importance of careful monitoring of trends to ensure the prudent use of natural resources. Given that it is not possible to precisely match resource conservation today with needs of the next generation, it is much less serious to err on the side of conservation than to err on the side of profligate use.

The United States met growth in domestic and foreign demand for food and fiber with essentially the same real volume of farm production inputs from 1920 to 1987. The source of increased output or new wealth was not natural resources or raw labor but increased productivity from creation of knowledge through human ingenuity, education, research, science, and technology. The result has reduced pressure on soil and water resources.

Sharply raising farm commodity prices or substantially curtailing soil erosion and conversion of cropland to urban uses are costly and not effective means alone to respond to emerging food and fiber needs. As noted above, there is merit in preserving agricultural land for option value and other reasons. But the lowest environmental cost and overall economic cost strategy is to invest in development of new agricultural technology to raise farm productivity to meet demand for food and fiber with minimal pressure on natural resources and the environment. Another alternative is to scale back living standards, an alternative only a few Americans are likely to pursue (CAST, 1988).

POLICY OPTIONS

Commodity programs have supported conservation by converting millions of acres to soil conserving uses. However, sometimes high price supports have encouraged conversion of grazing land and wetlands from soil conserving uses to cropland. The 1985 farm bill attempted to better serve conservation than did past programs. The *Conservation Reserve Program* (CRP) will include up to 45 million acres of highly erodible cropland. The CRP can reduce soil and water loss while contributing to wildlife habitat, recreation, reserve production capacity, and farm income. No more than 25 percent of the cropland in any one county can be placed under contract, although upon government approval of a formal request from counties the limit can be exceeded if the county is not "adversely impacted." Three other provisions in the 1985 Act — the Sodbuster, Conservation Compliance, and Swampbuster provisions — also relate to conservation and hence to long-term vitality of agriculture.

The *Sodbuster* provision applies to land not considered cropland prior to the 1985 Act. If a farmer breaks sod on that land, he must develop and implement an approved conservation plan that crop year to be eligible for any government price supports, Farmers Home Administration credit programs, or federal crop insurance and disaster payments. The *Swampbuster* provision applies to wetlands. Such wetlands are not eligible for government programs if their conversion to cropland began after enactment of the 1985 Act.

The *Conservation Compliance* provision applies to land considered cropland when the 1985 Act was enacted but which has a high erodibility rating. If the land is cropped, to maintain eligibility for government programs the farmer must have an approved conservation plan by 1990 and the plan must be fully implemented by 1995.

The CRP by late 1987 had enrolled only a small portion of the 80 to 100 million acres eligible for the program and needing conservation measures. Although the Food Security Act of 1985 has made significant strides in protecting the environment, selected changes would improve on the Act. The several possible changes or additions to the CRP listed below are designed to build on or augment the attractive multiple long-term benefits of CRP for a sustainable agriculture. They are options deserving of consideration but would require in-depth analysis before implementation (see CAST, 1988).

— Allow the Conservation Compliance provisions and CRP expansion to work together. Eligibility of erodible land for regular price support and diversion programs provides stiff competition for cost-effective implementation of the CRP. Combining the "carrot" of CRP with the "stick" of ineligibility for other programs raises cost-effectiveness of government funds used to convert the most erosion prone land to soil conserving uses. This option would bring more highly erodible cropland into the CRP.

— Increase the proportion of land eligible for the CRP in counties with severe erosion problems; at the same time allow grazing and haying of additional CRP land to reduce unfavorable economic impacts on local communities and to lower bid costs to the government. Increasing the size of the CRP would permit reducing regular acreage diversion and paid diversion programs. The potential unfavorable impact on cattle production and prices needs to be considered, however.

— An alternative to the above option would be *cropland easements* on highly erodible land. The owner could use the land for grazing, haying, wildlife habitat, recreation, or other uses under an approved conservation plan but could not crop the land. The government could obtain cropping rights by purchase or loan arrangements on a bid basis. The easement might be for an indefinite period. In the case of a loan, the owner could repay the loan with interest to regain cropping rights or the government could temporarily remove easement restrictions on cropping during a national emergency. An advantage of the easement would be permanent conversion of much highly erosive cropland to soil conserving uses while allowing beneficial uses of land that could help to keep local communities economically viable in the long run. The program, operated on a bid basis to reduce government cost, could remain in place serving multipurpose objectives in the absence of conventional cropland diversion programs.

— Reduce the risk to farmers of putting land in the CRP in the face of possible future inflation by indexing CRP payments to an index of cash rents, to prices received by farmers, or to the Consumer Price Index. Land would be offered to CRP at lower bid prices if protected from risks of inflation.

— Expand the CRP or easements to control use of nonrenewable groundwater supplies for irrigation on a countywide basis. Given high erosion rates and mining of groundwater, continued cropping in the Ogallala reservoir is difficult to justify in the face of current excess farm production capacity and possible need for food supplies in the future. Turning off a few scattered pumps in a county does little to halt the decline of groundwater use — nearby wells will continue to deplete the supply. Pumping needs to be halted over a considerable area but dryland cropping, haying, and grazing consistent with a conservation plan could be continued to help rural communities remain viable. In this and other cases where grazing and haying are permitted, attention needs to be paid to safeguard the livestock industry from too rapid an expansion of supply.

— CRP could place even greater emphasis on controlling erosion in criteria for acceptance of bids. A premium would be paid to obtain the most erosion-prone cropland in CRP. Holding reserve production capacity, wildlife habitat, and recreational lands would be important but secondary objectives. The CRP would focus solely on only highly erodable acres, whole fields and farms might rarely be in CRP. Other conservation programs also would narrowly focus on the most erodable cropland.

As concern over natural resource and environmental issues relating to agriculture intensifies, interest in applying environmental regulations to agriculture also intensifies.

Many of the environmental impacts of agriculture affect third party interests outside of agriculture. The traditional voluntary USDA approach to soil and water management may not be acceptable unless a high compliance rate can be demonstrated.

Hazards from agricultural chemicals directly affects farmers and chemical users. Applicators face risks from improper handling and disposal of chemicals. In some areas, farm families who depend upon private wells for drinking water can be exposed to agricultural chemicals through groundwater contamination. Farmers, like consumers, are exposed to residues in foods from pesticides, antibiotics, and other additives. If farmers are to play a constructive role in protecting their own as well as other environments, they as well as society must be well informed. Improved educational programs are needed for farmers about management practices to reduce the environmental and health risk of chemicals. Continued monitoring of natural resource and environmental trends is necessary so that agriculture and society can make necessary decisions and adjustments to conserve and use resources in a sound manner for this and future generations.

Other policy questions for resource conservation and development as follows (Tweeten, 1984):

1. To what extent can the effectiveness of soil and water conservation measures be enhanced by focusing limited public funds more narrowly on areas of greatest need?

2. To what extent is it feasible to provide all farm operators and landowners with measures of soil erosion rates in excess of tolerance levels on their land? Many farmers do not know how rapidly their land is eroding. Would it make sense to make these rates public knowledge to bring peer pressure for conservation?

3. To what extent is it feasible to expand the role of the market-price system in water allocation among all users by emphasizing negotiable water rights?

4. To what extent is it feasible to set up social experiments to evaluate soil conservation policies with a control group and experiment group? The latter would be divided into subgroups offered differing treatments of soil conservation education, subsidies, and technical assistance and regulation. Results would be monitored and analyzed to determine effective and acceptable conservation practices.

5. To what extent is resource conservation best served with limited public funds by (1) direct investments in resource conservation education, technical assistance, and subsidies, versus (2) tieing of conservation measure to commodity programs, versus (3) indirect investments in production technologies that improve crop and livestock production efficiency and reduce demand for soil and water? For example, is resource conservation furthered to a greater extent by investing $1 billion in genetic engineering or in cost-sharing to build land terraces, or in a long-term cropland easement?

6. To what extent would federal block grants for conservation to individual states (with states then establishing allocations and procedures) further resource conservation objectives?

7. To what extent is it desirable or feasible to tax or in other ways penalize landowners for erosion rates above soil tolerance levels?

REFERENCES

Adams, R.M., S.A. Hamilton, and B.A. McCarl. 1984. *The Economic Effects of Ozone on Agriculture*. Washington, D.C.: U.S. Environment Protection Agency.

Alt, Klaus and John Putman. April 1987. Soil erosion dramatic in places, but not a serious threat to productivity. Pp. 28-30 in *Agricultural Outlook*, AO-129. Washington, D.C.: ERS, USDA.

American Agricultural Economics Association Soil Conservation Policy Task Force. 1986. *Soil Erosion and Soil Conservation Policy in the United States*. Occasional Paper No. 2. Ames, Iowa: AAEA Business Office.

Carrying Capacity, Inc. 1986. Beyond oil. (Summary report). Washington, D.C.

Clark, E.H., J. Haverkamp, and W. Chapman. 1985. *Eroding Soils: The Off-Farm Impacts*. Washington, D.C.: The Conservation Foundation.

Cornucopia Project. 1981. *Empty Breadbasket?* Emmaus, PA: Rodale Press.

Council for Agricultural Science and Technology (CAST). 1988. Long-term viability of U.S. agriculture. Ames: CAST.

Council on Environmental Quality (CEQ). 1984. *Environmental Quality, 1984*. Washington, D.C.: U.S. Government Printing Office.

Council of Economic Advisors. 1987. *Economic Report of the President*. Washington, D.C.: U.S. Government Printing Office.

Crosson, Pierre and Ruth Haas. 1982. Agricultural land. Pp. 253-282 in Paul Portney, ed., *Current Issues in Natural Resource Policy*. Washington, D.C.: Resources for the Future.

Crosson, Pierre. 1986. Soil erosion and policy issues. In T. Phipps, P. Crosson, and K. Price, eds., *Agriculture and the Environment*. Washington, D.C.: Resources for the Future.

Crosson, Pierre. 1987. The long-term adequacy of land and water resources in the United States. Washington, D.C.: Resources for the Future.

El-Ashry, Mohamed T., Jan van Schilfgaarde, and Susan Shiffman. 1985. Salinity pollution from irrigated agriculture. *Journal of Soil and Water Conservation* 40:48-52.

Fantel, R.J., G.R. Peterson, and W.F. Stowasser. 1985. The worldwide availability of phosphate rock. *Natural Resources Forum* 9:5-24.

Flora, Newt. September 1987. Show us facts regarding pesticide usage. *Farm News and Views* 68:5 (Oklahoma City: Oklahoma Union Farmer).

Fraley, R.T., S.G. Rogers, and R.B. Horsch. 1986. Genetic transformation in higher plants. *CRC Crit. Rev. in Plant Sci.* 4:1-46.

Frederick, Kenneth D. 1982. Water Supplies. Pp. 216-248 in Paul Portney, ed., *Current Issues in Natural Resource Policy*. Washington, D.C.: Resources for the Future.

Gibbs, M. and C. Carlson, eds. 1985. Crop productivity — Research imperatives revisited. Proceedings of international conference held at Boyne Highlands Inn October 13-18, 1985 and Arlie House December 11-13, 1985.

Glotfelty, D.E., J.N. Seiber, and L.A. Liljedahl. 1987. Pesticides in fog. *Nature* 325:602-605.

Heck, W.W., O.C. Taylor, R.M. Adams, G. Bingham, J. Miller, E. Preston, and L. Weinstein. 1982. Assessment of crop loss from ozone. *Journal of Air Pollution Control Association* 32:353-61.

Hoar, S.K., et al. 1986. Agricultural herbicide use and risk of lymphoma and soft-tissue sarcoma. *Journal of the American Medical Association* 256:1141-47.

Hopper, David. 1978. Distortions of agricultural development resulting from government prohibitions." Pp. 69-78 in T.W. Schultz, ed., *Distortions of Agricultural Incentives*. Bloomington: Indiana University Press.

Huszar, Paul C. and Steven L. Piper. 1986. Estimating the off-site costs of wind erosion in New Mexico. *Journal of Soil and Water Conservation* 41:414-16.

Landsberg, Hans, John Tilton, and Ruth Haas. 1982. Nonfuel minerals. Pp. 74-116 in Paul Portney, ed., *Current Issues in Natural Resource Policy*. Washington, D.C.: Resources for the Future.

Lee, Linda K. 1984. Land use and soil loss: A 1982 update. *Journal of Soil and Water Conservation* 39:226-28.

Martin, Philip and Alan Olmstead. May 14, 1984. Sprouting farm machinery myths. *The Wall Street Journal*, p. 25.

McDonald, Kim. October 24, 1984. Attempts to halt genetic research anger scientists. *Chronicle of Higher Education*, pp. 7ff.

Nielsen, Elizabeth and Linda Lee. January 1987. Potential groundwater contamination from agricultural chemicals: A national perspective. Pp. 23-32 in *Agricultural Resources*. AR-5. Washington, D.C.: ERS, USDA.

Nielsen, Elizabeth G. and Linda K. Lee. 1987. *The Magnitude and Costs of Groundwater Contamination from Agricultural Chemicals: A National Perspective*. Agricultural Economic Report No. 576. Washington, D.C.: ERS, USDA.

Office of Technology Assessment (OTA). 1986. *Technology, Public Policy, and the Changing Structure of American Agriculture*. Washington, D.C.: U.S. Congress, OTA.

Randall, Alan. 1987. *Resource Economics*. New York: John Wiley.

Sivard, Ruth. 1980. *World Energy Survey*. New York: Rockefeller Foundation.

Sloggett, Gordon. 1985. *Energy and U.S. Agriculture: Irrigation Pumping, 1974-83*. Agricultural Economic Report 545. Washington, D.C.: ERS, USDA.

Strohbehn, Roger, ed. 1986. *An Economic Analysis of USDA Erosion Control Programs: A New Perspective*. Agricultural Economic Report No. 560. Washington, D.C.: ERS, USDA.

Tutwiler, Ann and George Rossmiller. Winter 1987. External events and the recovery of U.S. agricultural exports. Pp. 20-22 in *Resources*. Washington, D.C.: Resources for the Future.

Tweeten, Luther. 1984. Summary and synthesis. Pp. 568-583 in Burton English et al., eds., *Future Agricultural Technology and Resource Conservation*. Ames: Iowa State University Press.

U.S. Department of Agriculture. 1986. *The Second RCA Appraisal.* (Preliminary draft.) Washington, D.C.

Yeh, Chung J., Luther Tweeten, and C. Leroy Quance. February 1977. U.S. agricultural production capacity. *American Journal of Agricultural Economics* 59:37-48.

CHAPTER TEN

Poverty, Human Resources, and Rural Development

Progress in alleviating problems of farmers such as poverty is closely tied to rural and national economic progress and welfare reform. Farm people depend relatively far more on nonfarm income than rural areas or the nation depend on farm income. Farmers as a whole receive 60 percent of income from off-farm sources but families on small farms which account for most farms receive a much higher percentage.

Farmers account for only 2 percent of national income, thus farmers have only modest marginal effect on the nation's economy. A USDA study (Deavers and Brown, 1985, p. 9) classified 27 percent of all nonmetropolitan counties as farming oriented (20 percent or more of labor and proprietors' income from farming). These counties contained only 12 percent of the nonmetropolitan population. Approximately the same percentage of nonmetropolitan counties were classified as manufacturing dependent but these counties contained 40 percent of the nonmetropolitan population. It follows that far more rural people depend on manufacturing than on farming for an economic base.

In the long run, farm income keeps pace with income of nonfarmers. Thus nonfarm income ultimately determines farm income per capita. The short-run economic destiny of farmers and the family farm also depends heavily on off-farm jobs and hence on performance of the nonfarm economy. And the ability of farmers to perform well in farm or nonfarm activity depends heavily on human resource investments in education and training. Those who fall behind slip into poverty. In the 1980s the poverty rate among farm families averaged approximately 20 percent compared to the 14 percent poverty rate for the U.S. as a whole. With an improved farming economy in 1987 the farm poverty rate dropped below the nonfarm poverty rate. These rates might be less than half that indicated if in-kind payments and assets were considered. By any definition, however, rural poverty was severe in Appalachia, the Mississippi Delta, the Black Belt of Alabama, and on Indian reservations. Among farmers, poverty is largely confined to noncommercial farms, especially to full-time noncommercial farms.

Rural development is defined as improving well-being of rural people, wherever they eventually reside (Tweeten and Brinkman, 1976, p. 4). Well-being is improved by raising real income while reducing inequality and instability. The term "wherever they eventually reside" is included to avoid a fixation on place prosperity by solely trying to bring jobs to rural areas. For many rural people, the surest route to development is first to obtain a good education, then, if necessary, leave the community for another offering larger compensation for human resources. The low-income home community should not be penalized for investing at considerable sacrifice in human resources, that is, in schooling of people who leave.

CHARACTERISTICS OF RURAL AREAS OF IMPORTANCE TO AGRICULTURE AND RURAL AMERICA

Compared to urban counties, rural counties display several distinctive features:

1. Rural counties on the average have lower per capita income, higher poverty rates, higher dependency rates, and lower labor force participation rates. Nonmetro areas have lower proportions of school-age youth in school and lower proportions of persons in professional, technical, managerial, and administrative occupations (Nilsen, 1980).

2. Rural counties have lower rates of population and employment growth. In the 1970s, population and employment grew faster in rural than in urban counties. This departure from the historic pattern undoubtedly helped divert public attention from rural problems in the 1970s. With return to slower rates of growth in population and employment in rural than in urban areas in the 1980s, attention has returned to issues of rural economic policy. In the 1980s per capita income fell in rural areas compared to income in urban areas. The cause was transitory factors such as an overvalued dollar as well as long-term structural changes from deregulation of transportation, communication, and banking; the shift to high-technology and service industries; changes in technology which require less raw materials from extractive industries per unit of output; economies of size in provision of community services; and preferences of people for amenities of living found only in larger centers (see Mulkey and Clouser, 1987).

3. The single most distinguishing feature of rural areas is population dispersion. Sparsely populated areas offer environmental and other amenities sought by many Americans but pose unique problems in providing quality community services at low cost per capita. Economies of size characterize many manpower as well as other services. Determining an appropriate level of services and paying for them is a continuing challenge for many rural communities.

4. Industrial and occupational composition of rural counties is becoming more and more like that in urban counties (Tweeten, 1983). However, rural counties continue to depend more heavily than do other counties on extractive industries such as agriculture, lumbering, and mining. These industries compete for exports or with imports, hence were especially influenced by the international exchange value of the dollar. Like agriculture, rural areas in general are ever more closely tied to the global economy. Extractive industries along with manufacturing, now the largest single basic industry in rural areas, have been characterized by slow growth or decline, creating community adjustment problems.

5. In part because farming continues to be dominated by large numbers of family-sized operations while small towns are dominated by small family businesses, self-

employment is approximately twice as frequent in rural areas as in urban areas. Because of this and other unique rural characteristics, many national policies designed for urban areas do not work well in rural areas.

6. Principal economic problems of rural areas are *poverty* and *underemployment*. Each problem can require a different solution. Many of the poor are aged and disabled individuals not underemployed, and their incomes can be raised at least government cost by transfer payments. Earnings of the underemployed who are earning below their capabilities can be raised at least public cost by job development and by human resource development programs of eduction, training, job search assistance, and relocation allowances.

PAST PROGRAMS OF RURAL DEVELOPMENT

Federal government programs to promote rural development can be divided into five general approaches: (1) monetary and fiscal policies promoting national full employment, (2) community improvement through better local decisions that more fully utilize local resources, (3) bringing people to jobs through various human resource development and labor force programs, (4) bringing jobs to people through industrial development and related measures, and (5) transfer payments.

HUMAN RESOURCES AND FULL EMPLOYMENT

The Employment Act of 1946 committed the federal government to maintain full employment, a task that states or regions acting individually cannot perform. National full employment is a precursor of rural area development: Alleviating underemployment in rural areas by programs to upgrade labor skills and assist labor mobility is hampered if jobs are scarce; generating employment within reach of rural residents with special federal programs is of little or no net benefit if jobs are merely shifted from metropolitan communities where unemployment is already high. The term "full employment" here is consistent with 5-6 percent national unemployment and with pockets of unemployment in excess of that level but the natural unemployment rate will vary with demographic changes. A major purpose of structure policies is to reduce the natural unemployment rate through developing human resources and removing barriers to employment.

COMMUNITY IMPROVEMENT THROUGH SELF-HELP

Another approach to rural development is community self-help. Under this approach, federal assistance is minimal, and community development is viewed as a *process* whereby those in a community arrive at group decisions and take actions to enhance the social and economic well-being of the community. The role of the federal government is to provide education specialists from the extension service and technical specialists from other agencies to serve as catalysts, helping local people to identify problems and leaders and to organize for action (Tweeten and Brinkman, 1976, ch. 9).

Although broad public programs such as road and public utility construction began much earlier, the first programs designed specifically to promote rural development began in the 1950s (Table 10.1). The programs emphasized self-help. Problems of depressed areas in an affluent society had been well-identified by the mid-1950s, helping spawn the Rural Development Program under the Eisenhower Administration in 1955. The general approach was to form rural development committees composed of local leaders in selected counties for outlining and guiding the local program designed to develop rural resources.

Public research, extension, and technical assistance were focused to a greater extent than in the past on the problems and opportunities of low income rural areas. The number of private, cooperative, and government loans was increased, but emphasis was on local leadership and initiative to solve local problems.

Congress allocated only $2 million to operate the Rural Development Program, and it gave the Farmers Home Administration (FmHA) some additional lending authority to operate under the program. In 1960 it reported that 210 counties participated in development programs and that 18,000 new full-time jobs were created in the year as a result of industrial growth and new business. Nevertheless, in the judgment of Cochrane (1965, p. 202) "the Rural Development Program never really moved out of the pilot stage...In general [it] barely touched the hard-care underemployment-poverty problem in rural areas."

Under the administration of President John Kennedy, the Rural Development Program was reorganized and named Rural Areas Development in 1961 (Table 10.1). As with its predecessor, the new program assumed that the local community would provide the leadership and initiative for the development process. Like the former program, it was basically a planning and coordinating program, designed to focus some of the activities of existing agencies on alleviation of poverty. The Rural Areas Development effort from 1961 to 1966 entailed organization and promotion of an estimated 20,000 projects ranging from community facilities to industrial parks. The advent of the new program was not met with a significant budget increase; hence it was essentially a continuation of the former pilot project. Cochrane's statement about the Rural Development Program can also be applied to Rural Areas Development.

The Rural Community Development Service was formed in February 1965 to allow the U.S. Department of Agriculture to provide leadership in rural development (Table 10.1). Technical action panels operated at local levels. Federal representation on the panel usually consisted of local officers of the FmHA, the Soil Conservation Service, the Agricultural Stabilization and Conservation Service, and the Federal-State Cooperative Extension Service. The Rural Development Act of 1972 (discussed later in this chapter) continued efforts by the U.S. Department of Agriculture to develop jobs in rural areas but the Act was never fully funded. With the rural turnaround featuring more rapid rates of growth in rural than in urban areas in the 1970s, rural development efforts lagged.

The Rural Regeneration Initiative announced by the USDA in 1987 continued the tradition:

TABLE 10.1. Federal Programs for Rural Development

Presidential Administration and Year	Major Legislation, Agencies, Committees, Panels, etc.	Goals
Grover Cleveland		
1893	• Office of Road Inquiry (USDA)	• Work of a demonstration/educational nature
Theodore Roosevelt		
1905	• Office of Public Roads (USDA) organized; replaces Office of Road Inquiry (1893)	• Construct object lesson roads; test road-building materials
1908	• Country Life Commission appointed	• Major report on needs of rural population
Howard Taft		
1912	• Office of Public Roads (USDA) receives appropriations to supervise building of rural post roads	
Franklin Roosevelt		
1933	• Tennessee Valley Authority established	
1935	• Rural Electrification Administration (USDA) organized	• Bring electricity to farms
	Resettlement Administration organized	• Resettle farm laborers and disadvantaged rural resident in part-time farming communities
Harry Truman		
1949	• Rural Telephone Loan program begun	
Dwight Eisenhower		
1953	• Interstate Highway System receives first appropriations	
1954	• USDA committee asked to report on agricultural development	• Call attention to rural development problems
1955	• Rural Development Committees organized	• Aid local communities in establishing new training programs and other activities
1959	• President established interdepartmental Committee on rural Development	Coordinate all federal efforts in rural development
John Kennedy		
1961	• Office of Rural Areas Development (USDA) established; Rural Development Committees replaced by Rural Area Development Committees	• Eliminate rural underemployment
1962	• Rural renewal program authorized by Congress	
Lyndon Johnson		
1964	• Economic Opportunity Act (war on poverty) enacted	• End rural poverty
	• Job Corps organized	• Provide opportunities for disadvantaged youth
1965	• Housing and Urban Development Act passed	• Improve rural and urban housing
	• Rural Community Development Service (USDA) replaces Office of Rural Areas Development	• Coordinate USDA's rural activities
	• Interagency Task Force on Agricultural and Rural Life established	• Recommend legislation to improve rural life
1966	• National Advisory Commission on Rural Poverty	• Develop major program for attacking rural poverty
1967	• *The People Left Behind* published by National Advisory Commission on Rural Poverty	• Call Nation's attention to problem of rural poverty
Richard Nixon		
1969	• Presidential Task Force on Rural Development	• Recommend programs for public and private sector
1970	• Rural Community Development Service abolished; functions transferred to USDA Departmental Rural Development Committee	• Coordinate USDA rural development programs
	• USDA Committee for Rural Development set up in each State	• Coordinate USDA programs for rural development within States
1971	• Rural Development Service (USDA) organized	• Direct USDA rural development programs
	• Rural Telephone Bank organized	• Finance rural telephone cooperatives
	• First Regional Rural Development Center established	• Carry out regional extension and research for rural development
1972	• Rural Development Act signed into law	• Broad authority for rural development programs
1973	• Congressional Rural Caucus organized	• Emphasize needs of rural areas
Gerald Ford		
Jimmy Carter		
1978	• White House Rural Development Initiatives on health, water, sewers, communications, energy, transportation	• Secure cooperation in solving these problems
	• USDA's Rural Development Service merged into FmHA	• Emphasize rural housing problems
1980	• Rural Development Policy Act passed by Congress	• Extend authorizations for appropriations
	• USDA establishes National Advisory Council on Small Community and Rural Development	• Give varied groups opportunity to participate in policy and program planning
Ronald Reagan		
1981	• USDA established Office of Rural Development Policy	• Formulate policy and coordinate rural development efforts
1982	• National Advisory Council on rural Development established	• Identify rural problems and support rural development policies

SOURCE: Rasmussen (1985, pp. 6, 7).

1. The Cooperative Extension Service placed additional emphasis on rural revitalization. The feasibility of setting up rural technology centers at land-grant universities was explored.

2. Rural Enterprise Teams consisting of four or five specialists were organized to go on call to communities or counties to assist in business and job development.

3. A clearinghouse was made available so that all information about federal programs could be obtained at a single phone number and referred to the appropriate agency for follow-up.

4. All USDA agencies with a research mission were instructed to increase rural development efforts.

5. The Farmers Home Administration used the 1987 Business and Industry Guarantee Loan Program to create more jobs in rural areas.

6. Rural revitalization received greater priority by combining all coordination and direction in the office of the Deputy Secretary of Agriculture.

These initiatives were not unlike those of the Rural Development Program of 1955. Some of the efforts were worthy. Several features were similar to past efforts: they received very little funding, they lacked criteria to focus funds where needed, they have no significant impact on rural underemployment and poverty, and they raised unwarranted expectations for progress.

In summary, community improvement process efforts by catalysts from the extension service and other agencies by and large did not bring people to jobs or jobs to people. The efforts brought some community betterment and improved services but not much development in the form of higher incomes. Lack of follow-up with planning and adequately funded action programs shares the blame for failure to achieve much economic success. Policymakers failed to perceive the magnitude of rural development problems. The development process was viewed as a labor intensive activity: It was thought that sending enough personnel to help the local community would bring development. In fact, rural development is human and material capital intensive activity. Community improvement efforts got ahead of the research, planning, and capital required to make needed progress. The Cooperative Extension Service has performed and will continue to perform well in helping communities make better use of their resources. But even the most efficient use of local resources would leave many rural communities lagging seriously behind the rest of the nation.

Bringing Jobs to People

The Accelerated Public Works program and the Area Redevelopment Administration (ARA) established in 1961 during the Kennedy Administration signaled two important new directions. One was the federal commitment to create jobs expressly in depressed rural areas; the other was recognition of nonfarm rural problems as evidenced by placing major job-generating development programs under the U.S. Department of Commerce rather than the U.S. Department of Agriculture. The U.S. Department of Agriculture continued some development programs: Rural Renewal (a small land-oriented program administered by the FmHA) and Resource Conservation and Development (a water-oriented program administered by the Soil Conservation Service) established by the Food and Agriculture Act of 1962 continued to provide grants and loans to conserve and develop natural resources; the FmHA provided operating and housing loans to farmers and utility loans to towns; and the Agricultural Stabilization and Conservation Service provided farm commodity programs. But these activities, along with community improvement programs mentioned earlier, were not intended to fulfill the need for jobs in rural areas.

The Area Redevelopment Administration (ARA) established by the Area Redevelopment Act of 1961 provided (1) loans to support job-creating commercial and industrial enterprises, (2) grants and loans for public facilities, (3) technical assistance to bridge the knowledge gap, and (4) retraining programs to fit workers to new jobs. Over the entire period of ARA operation (1961-1965), the agency approved industrial and commercial loans for a total federal investment of $176 million, resulting directly and indirectly in an estimated 66,585 jobs, for an average investment of $2,645 per job.

The Accelerated Public Works program was designed to provide useful short-term employment to help communities improve their public facilities. The program, established by the Public Works Acceleration Act signed into law in 1962 and administered by the ARA, eventually entailed expenditures totaling $1.7 billion and created an estimated 200,000 on-site and off-site jobs — for a cost of $8,500 per job. These and previous estimates of jobs created and costs per job are not highly reliable, however.

The Public Works and Economic Development Act of 1965 created the Economic Development Administration (EDA) and supplanted legislation establishing the ARA and Accelerated Public Works program. Many of the features of the earlier programs were retained, but notable changes were made.

In recognition of the importance of a viable or functional economic area, the act of 1965 provided for creation of multicounty *economic development districts,* with each district to contain two or more *redevelopment areas* (usually low-income or high-unemployment counties), and with each district to be of sufficient size, population, and resources to foster economic development. In addition, the legislation stated that each district must have one or more *growth centers* of sufficient size and potential to generate economic growth necessary to alleviate the distress of the redevelopment areas within the district. These centers could be located in redevelopment areas or in nondesignated counties.

The majority of funds under the Public Works and Economic Development Act of 1965 went for grants to finance public works and development facilities. Pubic works provided water and sewer systems, access roads, and the like to encourage industrial development. Business loans were used to encourage private investment. Technical assistance helped distressed areas understand the scope of their problems as well as their economic potential. Technical assistance provided by EDA under the act also included management and operational assistance to private firms.

Title V of the Public Works and Economic Development Act of 1965 provided for the establishment of economic development regions. By the end of fiscal 1967, five such regions had been designated: Upper Great Lakes, New England, Coastal Plains, Four Corners, and Ozarks. In addition, the Appalachian Regional Development Act of 1965 had established that region as the focus for special development efforts. Commissions were established to administer programs in each region. The Appalachian Regional Commission emphasized programs for highways, health, and vocational education. The Ozarks Regional Commission emphasized human resource development and data systems, including community profiles used by prospective firms to locate plant sites and information systems to help local manufacturers determine the best source of inputs for their firms. Unfortunately, little data are available to judge which regional commission used the most efficient approach, a judgment that can be made only within the context of the goals of the region.

It is important to emphasize that like previous federal programs, the job creation efforts of the EDA entailed comparatively few funds; EDA expenditures for all purposes averaged only approximately $300 million annually for the whole country in the 1960s. Evaluations indicated that EDA job development programs were reasonably cost-effective in creating jobs per unit of program outlay but funding was too small and poorly focused to have a significant input on rural underemployment and poverty. Early "worst first" efforts of EDA focusing on the most depressed counties were reasonably cost-effective in creating jobs. However, funding of EDA as well as FmHA business and industrial loans was always too modest and diffused to have much impact (Tweeten and Brinkman, 1976, ch. 14).

The Rural Development Act of 1972 was intended to replace EDA and provided $367 million of new grant authorization and virtually open-ended loan authorization. The $50 million of grants authorized for development of industrial parks and other services and facilities to support industry could have been reasonably cost-effective in creating jobs, but the sewer and water development grants of $300 million and pollution abatement grants of $75 million would have drawn few permanent new jobs to rural areas.

Based on an optimistic estimate of only $5,000 required to create a permanent new job, the entire *new* grant authorization would have generated only 73,000 new jobs even if fully funded, which it was not. Accounting for the job cost-ineffectiveness of most of the grants reduces the estimate; accounting for the influence of development loans increases the estimate.

The Rural Development Act of 1972, the last major rural development effort, was never fully funded and was not effective. Chances for special substantive rural development legislation have declined since 1972. Meanwhile urban areas have growing political muscle and continue to push for urban enterprise zones providing federal tax breaks to industries locating in the inner city. Such legislation would repeat some of the failures of earlier rural development programs. A wise strategy would appear to be for rural people to team with urban people in devising a sound program serving needs of rural and urban areas.

Opposition to federal programs of industrial development in rural areas has come from urban unions contending that such programs "rob Peter to pay Paul" and from rural industries contending that such programs create unfair competition with existing firms. To allay such criticism, the Rural Development Act of 1972 precluded financial assistance in the form of loans and grants to private business enterprises "which is calculated to or likely to result in transfer from one area to another of any employment or business activity" and "which is calculated to or likely to increase production in an area when there is not sufficient demand to employ the capacity of such existing enterprises in the area." The solution is to locate new plants in areas void of like industry. This solution can unintentionally preclude obtaining external economies of scale that accrue because several firms from the same or related industry locate in the same area. In the long run, presence of such economies can be decisive in creating economic viability for industry in an area.

Section 601 of the 1972 act required heads of *all* executive departments and agencies to give "first priority" to the location in rural areas of all new offices, facilities, and installations. The Agricultural Act of 1970, which required similar action "insofar as practicable" was found to be weak and easily circumvented. The 1972 act also had an escape clause in that the Senate-House Conference Report made clear that a new structure need not be located in a rural area "if there is an overwhelming reason" for locating it elsewhere. The term "rural" here refered essentially to non-SMSA (standard metropolitan statistical areas), whereas in the Agricultural Act of 1970 rural was defined for this provision as "areas of lower population density," an ambiguous concept that made rigorous application difficult.

In short, the 1972 act completed a circle: (1) returning to the U.S. Department of Agriculture a program that had originated there but in the interim was in the U.S. Department of Commerce; (2) returning to vaguely defined criteria for assistance following a program (EDA) with meaningful, if somewhat misguided, criteria to focus its development efforts; and finally (3) returning to a program lacking a concept of critical mass of growth resources or growth center. Multicounty districts and administrative centers were maintained by other programs for planning purposes, however. Growth centers fell victim to metropolitan enthusiasts who claimed cities of under 250,000 were too small for economic viability and rural enthusiasts who claimed cities of that size were too remote, who saw no need to focus development efforts into a critical mass, and who saw every community as an opportunity for job growth.

A modest program of business and industrial loans was continued under the Farmers Home Administration; the Economic Development Administration maintained a small rural development effort in the 1980s. EDA, in its struggle for political survival, eventually spread its job development effort so thinly among areas that a critical mass of resources for development was seldom assembled. "Worst first" has long been abandoned. Urban areas disproportionately absorbed EDA efforts and farmers disproportionately absorbed modest FmHA efforts to create jobs for rural people. Meaningful targeting of job creating efforts of EDA, FmHA, or urban enterprise zones extended to rural areas on the basis of *underemployment* appears to be out of the question in the foreseeable future. It follows that public efforts to reduce underemployment in rural areas will largely come through manpower and other human resource development policies rather than through industrialization policies — bringing jobs to people.

Place prosperity continued to be a focus of federal rural development programs despite the weak case for it both in theory and application. For example in 1987 legislation (H.R. 1800) was introduced in Congress to expand government grants and loans channeled through local nonprofit private and public groups for stimulating growth and development in depressed rural ares. The bill was to make $20 million available for lending and $25 million in grants per year. Another bill (S.983) would establish rural enterprise zones to stimulate the creation of new jobs and to promote revitalization of economically distressed rural areas through tax relief, regulatory relief, and improved local services. Meanwhile, the Reagan Administration announced that the Economic Development Administration was redirecting its assistance to rural areas, providing 75 percent of its allocation to rural areas in 1986.

It is reasonable to conclude that rural areas have no federal development policy, only small piecemeal programs administered by diverse federal agencies with a narrow perspective. Rural development too often has narrowly focused on place prosperity, chasing smokestacks, and local job development as opposed to improving quality of life for rural people even if they go elsewhere. If rural communities and states wish to engage in boosterism to draw industry, that is their privilege. While the federal government can wish them well, it does not need to be a part of such effort. If a useful underemployment measure could be utilized as a criterion for tax or other incentives to attract jobs in rural enterprise zones, such effort might serve society by productively utilizing underemployed labor with little opportunity cost. But a satisfactory underemployment variable to guide federal efforts is not yet available. If it were, it would be ignored by politicians who want political rather than economic factors to guide job development. Past history provides no basis to conclude that special rural job generating efforts by the federal government can be equitable, efficient, or effective in ending rural poverty and underemployment. It is well to consider alternative policies.

Bringing People to Jobs — Human Resource and Equity Policies

Bringing people to jobs has been by far the largest and most effective, albeit de facto, program to raise income of rural people. It was largely guided by market forces. The massive migration to metropolis contributed to economic progress but, for the most part, federal programs did little to ease the burdens of migration by assisting people before or after the move. Schooling to improve general competence for a wide range of employment opportunities, and programs to improve the functioning of rural labor markets, to train people for jobs, and to assist workers in securing the best employment available (by providing cash assistance, labor market information, and counseling) can contribute much to the socioeconomic progress of rural people. Human resource investment programs, whether they result in local or distant employment for recipients, are a fundamental component of rural development efforts.

The Public Employment (Job) Service (ES). Provision of job information and a clearinghouse to match job seekers and employers has some properties of a public good which the private market operating alone will not provide in efficient quantity. Private employment agencies may have difficulty appropriating benefits of job information made available to all workers and employers. Only one private agency is able to operate efficiently to provide a comprehensive job clearinghouse at acceptable cost per unit in many local labor markets. Such a natural monopoly may require some regulation to act in the public interest. Nonetheless, arguments for public provision of job services are not overriding. It would be a mistake to locate a public employment service in every rural community without attention to benefit-cost ratios and appraisal of alternative, possibly private, job service delivery systems.

Operation of the Employment Service (ES, also known as Employment Security Offices or Job Services) in 2,600 locations throughout the country is the responsibility of state governments but with funding by federal grants. Federal legislation establishing the employment service in 1933 focused on overcoming labor market imperfections in matching workers to jobs and in overcoming skill shortages. With the "war on poverty" beginning in 1964, the emphasis shifted to labor initiatives targeted at minorities, welfare recipients, and low income youth. Fairly comprehensive programs providing training as well as job market information include the Manpower Development and Training Act (1962), the Economic Opportunity Act (1964), the Comprehensive Employment and Training Act or CETA (1973), and the Job Training Partnership Act (1982). In constant 1983 dollars, annual expenditures on these programs rose from approximately $3 billion in the late 1960s to a peak of around $14 billion in the late 1970s, and declining to approximately $4 billion in the early 1980s with further cuts programmed for the next several years (Burtless, 1984, p. 18).

As federal funding through grants to states has declined in recent years, responsibilities of the ES have increased. The ES, for example, administers work tests that determine eligibility for welfare and food stamp programs. Use of the ES to assist the poor in obtaining employment may have compromised the agency's effectiveness with other workers. According to the Congressional Budget Office (July 1982, p. 23), "The Service has acquired a reputation for dealing largely with economically disadvantaged job seekers with low levels of skills." Relatively few individuals seeking or firms offering better paid jobs use the Employment Service. A Department of Labor survey reported that only one-fourth of all employers representing 36 percent of all job vacancies listed the openings with their public employment service (U.S. Department of Labor, 1976). A much smaller proportion, about 5 percent, of job seekers find jobs through the ES.

Data on use of the public employment service in rural areas were obtained from family heads in the control and experimental groups of the rural income maintenance experiment in Iowa and North Carolina in 1970 (see Tweeten and Brinkman, 1976, ch. 4). Family heads were asked where they would refer someone looking for work. Two-fifths of the respondents were unable to suggest a place to get help. Twenty-three percent of respondents suggested the public employment service. Only 15 percent of the rural heads who had employment problems used the public employment service. Intensive surveys confirm that many who could benefit from labor services do not register, in part because the ES relies on referrals to local employers who cannot meet needs for employment.

The Department of Labor has had a modest interstate clearing system between state agencies. The system has attempted to match employees willing to relocate with employers willing to recruit out-of-area workers (Congressional Budget Office, July 1982, p. 51). An evaluation of the Job Service Matching System (JSMS), a computerized process matching workers to jobs and operating in 24 states, indicated that computerization had done little to improve the effectiveness of the Employment Service (Congressional Budget Office, July 1982, pp. 46, 47).

In short, the public employment service has many shortcomings. Interarea recruitment is minimal. When used at all, services are mostly to provide workers for existing or potential local employers. The service rarely refers potential workers to the best opportunities available anywhere.

The matching of workers and jobs, if done well, is complex. Computerized nationwide information systems; systematic job counseling beginning at the high school level; and well-staffed mobile employment teams for rural areas are a few of the potential improvements. These improvements should be monitored for efficiency. If social costs exceed benefits after a reasonable period required to become established, the programs should be changed or abandoned.

Training Programs. CETA training programs in 1980 had 360,000 participants in classroom training at an average cost of $2,700 per trainee, 100,000 in on-the-job training at $2,100 per trainee, and 300,000 in the work experience program at a cost of $2,200 per trainee. The latter program provided subsidized jobs that gave some training and encouraged favorable work habits and attitudes. CETA trainees were mostly the disadvantaged and included a high proportion of enrollees from families receiving public assistance (33 percent) and from minorities (44 percent). Most enrollees were youth — only 15 percent of trainees were over 44 years of age in 1980.

Previous studies and data systems provide few clues regarding how CETA or its successor, the Job Training Partnership Act, has influenced the rural labor force of over 30 million. Despite cutbacks in federal training programs for all sectors, the issue is of continuing concern.

State and local vocational-technical programs have grown in recent decades and are far more important than federal programs in providing training. The state programs serve broad classes of people and tend to have higher completion and placement rates than federal training programs for the disadvantaged. Many of the programs are of sufficient duration and quality to add much to income.

On the other hand, increasing evidence points to frequent cases of low-quality vocational-technical training or training for jobs which do not exist. A related problem is placement efforts focused narrowly on the local job market without sufficient attention to regional and national markets and to the projected supply-demand balance for various skills. Many local-state vocational-technical schools need to improve outlook, placement rates, and training quality. These issues including the private and social payoff from vocation-technical schooling as it relates to rural areas are a priority for research (see Tweeten and Brinkman, 1976, chapter 4, for earlier studies).

Off-farm wages of farmers increase with additional schooling. However, vocational training frequently has had a negative affect on off-farm wage rates (see Huffman, 1985). One explanation is that vocational training does little to improve earnings. An alternative and perhaps a more plausible explanation is that individuals who choose vocational training are less able and earn a lower wage than others, other things equal. Vocational training does increase the probability of farmers participating in off-farm work.

In general, programs that rank high in efficiency measured by cost-benefit ratios or rates of return on investment rank low in equity, defined as providing the most benefits to those with fewest resources. One of the lessons learned from the 1960s is that it is very expensive to compensate through remedial education programs for inadequate genes, family background, ill health, and discrimination that stifle the development of competence. A related lesson is that the schooling has a modest impact on economic outcomes compared to fate and the home. Rates of return on investment in general schooling have been found to be favorable, and providing equal educational opportunity is imperative. But additional efforts are needed to deal with equity problems. Later proposals for reform of welfare, school funding, and taxes treat this issue.

PRINCIPLES AND ROLE OF GOVERNMENT IN RURAL DEVELOPMENT AND PROVISION OF COMMUNITY SERVICES

Based on previous discussion and the conceptual framework developed in Chapter 2, several principles guiding the role of the federal government in rural development and particularly rural services are listed below.

1. Because of its capabilities to bring efficient resource and product allocations, the private sector is central to rural development. But there is widespread agreement that the public sector must play a role in provision of public goods, correcting externalities, and seeing to those who cannot provide for their own basic needs. Thus the government appropriately is involved in providing roads, schooling, public (welfare) assistance, and police and fire protection.

2. Public services ordinarily are most efficiently provided (funded and administered) by the most local unit of government within which the service benefits are realized. When the federal government subsidizes local utilities and services such as water, electricity, and school buses, the result is to encourage uneconomic sprawl of urban residents into high-service-cost rural areas because residents do not pay the costs while receiving the benefits of services.

For the most part, funding and administration of community services such as water, electricity, waste disposal, streets and country roads, community parks, and fire and policy protection are best left to local governments. An exception is education and welfare services. Prior to the 1970s, migration of millions of rural people transferred massive human capital in the form of local investments in schooling from rural areas to cities. With the migration turnaround in the 1970s, the direction of needed compensation for net transfers also turned around only to reverse again in the 1980s. Many rural communities continue to experience emigration not compensated by immigration of human capital. For many rural communities, economic benefits derived from education within their funding jurisdiction continue to fall well short of local costs incurred. This seeming disincentive to adequate local school funding does not show up in major underinvestment in education according to several studies (see Tweeten and Brinkman, 1976, pp. 139-43). One reason is rigid funding formulas imposed by states; another reason is parents wanting a good education for their children wherever they may eventually reside. However, a case can be made on equity grounds that local areas much less wealthy than the areas which receive the benefits of local investments in schooling should bear a smaller proportion of the costs of human capital formation.

Similar reasoning applies to welfare services. Areas least able to afford the costs of providing for the poor frequently have the highest proportion of the poor. Failure to provide adequate welfare in Mississippi, for example, spills over as costs to St. Louis or Chicago as the poor migrate to areas with more generous provisions for the disadvantaged. In the absence of outside funding, local communities face strong incentives to give the least possible welfare assistance so that the destitute will leave the community to take advantage of communities offering more generous welfare benefits. Thus urban Illinois has a stake in the welfare programs made available in rural Mississippi.

Less than full local funding of local schools is justified by externalities or spillovers of benefits from the local district into other districts and states. The externality would be eliminated by a federal takeover of all public common school funding because all benefits and costs tend to be realized in the nation. However, program operation is often superior with maximum equitable local or state funding and administration so that local initiative, control, and accountability are retained to the extent possible.

3. It is ordinarily appropriate to dispense rural development programs on the basis of equity and efficiency rather than on the basis of sector or occupation. Because individuals and families rather than communities are poor, it is wise to target transfer payments to individuals and families and not to communities. This principle rules out general revenue sharing and block grants. Following the principle would not necessarily rule out reimbursement to states for administering programs which fulfill federal guidelines.

This principle also does not rule out aid to communities where externalities (spillovers) are involved as noted in principle 2 above. The federal government has reason to pay for part of local services where the benefits are national and performance criteria are federally mandated. For example, wastewater treatment facility capital grants may be justified by the federal government to help communities comply with the Clean Water Act.

A corollary of principle 3 is that a solely rural or solely urban federal development program is inappropriate. A domestic development program to foster economic efficiency and distributive justice will treat metropolitan and nonmetropolitan areas in similar socio-economic circumstances alike.

4. A domestic development program ideally utilizes allocation criteria requiring minimal political or administrative discretion and utilizes accurately estimated and unbiased data. Examples of past federal efforts violating this principle are the Appalachian Regional Development Program which favored one region over all others for political reasons, and the early Economic Development Administration program which allocated resources on the basis of unemployment, a criterion biased against rural areas. Because it measures underutilized (not just unemployed) human resources, *underemployment* could be useful for allocating a job development program. Underemployment, an alternative to unemployment to allocate development efforts, would not be biased against rural areas. Underemployment has not been estimated with sufficient timeliness and reliability to be operational (Gilford et al., 1981).

Debate has long raged over the benefits of a *growth center* development strategy. Growth centers of say 25,000 or more population would be within easy reach of the vast majority of current and potential rural workers in the trade-commuting hinterland. A growth center is defined as an economically viable community capable of self-sustained growth because it provides the critical mass of financial, communication, and transportation infrastructure and the pool of skilled workers and industry diversification necessary to realize internal and external economies of size for industry. Although conceptually appealing, the idea has never been politically acceptable because every rural community wants to be eligible for federal programs attracting industry.

5. A development program allocating on the basis of actual performance is preferred to one allocating on the basis of possible performance. Communities may realize very large net benefits from attracting a new industrial plant of the right kind, and it is not surprising to see communities and states lavishly expending time and money to compete for jobs. Communities seeking plants far outnumber available plants, and most communities will be disappointed. The nation as a whole gains little whether the industry locates in one area of high underemployment versus another, and the firm itself rather than the federal government might best decide where to locate. Millions of dollars have been wasted in unutilized industrial parks constructed to attract industry. Rather than build an industrial park which stands a high chance of going unused, a more appropriate domestic development strategy for the federal government (if one is to be pursued at all) would be to subsidize wages or earnings of low income workers. Such a program would automatically tend to target areas of high underemployment in rural or urban areas.

6. It is well to maintain the integrity of the tax and expenditure structure by maintaining their separability. The notable case in point is tax exempt bonds used by local governments. The tax free bond provides a tax loophole to the super rich while subsidizing government activities often in direct competition with the private sector which could perform the task more efficiently. The result is to increase tax burdens on others while creating unfair competition to the unsubsidized private sector. If the federal government wishes to subsidize loans for local businesses including beginning farmers, that subsidy could be made explicit and financed directly from taxes. The tax system fairness would be better maintained and tax receipts enhanced if worthy local government investments (e.g.,

building schools) were subsidized directly (if at all) by the federal government rather than by a tax shelter.

FUTURE DIRECTIONS FOR FEDERAL DEVELOPMENT EFFORTS

The core rural development efforts must center on human resource development: general and vocational-technical education, public assistance, and employment counseling, job information, and mobility assistance.

Few if any rural communities retain all local students after they complete school. Youth deserve the chance to develop adequate skills to earn favorable incomes elsewhere if not at home. General education investment has been found to earn a 10 percent real rate of return but the rate is lower if people are confined to a limited-resource community (see Tweeten and Brinkman, 1976, chs. 4 and 5). Vocational-technical training rates of return have been found to average about 15 percent but vary widely. Returns are negative when people are trained to work in industries without job opportunities.

IMPROVING RURAL COMMON SCHOOLS

Improvements are needed in the quality and funding of common schools attended by rural youth. The formula for improving schools is simple but difficult to apply. Teachers are needed who demand much from students especially in the basics, and provide discipline and motivation for the student to deliver. Teachers able to achieve these objectives must be rewarded if schools are to stimulate and retain them. If monetary and nonmonetary rewards such as respect and status are low, poorly trained students will become poor teachers who expect and receive mediocre performance from students. Teachers rewarded only according to age and experience under a rigid salary scale cannot be expected to raise standards; merit must be recognized and rewarded to achieve excellence.

In determining overall funding (as opposed to within-school allocation) per pupil, it is first useful to determine an overall socially *efficient* level which can differ among areas by cost of living and other factors. Such an overall level might strive for a say 10 percent real rate of return on investment but also might be increased to recognize positive externalities, defined as benefits of schooling that do not show up solely as income to the individual. The benefits to society from more education are more than the simple sum of income gains to individuals.

A second adjustment is to reward funding *effort* as opposed to ability to pay. Suppose half of local-state school funding is from local property taxes and the other half is from state income and sales taxes. Suppose also that the state average value of property assessed at current market value is $20,000 per student. Suppose the local contribution is judged to need to average $1,000 per student or 5 percent of assessed value. Under the *power equalization* formula, each district would be assigned a hypothetical tax base of $20,000 per pupil regardless of the actual tax base. Each school would decide for itself its tax effort. If it elected to spend $2,000 per pupil, it would be assigned a tax rate of $2,000/$20,000 = 10 percent and if it elected to spend $1,000 per pupil its tax rate would be $1,000/$20,000 = 5 percent.

Each district's chosen tax rate would then apply to its actual tax base. All property taxes would go into the state pool of funds from which payments would be made to each district based on its chosen level of local expenditures per student. High wealth districts would help finance low wealth districts. *The contribution from the pool of local property taxes would be matched by a state contribution from income and sales taxes up to $1,000.*

A final adjustment could be made for *spillover* of benefits. One way to make such adjustment would be to calculate from census data net spillout of benefits across geographic boundaries (for an example see Hines and Tweeten, 1972). States with net spillin would compensate states with net spillout. Such calculations are difficult to make in practice and states experiencing spillins would resist compensating states experiencing spillouts of benefits. A more practical but not entirely satisfactory solution is to recognize that 30 percent of educational benefits spill over state boundaries on the average. The federal government accordingly would supply 30 percent of common school funding.

How the modified power equalization formula would operate in a hypothetical situation of funding per pupil is illustrated below:

	Dist. Wealth Low ($10,000/pupil)			**Dist. Wealth High ($30,000/pupil)**		
	Effort			**Effort**		
Item	**Low (2%)**	**Medium (5%)**	**High (10%)**	**Low (2%)**	**Medium (5%)**	**High (10%)**
Taxes Paid (effort percentage multiplied by actual base for pool)	$200	$500	$1,000	$600	$1,500	$3,000
Funds Received (effort percentage multiplied by $20,000 state base from pool)						
From Pool	400	1,000	2,000	400	1,000	2,000
From State	400	1,000	1,000	400	1,000	1,000
From Federal	850	850	850	850	850	850
TOTAL	$1,650	$2,850	$3,850	$1,650	$2,850	$3,850

The low wealth district with $10,000 of assessed value per pupil deciding to make a high effort of 10 percent of its base contributes $1,000 to the state pool per pupil and receives back $2,000 from the pool (10 percent of the state average property value of $20,000 per pupil). It receives $1,000 of state funds plus $850 of federal funds for a total spending of $3,850 per pupil. If it put forth a medium *effort* the state would have matched its $1,000 pool allotment to bring overall spending to $2,850 per pupil.

The high wealth school district desiring to make medium effort would pay $1,500 into the state pool (5 percent of its actual wealth of $30,000 per pupil). It would draw out $1,000 per pupil from the pool (5 percent of the average value of all state property per pupil). Wealthy districts would assist school funding of poor districts. But wealthy and poor districts making the same local funding effort would receive equal overall funding per student.

Under the medium effort scenario, average local districts would provide 35 percent of the funding, the state would provide 35 percent, and the federal government 30 percent. These percentages compared to average historic levels are much higher for the federal government and lower for the local district. Such rates would help to correct for spillovers, tend to bring low wealth district outlays to needed levels, would tax more heavily those with ability to pay, and yet leave major control of common schools in the hands of local districts and states. It would free some local taxes to pay for water, road, waste disposal, and other local services. Federal and state funds might focus on improving basic cognitive skills while local funds would be used to fund sports and other extracurricular activities. Administrators might be paid according to achievement test performance of students adjusted for socio-economic backgrounds.

Improving Vocational-Technical Schooling

Vocational-technical schooling is also important to the well-being of rural people. Many states have sharply expanded facilities for vocational schooling in recent years, and many such schools are within commuting distance of rural students. Several points need to be kept in mind regarding vocational schooling:

1. Although the social rate of return has averaged a respectable 15 percent based on some past studies (see Tweeten and Brinkman, ch. 4), expansion of output has sometimes been poorly planned, training students for occupations without favorable job opportunities. More attention needs to be paid to supply-demand prospects and placement.

2. Private firms tend to underinvest in training employees because benefits often are not appropriable. The firm first pays for training and then pays a higher wage approaching the value of increased productivity or the trained worker will go elsewhere. Still, on-the-job training tends to be efficient because it prepares workers for existing jobs. The federal government often saves training costs by paying private firms to train workers. In many instances, the firms subsequently employ the trained workers.

3. Special programs are needed for the disadvantaged. In many instances where federal involvement is warranted, persons can be trained in state facilities with partial federal funding of trainees. In other instances federal training facilities may be needed. The Job Corps provides a campus-like program of comprehensive training, counseling, and job search assistance for the hard-core disadvantaged. Costs averaged $14,000 per trainee in 1984. Although expensive, the program has had positive results. Some estimates indicate the program returned $1 for each $1 invested (Antipoverty policy: Past and future, Summer 1985, p. 15). If these benefits and costs were adjusted for the utility

of income for those who pay the costs and receive the benefits, the program would be rated as one of the more successful of a generally disappointing array of federal manpower programs.

MAJOR WELFARE REFORM

Current welfare programs have been criticized for discriminating against intact poor families and for causing family disintegration, for discouraging employment, for encouraging welfare dependency, and for being an administrative nightmare with problems of coordinating multiple income-conditioned programs. Three major welfare reform proposals have been advanced. Each has advantages and disadvantages as noted briefly below.

Negative Income Tax. The negative income tax is an extension of the positive income tax concept. The government is paid a positive tax by individuals or families whose income is sufficiently high; the government pays a negative tax or transfer to the individual or family whose income is sufficiently low.

The basic form of the plan is:

$$P = G - rE \qquad P = 0 \text{ for } E > \frac{G}{r}$$

$$Y = P + E$$

where:

P = government payment,
E = earnings,
Y = total income,
G = guarantee — the government payment to the person who has no earnings, and
r = tax rate, the sacrifice of government payment per additional dollar earned.

The breakeven level, the earnings at which government payments cease, is where $P = G - rE = 0$ or $E = G/r$. If the tax rate is 50 percent and G is $6,000, then breakeven earnings are $12,000. An illustrative schedule of earnings, payments, and total income for these parameters are as follows for the negative income tax:

Earnings	Payment	Income
	(dollars per year)	
0	6,000	6,000
4,000	4,000	8,000
8,000	2,000	10,000
12,000	0	12,000

Decisions would have to be made regarding guarantees by size of family, work requirements, and tax rates. The plan received serious consideration from several past administrations and from Congress and has been tested in actual trials in rural and urban areas.

The Rural Income Maintenance Experiment conducted from 1970 and 1972 in Iowa and North Carolina estimated that a negative income tax with a 45 percent tax rate and income guarantee at 80 percent of the poverty threshold would reduce hours worked by families as a whole by 13 percent (Bawden et al., 1976, p. x). Responses differed greatly among family members: Husbands reduced hours worked very little while wives reduced hours worked for wages by 27 percent and dependents by 46 percent. Labor supplied by farmers was not reduced by the negative income tax.

The Seattle-Denver negative income tax experiment, initiated in the early 1970s and considered to be the best run of several such efforts, included 4,800 families. Prime age males in the 5-year negative income tax program reduced annual hours of work by 9-10 percent, their spouses reduced work by 17-20 percent, and women heading single-parent families reduced annual hours by as much as 32 percent (Burtless and Haveman, 1984, p. 108).

These work reductions for urban families in the Seattle-Denver study and for rural families in the Iowa-North Carolina study were sizable and much larger than for farm families. In the words of Burtless and Haveman (1984, p. 108), these reductions

> ...are large enough to cause alarm among conservatives already opposed to a NIT [negative income tax] and even among centrists with no strong opinions about the desirability of a NIT.

Demogrant. A second major welfare reform proposal is the demogrant which provides a flat grant to every person in the nation. No means or work eligibility test is required, hence the program is easily administered. Suppose the demogrant is D = $1,500 per person. Suppose overall federal income tax rates are raised 10 percentage points to pay for the program. A schedule of earnings, payments, and income for a family of 3 is therefore:

Earnings	Payment	Income (after special taxes)
	(dollars per year)	
0	4,500	4,500
6,000	4,500	9,900
12,000	4,500	15,300

The demogrant itself would not be taxed. The payment would continue for all income levels. However, there is a breakeven level. Those with earnings of over $15,000 would pay more taxes than they receive from the demogrant, hence would be transferring income to those with lower earnings.

Wage or Earnings Supplement. The final major welfare reform proposal is the wage or earnings supplement. Under the wage supplement, the government provides a payment P equal to some proportion s of the difference between a target wage T and whatever wage w the individual can earn in the market or

$$P = s(w - T) \qquad P = 0 \text{ for } w \geq T$$
$$Y = W(P + w)$$

where Y is income, W is hours worked, and other variables are as defined above.

If the target wage T is $6 per hour, if the wage tax rate is 60 percent (s = .6), and the individual works 2,000 hours per year, results for alternative wage rates w of $1 or greater are as follows:

Wage	Payment	Earnings	Income
(dollars/hr.)		(dollars/yr.)	
1.00	3.00	2,000	8,000
3.00	1.80	6,000	9,600
6.00	0.00	12,000	12,000

The total annual supplement is $6,000 for a wage of $1 per hour, is $3,600 for a wage of $3 per hour, but drops to zero for a wage of $6 per hour.

An earnings supplement would be the counterpart of the wage supplement for self-employed persons such as farmers. For example, the earnings supplement would match earnings $1 for $1 up to $4,000 after which 50¢ of supplement would be subtracted for each dollar earned. Results would be as follows:

Earnings	Payment	Income
	(dollars per year)	
0	0	0
2,000	2,000	4,000
4,000	4,000	8,000
8,000	2,000	10,000
12,000	0	12,000

Figure 10.1 illustrates the earnings-leisure tradeoffs for the three plans presented above. Given wage rate w_0, a person employed full time works W hours for Ww_0 earnings. The attainable earnings-leisure tradeoffs follow the straight line Ww_0L, and the optimal combination is where indifference curve I_0 is tangent to the earnings leisure line at A.

With no work, the demogrant provides income D, the negative income tax the guarantee G, and the wage supplement provides no payment. The line of attainable combinations of income and leisure with wage supplement P per hour is the straight line $W(w_0 + P)L$ tangent to the highest indifference curve attainable I_1 at WS.

Assuming B is breakeven earnings for the negative income tax, the line of attainable income-leisure tradeoffs is Ww_0BG in Figure 10.1. The slope of that line is flatter than that for the wage supplement, hence is tangent to indifference curve I_1 in Figure 10.1 at greater leisure and lower income than WS. The demogrant lies between the two above welfare reform proposals in earnings-leisure tradeoffs. Aside from the modest tax rate, the demogrant income-leisure tradeoff line lies above Ww_0L by a constant D equal to the demogrant. It is tangent to I_1 at Dem. Whether points Dem, WS, or NIT entail more or less leisure than without welfare reform at A depends on income and substitution effects in Figure. 10.1

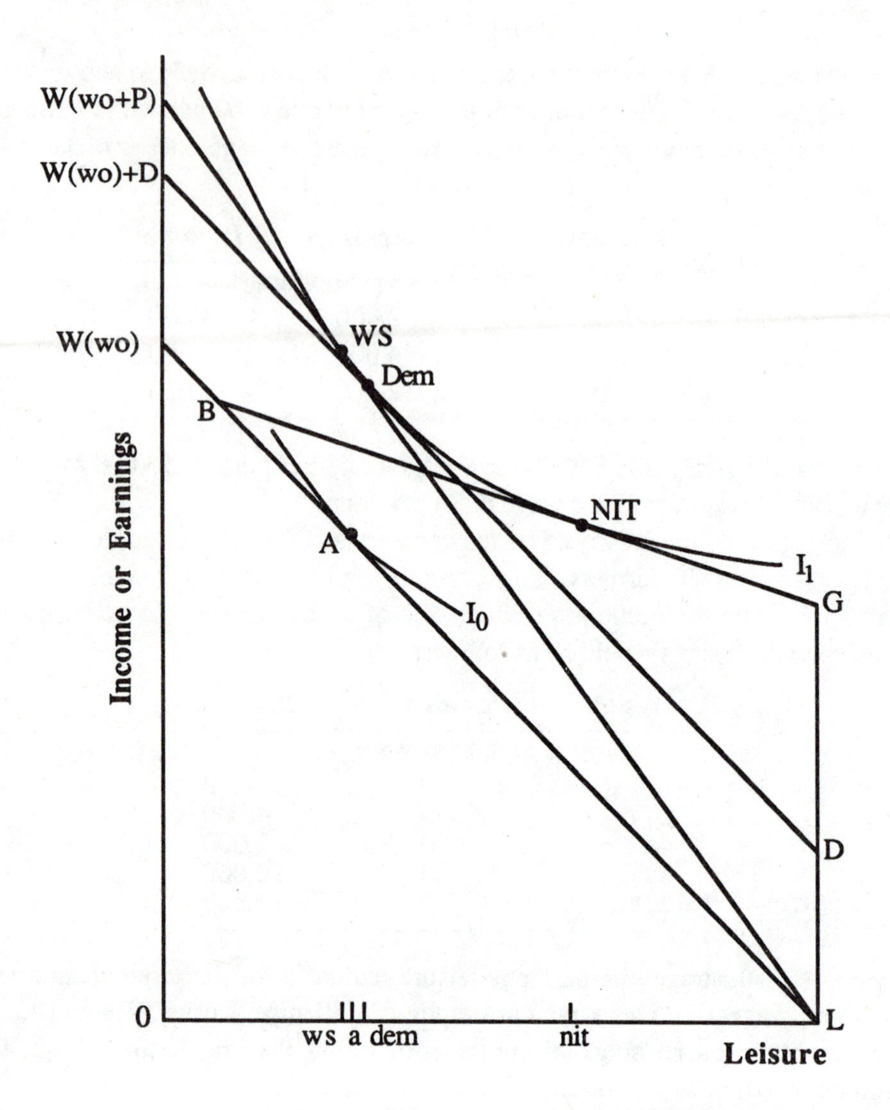

FIGURE 10.1. Earnings-Leisure Impact of Wage Supplement, Demogrant, and Negative Income Tax, and Where *a* Is Leisure Without Intervention

Each of the three reform proposals may reduce national income and raise social cost. The wage supplement entails a social cost because it induces more work and less leisure than would be optimal in a well-functioning economy. The opposite holds for the negative income tax; the demogrant ranks in between. However, given actual institutional and other constraints to employment, given the strong tendency of the wage supplement to target those with low income, and given its strong inducements to work, a case can be made that a wage supplement would increase full real national income and provide a net social gain. Given the net transfers to low income people of the demogrant and negative income tax, those reforms too might raise national well-being but national money income would be reduced. That is, national income adjusted for utility as shown in Chapter 2 might be increased by major welfare reform. The strong work disincentive of the negative income tax and the high Treasury cost of the demogrant make them less serious contenders than the wage or earnings supplement for welfare reform. The negative income tax and demogrant might receive less national popular support and funding than the wage supplement because they induce leisure as noted in Figure 10.1.

Modification of Welfare

Modification rather than major overhaul of the current welfare system seems likely. In 1988, the Aid to Families with Dependent Children program was modified to make a family with children and two parents in the home eligible for welfare in all states. If applied to underemployed families where parents are working but at low earnings, it could benefit the working poor rural self-employed families disproportionately found on small full-time farms in the South.

Workfare federal welfare reform will require of welfare recipients comprehensive remedial education, job-training, or work. Some reformers have proposed making available public employment as last resort. Pay would be low enough to encourage welfare recipients to exhaust private employment options before turning to public employment. Public employment has not had a favorable record of success when practiced on a large scale. Subsidies to private firms to employ those on welfare have more merit. Child care and Medicaid benefits might be retained for some prescribed period after recipients leave welfare roles to reduce incentives to remain on welfare and out of work. Automatic pay deductions from earnings of parents absent from the home where children are being raised would reduce public welfare costs. Under workfare, eligibility of more families for assistance plus additional costs for improved and expanded child care could raise overall welfare costs. The main advantage of workfare is to reach more needy families while discouraging long-term dependence on welfare.

Welfare reform offers no quick solutions. A study by the Manpower Demonstration Research Corporation found that state workfare and training programs raised the employment rate for welfare mothers by only 3 - 9 percent (Shapiro, 1987, p. 19). Less than one-sixth of all welfare mothers receive any child support from fathers of their children. The Congressional Budget Office estimated that a nationwide plan to identify fathers and garnish their wages would raise only $374 million over five years from absent parents (Shapiro, 1987, p. 19).

Welfare reform is needed, but the proposed modifications listed above (however desirable) would not have a major impact on farm, rural, or urban poverty and underemployment.

Programs to Assist Displaced Workers

Some farm operators and their families will be unable to maintain the efficient farm size, management skills, and technological pace necessary to succeed in a dynamic economic environment. Other farm families will be unable to form or maintain an economic unit and find satisfactory means of livelihood on the farm because of misfortune such as illness, disability, an unexpected turn of weather or prices, and a host of other factors simply labeled bad luck. The pace of technological change almost certainly will leave some farmers redundant. The public has a role in providing a safety net for such persons. Commodity programs and credit programs alone do not meet the needs of families left behind. Farm activists notwithstanding, the nation is no more likely to halt technological change or guarantee tenure to farmers than it is for cobblers, miners, or the local mom-and-pop grocery employee.

Farmers have a more difficult time adjusting to alternative employment than others because job, family, home, status, and worth are closely integrated. Farmers facing financial failure typically progress through successive stages of disbelief, anger, and resignation. With time, subjective indictors (what they say) and objective indicators (income, housing, access to community services, etc.) indicate most are better off for having made adjustments (see for review Tweeten and Brinkman, 1976, pp. 88-92). But public programs can ease trauma during the difficult adjustment period. With proper assistance, many of displaced farmers can find satisfactory jobs outside of agriculture. Farm families can benefit from several forms of help to make a successful transition to a nonfarm job (Mazie and Bluestone, 1987, p. 1):

— *Personal support.* Such support can include counseling, help in assessing their financial condition, and legal and technical information to help them adjust to new circumstances and make decisions in selling their farm assets.

— *Financial bridges.* Displaced farm families need a source of income until they can obtain work in the nonfarm sector.

— *Help to find work.* Skills assessment, classroom and on-the-job training, and job search and relocation assistance can help them find new work.

The main federally funded source of help is the dislocated worker programs (Title III) authorized by the Job Training Partnership Act (JTPA). These programs, developed and administered by state and local institutions and supported by JTPA block grants, service all eligible dislocated workers — urban or rural, wage earner, or self-employed. Most federally funded programs providing such help for displaced workers have been designed for wage earners, not self-employed persons like farmers. Such programs often are not suited to farmers. Innovative state and local institutions are using federal programs and creating new programs tailored to farmers' special requirements.

Farmers displaced by structural changes in agriculture are eligible for these programs, but not many participate. Those who do participate are often disappointed because the programs are geared to urban industrial workers. Because dislocated farmers are often long on skills but short on recognizing them and how they can be used outside of farming, the programs emphasize building confidence, assessing skills and interests, and developing job search skills. Personal and career counseling is included. Many states will reimburse employers for half the wages paid while providing on-the-job training for six months. Stipends are not available to those in classroom training, but funds for tuition and books are available for up to one year. Supportive services such as child care, transportation, and counseling are also available. Organizations linked to the community, such as the Cooperative Extension Service, are often most effective at encouraging farmers to learn about and take advantage of available programs.

The principle of providing assistance under the Trade Adjustment Assistance (TAA) program enacted in 1962 and the Job Training Partnership Act is to assist workers displaced for reasons beyond their control. This criterion is unworkable because it is impossible to judge in most instances whether the circumstances were within or outside of the individual's control. Job displacement in farming occurs because of changes in macroeconomic and trade policy, weather, unfortunate timing of asset purchases, poor management, and host of other reasons impossible to sort out. It makes sense to not attempt to assess reasons for displacement before providing assistance for displaced farmers.

With greater funding and recognition that training can increase earnings and GNP, training need not be confined to workers dislocated by exports or other reasons beyond their control. If a strong program of training and mobility assistance is available, it is feasible to more aggressively enhance productivity growth in agriculture while minimizing adjustment problems for those impacted. Farmers do not need a separate program but need special attention in a nationwide program designed for workers in all occupations.

Vocational Agriculture

The vocational agriculture program in high school has served youth well over the years, and is especially important in establishing a sense of pride, responsibility, citizenship, and leadership. Changes could improve the programs, however (CAST, 1988). The programs are costly per student, can unduly encourage students with bleak prospects for farming success to become operators, and can detract from needed training in science, mathematics, foreign languages, and English. The modern-day commercial farm requires a high level of business acumen, portfolio-management capabilities, and risk-management skills comparable to those required to run a sizable nonfarm business. Owners and operators of such nonfarm businesses often recognize the great risk and very high level of managerial ability required. A college education, even a Master of Business Administration (MBA) degree, is viewed as extremely useful.

On the other hand, those who would become modern-day commercial farm operators all too often underestimate the requirements to be a successful farm operator. Being the son or daughter of a farmer is not enough. It is important for prospective operators to be candidly informed of the high level of managerial competence, ability to handle risk, and resource requirements for an efficient, viable commercial farm. Imparting a romantic-nostalgic image of a farm way of life engenders expectations which inevitably will clash with reality, leaving the unprepared operator disillusioned, financially crippled, and prone to turn to taxpayers for maintaining the farming operation. Adequate training, experience, and counseling are essential to minimize unfavorable outcomes.

The current preparation of farm operators and mangers could be improved by

— More intensive training in science, mathematics, English, computers, communication, and other basic skills in high school while minimizing training in vocational courses — the latter reserved for post-high-school programs.

— Strengthened 4-H programs to teach citizenship, responsibility, leadership, and related topics in programs outside the school and with projects in agriculture and other fields available to youth from all walks of life, including the urban ghetto. Career counseling is an important component with students realistically appraised of requirements for and consequences of entering agricultural as well as other occupations.

— Strong post-high-school vocational agricultural training programs in area vocational-technical schools and junior colleges. Intensive training would take place in management as well as technical aspects of farming.

— Strengthened programs of training for commercial farm operators in four-year programs in colleges of agriculture. These programs would include not only rigorous training in traditional areas such as animal science, crop and soil science, and agribusiness management, but also in liberal arts, social sciences, humanities, mathematics, and basic science. Some universities currently have such programs and these can be strengthened. Many new farm operators will not come from a farm background; for them internships are important.

In short, the apprenticeship of growing up on a farm and being the son of a farmer is alone no longer adequate preparation for the successful operator and manager of tomorrow's commercial farm in a viable agriculture. For the foreseeable future, agriculture will be troubled by having too many rather than too few farms and operators. There will be more than enough farm families and farms to provide adequate food supplies and a competitive economic environment. Special public policies such as subsidized credit or related income supports are not needed now nor in the foreseeable future to assist young people to start farming. Subsidized entry only exacerbates the problem of more resources in agriculture than market demand will support and unfairly competes with established farmers for resources and markets.

FOOD ASSISTANCE

Food stamps are the most widely available welfare program. Food stamps are available to virtually anyone who can show need and is the only welfare program widely available to able-bodied adults in poverty and without children. Food stamps are available to working poor adults in intact families with able-bodied parents in the home, a family type disproportionately found in rural areas. The program reached 20 million persons in 1985 at a cost of $20 billion.

The food stamp program has been criticized for not ending hunger, for making payments essentially in cash rather than in kind, and for other reasons. Before examining proposals for reform, it is necessary to review briefly the conceptual base.

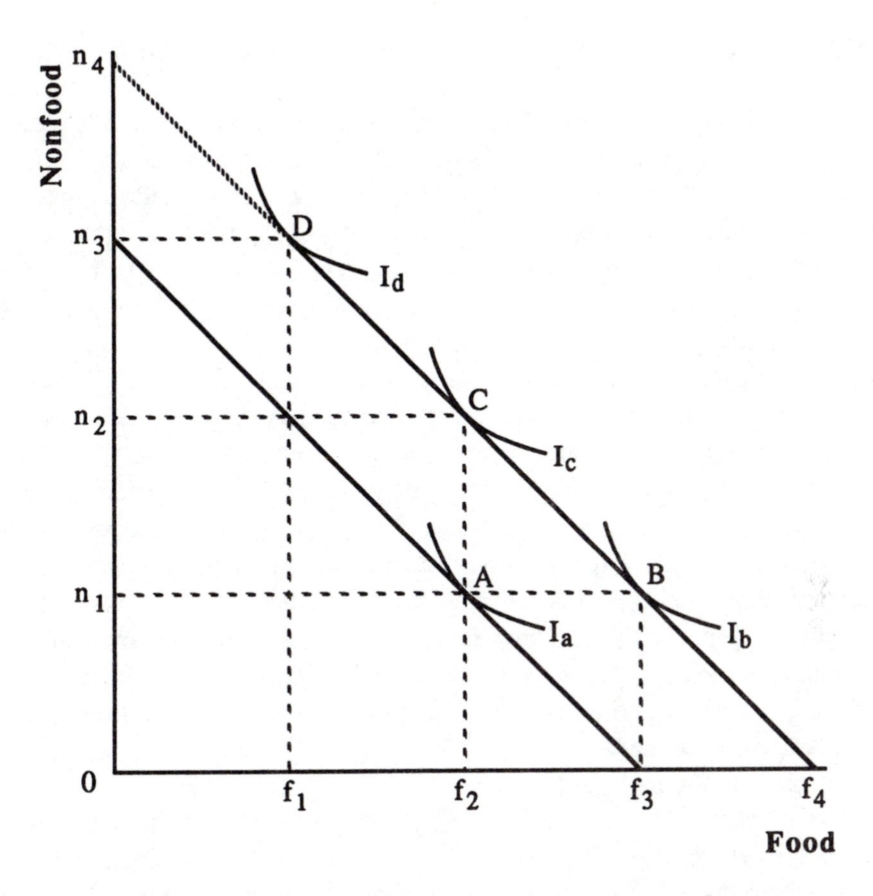

FIGURE 10.2. Illustration of Food Subsidy Benefits

Figure 10.2 illustrates the elementary concept that commodity distribution contributes less to well-being of consumers than does payment of a similar value in cash. The isocost line n_3f_3 of a given budget cash transfer from the government to the consumer with no earnings allows reaching the highest possible indifference curve I_a at quantity n_1 of nonfood and f_2 of food. If the government instead provides a quantity f_3 of food only, the cost to the government remains unchanged but the consumer with no other income and no ability to trade one commodity for another will be unable to reach indifference curve I_a. Surveys show that the public prefers to give food rather than cash assistance to the needy. If the public provides much more direct commodity distribution than value of food stamps, the family may be better off with commodity distribution especially if food can be bartered for other needed goods and services.

The Food Stamp Act of 1977 eliminated the existing purchase requirement. Previously, households receiving food stamps were required to spend approximately 30 percent of their income to purchase the stamps deemed adequate to provide a thrifty but adequate diet. This new provision removed the potential for vendor fraud in the handling of cash, simplified program administration, and lowered program administrative costs.

The overall face value of food stamps was less, however, reducing incentives to purchase food.

Figure 10.2 also illustrates why food stamps designed to provide an adequate diet do not do so if participants are rational. (It has been proposed since that the food stamp bonus be replaced with cash.) Suppose the consumer budget line from earned income is n_3f_3 so that the consumer optimally purchases n_1 of nonfood and f_2 of food to attain the highest possible indifference curve at A. Suppose society deems food quantity f_3 rather than f_2 to be adequate for a proper diet. A supplement $f_1 = f_3 - f_2$ of food or food stamps is provided, shifting the budget line to the right by the amount f_1. The new budget line of attainable combinations is n_3DCBf_4. The consumer with indifference curve I_b will purchase n_1 nonfood as before and the quantity of food f_3 providing an adequate diet. Consumption of food is increased by the amount of the food stamps. In fact, the indifference curve is likely to lie above and to the left of I_b. If the indifference curve is I_c tangent at C, then food consumption is unchanged at f_2 but nonfood consumption increases from n_1 to n_2 despite the gift of food. Another alternative is indifference curve I_d tangent at D, in which case total food consumption drops to f_1 from f_2 despite provision of free food or food stamps. In general, a food stamp subsidy allowing the consumer to increase food purchases from f_2 to the appropriate food need level f_3 will in fact result in consumption of less than f_3 because rational consumers will spend part of the additional buying power for clothing, shelter, and other needs.

The outcomes in Figure 10.2 depend on the position of the indifference curves and on the fungibility of the food assistance. Inability to convert food assistance into nonfood purchases flattens the budget line n_4DCBf_4, hence tends to move the tangency of the indifference curve along the budget line to a higher ratio of food use relative to nonfood use. But such a move lowers the highest indifference curve attainable.

The most widespread malnutrition problem in the United States is chronic excessive intake of calories and too much fat, cholesterol, salt, and sugar and not enough fiber and complex carbohydrates in diets. Food stamps do not treat these problems. Malnutrition can be viewed as primary or secondary where *primary real malnutrition* refers to those who do not have or cannot gain access to resources required for a nutritious diet adequate in calories. Because of the widespread availability of food stamps (even for indigents), and because food stamps provide sufficient resources to purchase a nutritionally adequate diet (although palatability may leave something to be desired), primary malnutrition is very rare indeed in the United States. *Secondary malnutrition* is widespread, however, and arises because people are not informed, motivated, or competent to make proper use of resources available to them and because food is only one of the essentials for people in poverty.

The majority of Americans experience some form of secondary malnutrition — eating or drinking in ways detracting from good health. Short of force feeding, which is unthinkable, secondary malnutrition will continue to be commonplace. At issue is how best to deal with it. Some suggestions are offered below:

1. The Thrifty Food Plan underlying food stamp allotments is adequate for a nutritious diet with good planning and food preparation and minimal waste. But even normal waste is at least double the assumed 5 percent in the Thrifty Food Plan used by

USDA to set food stamps allowances, and low income consumers frequently are unable to manage as well as other consumers. Many feel the basic food stamp allotment should be increased to assume 10 percent waste. Allowances could be adjusted among regions for cost-of-living differences.

Providing more food stamps will not solve the problem of secondary malnutrition, however. Having the means to purchase an adequate diet is a necessary but by no means sufficient condition to alleviate malnutrition.

2. A review of numerous studies indicates that greater buying power, general schooling, and nutrition education especially on the part of the homemaker raises family nutritional status (Morgan, 1986, pp. 1242, 43). The Expanded Food and Nutrition Education Program (EFNEP) has raised nutritional status particularly when operated in conjunction with the food stamp program. Greater buying power alone, whether from greater income or food stamps, helps (the marginal propensity to spend on food is well above zero for low income people), but the relationship between nutrient intake and income or buying power is not strong (see Davis, 1982, p. 1022).

3. Food assistance programs were initiated to alleviate hunger; welfare programs were initiated to alleviate poverty and deprivation. Poverty and hunger are inseparable, however, and attempts to segregate them in public programs inevitably causes conflicts for recipients, supporters of programs, and taxpayers. The public views food as a merit good — elimination of hunger benefits society as well as individuals. But food aid recipients need shelter, clothing, and other goods and services as well as food and hence continually press to reach higher indifference curves by receiving food in cash rather than in kind. Figure 10.2 illustrated that families eligible for food stamps capable of purchasing a nutritious diet will obtain neither adequate food nor other necessities if they are rational. General welfare reform along lines indicated earlier will help to reduce malnutrition among the poor.

4. Large numbers of eligible food program recipients do not participate. Many of those are near income levels entitling them to small allotments and many do not feel the benefits of food stamps are worth the hassle and stigma of obtaining them. However, an estimated 43 percent of eligible nonparticipants were not aware they qualified for the program (Allen and Newton, 1986, p. 1249). An effective outreach program is one way to reduce malnutrition and a case can be made to reinstate the outreach program terminated by the Reagan Administration in 1982.

5. Because the public sector does not meet all needs for food assistance, the private sector plays a key role through for example, "soup kitchens," "food banks," and a wide array of the charitable efforts by churches and other groups and individuals. Private charitable groups fill local nitches not met by public agencies. Private efforts are essential to supplement public programs for the indigent. The food stamp system works best for persons who have addresses. Renewed efforts in recent years to serve those who do not have an address have met with only partial success. Many of the indigent suffer from mental disorders, alcoholism, or drug addiction. Personal outreach is needed to serve even their minimum food needs — a job better performed by the private than public agencies. Establishing the appropriate public-private system to meet the needs of these people

remains a most serious challenge for food programs. The public sector must fill nitches including shelter for "street people" not filled by the private sector.

Many of the elderly poor living alone are incapable of providing adequate diets for themselves. Congregate meals and Meals-on-Wheels programs supported by public funds have improved their diets. These programs might be expanded to permit use of food stamps to purchase prepared meals. In some instances, it may be necessary to replace food stamps with cash to provide flexibility and access to the elderly with mobility problems. Similarly, indigent younger persons lacking permanent dwellings or addresses and unable to prepare meals would benefit from being able to use food stamps to purchase meals.

SUMMARY, CONCLUSIONS, AND DATA NEEDS

When most people think of "rural development" they think of local job creation and place prosperity. Federal efforts to promote such development have a dubious rationale and a mixed record of success at best. Individual rural areas and states may wish to pursue industrial plants to locate in rural areas but it is difficult to build a case for federal involvement except for technical, statistical, and informational support such as provided by the Cooperative Extension Service.

HUMAN RESOURCE PROGRAMS

No public policy is more essential to the well-being of farm and rural people than human resource investments. Rural areas have invested huge sums in schooling of youth. That enabled some of those youth to become competent farmers in a modern society and others to take their place with pride and dignity in nonfarm society. Children of farmers were outranked only by children of the managerial, professional, and supervisory (the top occupational group) in socioeconomic status in nonfarm jobs (Heady, 1962).

Problems remain, however. Rural common schools are frequently inadequate and inequitably funded. Numerous reforms were suggested herein to improve equity and efficiency.

Poverty remains a serious concern in noncommercial farming and in rural areas in general. Farm people have not been well served by welfare programs in part because the farm poor are frequently in families with both parents in the home. Options for reform of the welfare system and food programs were noted.

Transfer payments are the most cost-effective means to raise well-being of the elderly and other nonsalvageable poor. For others, human resource development programs have potential. Past federal programs to salvage the poor mostly have increased hours of work rather than pay per hour. Greater reliance on state vocational-technical programs offers the potential for longer, in-depth training necessary to raise earnings per hour. Of concern is whether programs will be evaluated for equity, efficiency, and impacts by sector, age, and other dimensions. Here it is well to recall the conceptual framework outlined in Chapter 2.

Manpower programs such as the Jobs Corp for the disadvantaged yielding a conventional benefit-cost ratio of only 1.0 can have a significantly greater *social* payoff than programs for more wealthy groups yielding higher conventional benefit-cost ratios.

Regarding work force programs, the situation as reported by Ray Marshall (1974, p. 119) in the 1970s has changed little:

> By whatever standard we judge manpower experiences, the evidence seems to support the conclusions that rural areas have been shortchanged in manpower efforts. An issue paper prepared by the Labor Department concluded that rural areas, with 22 percent of the population in 1969, received about 6.9 percent of labor and manpower efforts.

Labor force services have had a strong urban bias. While a case can be made to extend more work force services to rural areas, it is difficult to be precise without more hard data and analysis showing the payoff from such services to rural people. Such data are sparse indeed.

Data and Analysis Needs

Three general areas of data and analysis needs for rural areas are discussed below:

1. *Data measuring poverty, underemployment, and other dimensions of labor market performance and needs for public human resource development services are of high priority.* Some excellent descriptive statistics are available regarding the hired farm working force and farm operations. But relatively little is known about poverty and underemployment by economic classes of farms and even less is known of the characteristics of the 30 million worker nonmetropolitan labor force.

2. *Analysis of human resource development program payoffs.* On the one hand, the long list of past work force programs (see Tweeten and Brinkman, 1976, ch. 4) depicts the image of a federal-state labor force effort that is alert, imaginative, innovative, and bold. On the other hand, the image is a labor policy that is barren of proved, solid, comprehensive delivery systems integrated with other socioeconomic development efforts.

For the nonsalvable poor, such as the disabled and elderly, there is little alternative to income transfers. Transfers are more cost-effective than human resource development programs in using limited public funds to raise income because the latter programs devote considerable resources to administrative and training costs which produce little or no earnings for the target group.

For the underemployed and unemployed poor, a negative income tax has large disincentive effects. Public service employment programs are expensive to taxpayers and large programs are difficult to direct at producing valuable output because of political favoritism and objections of the conventionally employed. Federal efforts have been disappointing to bring jobs to people through industrialization incentives.

This narrows the list of successful federal human resource development programs to education, training, relocation assistance, and placement efforts. Monitoring performance of local-state institutions will be central because of their large role in human resource development.

3. Basic parameter. Predicting impacts of changes in human resource policies requires reliable estimates of parameters such as labor supply and demand elasticities and payoffs from programs including level and distribution of benefits and costs. If wages of farm and other workers were changed by higher minimum wage rates, by greater labor union coverage, or by higher payroll deductions for social security, disability insurance, and other social programs, what would be the impact on employment, payrolls, and output?

The economic payoff from *both* human resource and industrial development programs may be enhanced with balanced growth providing jobs and skills in depressed rural areas. Basic parameters relating population characteristics to public programs and, in turn, public programs to socioeconomic payoffs are required to model and evaluate alternative rural development strategies. What is the least-cost mix of public manpower, welfare, and job development policies to alleviate poverty and underemployment in rural areas? Better data and analysis are needed not only to determine program successes and failures but also to gauge how far prospective government programs can be expected to go to correct market failure while generating positive social benefits.

REFERENCES

Allen, Joyce and Doris Newton. Existing food policies and their relationship to hunger and nutrition. *American Journal of Agricultural Economics* 68:1247-52.

Antipoverty policy: Past and future. Summer 1985. *Focus* 8(2):13-15. (Institute for Research on Poverty, University of Wisconsin-Madison.)

Bawden, D. Lee et al. November 1976. The rural income maintenance experiment: Summary report for U.S. Department of Health, Education, and Welfare. Madison: Institute for Research on Poverty, University of Wisconsin.

Burtless, Gary. Fall 1984. Manpower policies for the disadvantaged: What works? *Brookings Review* 3(1):18-22.

Burtless, Gary and Robert Haveman. 1984. Policy lessons from three labor market experiments. Pp. 105-33 in R. Thayne Robson, ed., *Employment Training R and D, Lessons Learned and Future Directions*. (Conference proceedings of the National Council on Employment Policy, January 26, 27, 1984.) Kalamazoo, Michigan: Upjohn Institute.

Cochrane, Willard. 1965. *Cityman's Guide to the Farm Problem*. Minneapolis: University of Minnesota Press.

Congressional Budget Office. July 1982. Dislocated workers: Issues and federal options. Background paper. Washington, D.C.: U.S. Government Printing Office.

Council for Agricultural Science and Technology (CAST). 1988. Long-term viability of U.S. agriculture. Ames: CAST.

Davis, Carlton. 1982. Linkages between socioeconomic characteristics, food expenditure patterns, and nutritional status of low income households: A critical review. *American Journal of Agricultural Economics* 64:1017-25.

Deavers, Kenneth and David Brown. 1985. Natural resource dependence, rural development, and rural poverty. Research Report No. 48. Washington, D.C.: Economic Development Division, ERS, USDA.

Gilford, Dorothy, Glenn Nelson, and Linda Ingram, eds. 1981. *Rural America in Passage: Statistics for Policy.* (Report of the Panel for Statistics for Rural Development Policy, National Research Council.) Washington, D.C.: National Academy Press.

Haveman, Robert and Gregory Christainsen. November 1978. Public employment and wage subsidy in western Europe and the U.S. Discussion Paper 522-78. Madison: Institute for Research on Poverty, University of Wisconsin.

Huffman, Wallace. 1985. Human capital for agriculture. (Paper presented to conference *Agriculture and Rural Areas Approaching the 21st Century* held at Ames, Iowa, August 7-9, 1985.) Ames: Department of Economics, Iowa State University.

Hines, Fred and Luther Tweeten. September 1972. Optimal funding of elementary and secondary schooling. Research Report P-669. Stillwater: Agricultural Experiment Station, Oklahoma State University.

Marshall, Ray. 1974. *Rural Workers in Rural Labor Markets.* Salt Lake City: Olympus Publishing Company.

Mazie, S.M. and Herman Bluestone. 1987. Assistance to displaced farmers. Agricultural Information Bulletin No. 508. Washington, D.C.: Economic Research Service, USDA.

Morgan, Karen. December 1986. Socioeconomic factors effecting dietary status: An appraisal. *American Journal of Agricultural Economics* 68:1240-46.

Mulkey, David and Rodney Clouser. 1987. Rural development: Renewed commitment or passing fancy? (Paper presented at annual meeting of Southern Extension Public Affairs Committee, Asheville, North Carolina.) Gainesville: Food and Resource Economics Department, University of Florida.

Nilsen, Sigurd. 1980. Employment and unemployment statistics as funding allocators for non-metropolitan areas. Pp. 501-23 in National Commission on Employment and Unemployment Statistics, *Data Collection, Processing, and Presentation: National and Local.* Washington, D.C.: U.S. Government Printing Office.

Nilsen, Sigurd. March 1981. Nonmetro youth in the labor force. Rural Development Research Report No. 27. Washington, D.C.: ESS, USDA.

Rasmussen, Wayne. October 1985. 90 years of rural development programs. *Rural Development Perspectives* 2:2-9.

Shapiro, Walter. August 3, 1987. Tough, tightfisted, and traditional. *Time,* p. 19.

Tweeten, Luther. October 1983. Past and prospective economic development of rural communities. Pp. 174-197 in Congressional Research Service, Library of Congress, *Agriculture Communities; The Interrelationship of Agriculture, Business, Industry, and Government in the Rural Economy.* (Committee on Agriculture, U.S. House of

Representatives, 98th Congress, 1st Session.) Washington, D.C.: U.S. Government Printing Office.

Tweeten, Luther and George Brinkman. 1976. *Micropolitan Development*. Ames: Iowa State University Press.

U.S. Department of Health, Education, and Welfare. 1978. *Summary Report: Seattle-Denver Income Maintenance Experiment*. Washington, D.C.: Office of Income Security Policy/Research, HEW.

U.S. Department of Labor. 1976. *Recruitment, Job Search, and the United States Employment Service*. Employment and Training Administration, Research and Development Monograph No. 43. Washington, D.C.: U.S. Government Printing Office.

CHAPTER ELEVEN

Commodity Programs: A Short History

To understand the current structure of farm programs and what form legislation is likely to take in the future, it is well to examine history. This chapter reviews briefly the major commodity legislation since 1933. Sometimes programs that were advanced but not enacted into law are as instructive in understanding farm policy as are the programs in effect. Thus this chapter will include a discussion of the McNary-Haugen, the Brannan, and the Cochrane proposals. The chapter deals primarily with commodity programs of the federal government to alleviate commercial farm problems described at some length earlier in Chapters 1, 4, and 5. A Glossary at the end of Chapter 12 explains terminology.

PROGRAMS OF THE 1920S: A HISTORY OF FAILURE

The McNary-Haugen plan was originated by George Peek and Hugh Johnson of the Moline Plow Company. The plan was publicized by the slogan "Equality for Agriculture." It was a two-price plan, relatively simple in principle. A government export corporation would "buy" farm commodities not selling on the domestic market at a price (later called "parity") that bore the same relationship to prices paid by farmers as those in the pre-World War I period. The government corporation would dispose of excess supplies in the foreign market at the world price. Tariffs would be maintained to keep the commodities from being imported into the United States at the higher domestic price.

The farmer would receive an average price made up of the parity price for the domestic portion and the world price for the excess. The plan originated the concept of parity prices under the term "fair exchange value." The latter was defined as the same current ratio of farm prices received to the general price index as the ratio that prevailed in the 1905 - 14 period. If the ten-year prewar base period (1905 - 14) is cut to a five-year base (1910 - 14), and if the general price index is replaced by the index of prices paid by farmers, the "fair exchange value" becomes the "parity price" of the 1933 Agricultural Adjustment Act (Dubov and Rawls, 1974, p. 5). A historical review and calculations of agricultural price parity are found in Teigen (1987).

Five Defeated Bills

The first McNary-Haugen Bill was introduced in 1924 and was defeated in the House of Representatives. Southern and Eastern sections were unified in their opposition. the second McNary-Haugen Bill in 1925 never came to a vote. The third, even with new support from the American Farm Bureau Federation and amendments to make the bill more palatable, was defeated in both House and Senate in 1925. A fourth bill was passed by both houses of Congress in 1927 but was vetoed by President Coolidge with strong objections that the bill would aid farmers in certain sections at the expense of other farmers, that it was price-fixing, that it was impossible to administer, and that some portions of the bill were unconstitutional. A fifth bill was passed by Congress and vetoed by Coolidge in 1928. The principal features of the McNary-Haugen Bills were not dead, however, but became part of the New Deal legislation of the mid 1930s.

The Federal Farm Board

President Calvin Coolidge vetoed the last two McNary-Haugen proposals on the advice of his secretary of commerce, Herbert Hoover. Hoover took office as President in 1929. The Agricultural Marketing Act of that year established the Federal Farm Board. With a revolving fund of $500 million, the Board was to make loans to cooperatives to control surpluses by acquiring excess supplies, by constructing new storage facilities and other buildings, and by making advances to growers for their crops. The Board attempted to stabilize prices by purchasing stocks of wheat, cotton, and other commodities. Its limited finances were soon exhausted, and farm prices continued to fall. In 1932 the Federal Farm Board stated that its efforts had failed and recommended to Congress that legislation was needed to control production or quantities going to market.

Much effort of the Board was predicated on the erroneous assumption that the problems of agriculture could be solved with a more efficient marketing system. When the woes of a farm economy that had been depressed throughout the 1920s were compounded with the slack demand evolving from nationwide depression and lagging foreign demand for farm commodities in the early 1930s, the farm problem obviously became much larger than the Farm Board could handle. The Board was unable to support market prices and cope with existing excess production capacity without production controls. In May 1933 President Franklin Roosevelt abolished the Board by Presidential order.

PROGRAMS OF THE 1930S

The Agricultural Act of 1933

The first comprehensive farm price and income stabilization effort was the Agricultural Act of May 1933, creating the Agricultural Adjustment Administration and authorizing strong measures to raise the level of farm prices. Among these were authorizations (1) to enter into voluntary agreements whereby farms were paid to reduce acreage of "basic" crops, (2) to store crops on the farm with advance payments to producers for the crop, (3) to enter into voluntary marketing agreements with farmers and middlemen, and (4) to levy processing taxes to pay costs of adjustment operations and to expand markets (Benedict, 1953, p. 283). The "basic" commodities were cotton, wheat, corn, rice, tobacco, hogs, and milk. Acreage allotments were applied to the five crops listed above plus sugar and peanuts. Portions of cotton and tobacco acreage were plowed down as an emergency measure in 1933.

Nonrecourse loans were made for corn and cotton in 1933. The term "nonrecourse" meant that the advance loan made to farmers on collateral of corn at 45¢ per bushel (60 percent of parity) and on cotton at 10¢ per pound (69 percent of parity) need not necessarily be repaid. If the market price were above the loan price, the farmer could sell on the open market and repay the loan. If the loan rate were above the market price, the farmer could turn in the commodity (as full payment of the loan) to the Commodity Credit Corporation, a government-financed corporation created by executive order in October 1933.

Marketing agreements or orders were intended to raise prices by controlling the timing and volume of marketing fruits, vegetables, milk, and other commodities. The first marketing agreement covered handling of milk in the Chicago market and became effective in August 1933.

The production control features, including payments for taking land out of production, coupled with dry weather, sharply reduced production. Farm prices and incomes rose markedly from 1932 to 1936. The Agricultural Adjustment Administration program was halted by the Hoosac Mills decision of 1936 in which the production control and processing tax features of the program were declared unconstitutional by the Supreme Court.

Although not done under the Agricultural Adjustment Act of 1933, an international wheat agreement or cartel was formed by the United States, Australia, Canada and Argentina in 1933. The agreement committed major wheat exporters to cut acreage 15 percent in 1934. Only Australia honored the agreement. Argentina exceeded its export quota and the agreement collapsed in 1934.

The first known attempt at mandatory controls was a quota on tobacco production in Virginia in 1621. But the first workable mandatory controls were introduced by the Bankhead Cotton Control Act and by the Kerr Tobacco Control Act, both passed in 1934. Two-thirds of the producers of cotton had to approve the marketing quotas before they would go into effect. The required proportion was three-fourths before the tobacco allotment program would become effective. The quota was introduced by allotting to each producer certificates which, when accompanying the crop, would exempt it from "sales" tax. The tax was set high to prohibit sales not accompanied by certificates. Other processing tax proceeds were used to pay farmers to reduce production and to purchase commodities for distribution to persons unemployed and on relief.

Agricultural Adjustment Act of 1935

The Agricultural Adjustment Act of 1935 was not major or comprehensive legislation but gave the President authority to impose import quotas when imports interfered with agricultural price support programs. The act assigned 30 percent of customs receipts to promote agricultural exports and domestic consumption and to help finance adjustment programs.

Soil Conservation and Domestic Allotment Act of 1936

Commodity program supporters sought to shift program ties from processing taxes which were unconstitutional to soil conservation which was constitutional. The Soil Conservation and Domestic Allotment Act of 1936 combined conservation with production controls. Farmers were paid to voluntarily shift acreage from soil-depleting crops (which just happened to be those in excess supply) to soil-conserving legumes and grasses. For the first time the goal of income parity rather than price parity was introduced into the legislation. Although income parity was the stated goal in this and subsequent programs, price parity remained the operating concept. Income parity was defined as the ratio of purchasing power of the net income per person on farms to that of income per person not on farms which prevailed during the August 1909 to July 1914 period, a very favorable one for farmers.

Production was down because of the severe 1936 drought, but the large crop in 1937 showed that the program inadequately controlled production.

Agricultural Marketing Agreement Act of 1937

The Hoosac Mills decision had not invalidated the marketing agreement provisions of the 1933 legislation, but Congress saw fit to reaffirm certain provisions in the Agricultural Marketing Agreement Act of 1937. The 1937 act disclaimed authority to levy processing taxes and redefined the machinery for establishing marketing orders. Under provisions of the 1937 legislation and subsequent amendments, a large number of marketing orders have been established under the authority of the Secretary of Agriculture.

The act specified the terms for example for milk-marketing orders. Each order established minimum prices which handlers must pay for milk purchased from producers or associations of producers. Generally a minimum price was established for class II milk to be used to produce butter, cheese, evaporated milk, and other manufactured products. A higher minimum price was usually established for class I milk to be used in fluid products, principally to be sold to consumers as fresh milk. The terms of the order were developed through public hearings prior to issuance of the order. The role of the federal government was to hear all arguments and proposals, to evaluate proposals and resolve differences in the public interest, and to enforce provisions of the order.

Federal milk orders do not control production but increase farm income by the multiple price features. Because production is not controlled, prices can be established above equilibrium levels, causing overproduction and attendant need for the government to dispose of the excess outside the market. Due in part to overproduction generated by marketing orders, the government purchased, for example, 8.4 percent of all milk produced in the United States in 1953, 8.4 percent in 1961, and 8.0 percent in 1986. Percentages have fluctuated widely for other years, however. The tendency to overproduce was caused to come extent by the use of a "blend" price, giving producers an average of the prices for class I and class II milk. This practice obscured and distorted the marginal price in the two-price plan and gave the wrong economic signals to producers. Had farmers actually received the manufactured milk price for their added production, they would have been restrained from supplying more than the market would absorb.

Agricultural Adjustment Act of 1938

The Agricultural Adjustment Act of 1938 became the pattern for subsequent farm commodity programs. It contained the conservation provisions of the 1936 legislation. New features included (1) flexible nonrecourse loans for producers of corn, wheat, and cotton under specified supply and price conditions if marketing quotas were approved in referendum; (2) crop insurance for wheat; and (3) payments, if funds were available, to producers of corn, cotton, rice, tobacco, and wheat in amounts which would provide a return as nearly equal to parity as the available funds could permit.

The Secretary of Agriculture had discretion in establishing the nonrecourse loan rate between 52 and 75 percent of parity for wheat and cotton. The specific concept of parity price was used for the first time in legislation. A specific formula regulated loan rates on corn. The loan rate was to be 75 percent of parity if the supply was not expected to exceed a year's domestic consumption and exports and 52 percent of parity if supply was expected to exceed by 25 percent a year's domestic consumption and exports. There were several gradations between these extremes. The plan was to maintain an "ever-normal granary" of reserve stocks for possible emergencies.

Rice and tobacco referendums failed to obtain the necessary two-thirds approval for marketing quotas for the 1939-40 marketing year. Large crops were produced. The increased supply coupled with a decline in foreign demand brought lower prices, and growers approved tobacco marketing quotas for the 1940-41 crop year. Marketing quotas were in effect for cotton, peanuts, and wheat in 1941. Legislation also permitted mandatory controls on feed grains but the administration chose to avoid the referendum and rely on voluntary programs. Payment limitations were first introduced in 1938 legislation at $10,000 per recipient.

Marketing quotas and acreage allotments for wheat, tobacco, and cotton reduced acreage in the years they were in effect. Wheat acreage for example was 81 million in 1937 and dropped to approximately 63 million acres from 1938 to 1944 under the influence of controls. However, yields rose to partially compensate for the reduction in crop acreage.

Acreage allotments reduced corn acreage, but hybrid seed contributed to the sharp rise in yields. Surpluses accumulated, and farm prices fell 20 percent from 1938 to 1940. The government programs helped to avert a drastic drop in farm income, however.

Meanwhile efforts were made to expand demand for farm commodities. Section 202 of the 1938 legislation provided for establishment of four regional laboratories for scientific research to develop new uses and outlets for farm products. The distribution of surplus commodities at no cost to needy consumers began with pork in 1933, but the major food programs of today began with 1938 legislation. In addition to direct distribution to the needy of surplus farm commodities, a school-lunch program, low-cost milk program, and a food-stamp plan were initiated by the government. The food-stamp program reached 4 million people in 1941, was discontinued in 1943, and was revived later as discussed in Chapter 10.

FEDERAL CROP INSURANCE

A federal all-risk crop insurance bill was introduced in Congress in 1922. The proposal did not become reality until 1938 when the first Federal Crop Insurance Act was passed. It was signed by President Roosevelt as Title V of the Agricultural Adjustment Act of 1938. The original intent was to make the crop insurance bureau self-supporting. This feature was later changed to the requirement that premiums need only be sufficient to pay the indemnities. Coverage was closely associated with yields, although the intent was to cover cash production cost or investment in the crop. Coverage was to be low enough to avoid the "moral hazard" of deliberate neglect of crops by farmers to collect insurance. Coverage was first extended only to wheat, but in 1939, after one year of wheat crop insurance, a bill was passed to extend the program to the 1942 cotton crop.

Weaknesses in the program were apparent by 1943. Over $17 million had been lost on capital appropriated to operate the wheat program, and administrative cost had been high. Only one-third of wheat farmers and one-tenth of cotton farmers were participating. It is of interest that the largest net losses had accrued in the Midwest, while net surpluses had built in the Great Plains. The program was temporarily suspended in 1943 (Myrick, 1967, p.14).

The program was revived in 1945. The number of counties in the program was reduced from 2,400 in 1947 to 375 in 1948. The crop coverage has gradually extended to more countries and to a large number of grains, vegetables, and fruits. Through 1947, the ratio of losses to premiums was 1.48. From 1948 through 1963 the ratio was .93, hence the program was on a sounder financial basis. In 1980, Federal Crop Insurance replaced disaster payments, a component of commodity programs that was costing several billion dollars per year.

The Multiple Peril Crop Insurance (MPCI) Program in 1980 legislation redirected Federal Crop Insurance to place emphasis on (1) working through private insurance vendors; (2) allowing MPCI coverage electively at 50, 65, or 75 percent of growers' individual yield history, not county yield as earlier; (3) reducing premium rates with yield, i.e., a grower with top yields pays only about one-third the premium rate of the grower with below-average yields; and (4) integrating with commercial hail/fire insurance to form a comprehensive crop insurance package. In 1986, farmers insured a record 61 million acres (18 percent of cropland harvested) and received $6.2 billion in protection (approximately one-tenth of crop receipts). Loss payments to farmers totaled $595 million compared to premiums paid of only $285 million in 1986.

Thus all-risk crop insurance, though useful for many farmers, is not a large enough program to materially reduce risk in the farming industry. The program is most heavily concentrated in the Great Plains, Midwest, and Southeast. The program has reduced the risk from natural hazards such as drought, wind, and hail but does not protect farmers from market uncertainties such as unstable prices.

AGRICULTURE IN A WAR ECONOMY

The burdensome "ever-normal granary" stocks accumulated by the Commodity Credit Corporation proved a blessing in World War II. The emphasis changed from restraining production to encouraging production with high price guarantees that were to dominate farm legislation until the 1950s.

Congress passed legislation in 1941 to raise loan rates for cotton, corn, wheat, rice, and tobacco to 85 percent of parity on the 1941 crop. This legislation was followed by the Steagall Amendment later in 1941, which directed the Secretary of Agriculture also to support the prices of nonbasic commodities at 85 percent of parity if he found it necessary to increase their production.

An amendment to the Emergency Price Control Act of October 1942 raised the support rate to 90 percent of parity for corn, cotton, peanuts, rice, tobacco, and wheat, and for the Steagall nonbasic commodities. However, the rate of 85 percent of parity could be used for any commodity if the President deemed that a lower rate was in the interest of reduced feed and livestock costs for the national defense. Section 8 of the 1942 legislation provided that prices of basic commodities would be supported at 90 percent of parity for two years immediately succeeding the first day of January following a Presidential or Congressional declaration that hostilities had ceased. Loans were to be limited to cooperators, i.e., farmers operating in accordance with acreage or marketing quotas announced by the Secretary of Agriculture and accepted by the growers. Section 9 extended the two-year postwar guarantees to the Steagall nonbasic commodities (Benedict, 1953, pp. 415, 416).

The price support on cotton was raised to 92.5 percent of parity by a June 1944 law and was raised to 95 percent of parity by an October 1944 law. The Commodity Credit Corporation paid 100 percent of parity for cotton in 1944 and 1945.

Marketing quotas were retained throughout the war on barley and flue-cured tobaccos. Marketing quotas were retained to February 1943 on wheat and to July 1943 on cotton (Rasmussen and Baker, 1966, p. 74).

THE PAINFUL TRANSITION

The provisions of the Emergency Control Act extending high price supports were to terminate December 31, 1948. Price supports for basic commodities would then drop back to the range of 52 to 75 percent of parity as provided by the Agricultural Adjustment Act of 1938. However, new legislation, the Agricultural Adjustment Act of 1948, was approved. It established mandatory price supports at 90 percent of parity for 1949 crops of wheat, corn, rice, peanuts, cotton, and tobacco if producers had not disapproved marketing quotas. Price supports at 90 percent of parity were also provided for a substantial number of other crops and livestock, including hogs and chickens marketed before December 31, 1949.

Title II of the 1948 legislation was an amended version of the Agricultural Adjustment Act of 1938 and contained provisions for a new parity formula and sliding scale of price supports between 60 and 90 percent of parity. The sliding scale slated to begin in 1950 never became effective because the 1948 legislation was superseded by the Agricultural Act of 1949.

If producers approved necessary quotas and allotments, basic commodities were to be supported at 90 percent of parity for 1950 and from 80 to 90 percent for 1951 crops under the 1949 act. Cooperating producers were to receive price supports of 75 to 90 percent of parity, depending on supply, for 1952 and succeeding years if producers approved marketing quotas. Price support for tobacco was to continue at 90 percent of parity. Price supports generally from 60 to 90 percent of parity also were made available for wool, mohair, tung nuts, honey, Irish potatoes, whole milk, and butterfat.

Price supports were in fact maintained for basic commodities at 90 percent of parity through 1950. Supports for nonbasic commodities were maintained at lower levels where possible under the 1949 act. In 1950, price supports were discontinued for hogs, chickens, turkeys, long-staple cotton, dry edible peas, and sweet potatoes (Rasmussen and Baker, 1966, p. 75). Prices of chickens, turkeys, and sweet potatoes fell, but not so far that severe economic hardship resulted for producers. Any unfavorable long-term economic impacts of removing price supports were obscured by the Korean War.

THE BRANNAN PLAN

Secretary of Agriculture Charles Brannan in 1949 proposed a considerable departure from past programs. A key feature of the Brannan Plan was a shift to an income standard rather than a price standard as a measure of a "fair" return to farmers. The moving base period for income would initially be 1939 - 48, with the earliest year to be dropped and a new one added each year. The price support standard would be the level of prices for individual commodities necessary to raise farm income to that in the base period. The same relationship would be maintained among prices as in the base period. A second feature was that prices of perishable commodities would not be supported in the market. All commodities would be sold at prices that would clear the market, and the difference between the market price and the support price would be made up by compensatory (direct grant) payments to farmers. A third important feature was that supports would be limited to 1,800 units; a unit was defined as 10 bushels of corn, 8 bushels of wheat, or 50 pounds of cotton. The unit limit would exclude an estimated 2 percent of the farms from full coverage. The program was strongly opposed by all major farm organizations except the National Farmers Union on grounds that it discarded the "comfortable" price parity tradition, that it would entail great government expense, that it would obtain farm income through the Treasury rather than through the market, that it would be too difficult to administer, and that if Treasury costs were to be held down then production and marketing would have to be stringently curtailed to raise market prices to proposed levels. The Brannan Plan received little support from the Congress and was not enacted.

Secretary Brannan maintained price supports on basic commodities at 90 percent of parity to 1952. The Korean War strengthened demand for farm products and continued efforts to maintain fixed supports at high levels while postponing flexible supports. Legislation in 1952 stipulate that price-support loans for basic crops be at 90 percent of parity or higher through April 1953 unless producers disapproved marketing quotas. Mandatory supports at 90 percent of parity were extended to 1953 and 1954 crops by legislation approved in July of 1952.

Toward Lower Prices and Higher Surpluses

Secretary of Agriculture Ezra Taft Benson continued high price supports and marketing quotas to 1954, although quotas were not imposed on corn. Surpluses began to mount after the end of the Korean War in 1953. The Agricultural Trade Development and Assistance Act, Public Law 480, was approved in 1954. This Food for Peace Act provided substantial financial assistance for farm exports and was of major importance in providing for disposal of farm surpluses in foreign countries.

The Agricultural Act of 1954 established flexible supports for basic commodities. Supports were to range from 82.5 percent of parity to 90 percent in 1955, and from 75 to 90 percent thereafter, except for tobacco, which was to be supported at 90 percent of parity. Flexible price supports finally arrived in 1955 after repeated postponements since World War II.

The Agricultural Act of 1956 established the *Soil Bank,* the first large-scale effort since the 1930s to bring production in line with the utilization. Under one provision, the *Acreage Reserve*, farmers were paid to reduce below allotment levels the plantings of allotment crops — wheat, cotton, corn, tobacco, peanuts, and rice. Under a second feature of the 1956 act, the *Conservation Reserve*, farmers were paid to divert all or part of their cropland to soil-conserving uses under long-term contracts. The Conservation Reserve was a general cropland retirement program and was not directed at specific crops.

The Agricultural Act of 1958 made price supports for most feed grains mandatory. Changes were made in the cotton and corn programs. Farmers producing corn were given the option in a referendum of either (1) terminating allotments for the 1959 and subsequent crops and receiving supports at 90 percent of the average price of the preceding three years, or (2) keeping acreage allotments and receiving supports at 75 to 90 percent of parity. They chose the former alternative in a 1958 referendum.

Two basic approaches to commercial farm policy are workable: one is to adequately control production and support prices, the other is to leave production uncontrolled and let prices fall to levels that will clear the market. A Democratic Congress and Republican Secretary of Agriculture Benson concocted an unworkable combination: Price supports with ineffective controls at an inopportune time when the technological revolution had struck agriculture full force. The result was unconscionable levels of stocks. The carryover of wheat in July 1960 was 1.4 billion bushels and of corn in October 1961 was 2.0 billion bushels. A major shift in policies was clearly needed.

COMMODITY PROGRAMS IN THE 1960S

Soon after John Kennedy was inaugurated President in 1961, Secretary of Agriculture Orville Freeman was directed to expand the distribution of food to needy persons. Other measures taken to increase utilization of food included initiation of a food stamp plan and expansion of the school-lunch and Food for Peace programs.

The emergency Feed Grain Act was approved in March 1961. It was designed to divert corn and sorghum acreage to soil-conserving uses. Producers were eligible for price supports at 74 percent of parity if in 1961 they diverted to soil-conserving uses 20 percent of the average acreage they had devoted to corn and sorghum in 1959 and 1960. The national average support rate for corn was $1.20 per bushel and for grain sorghum $1.93 per hundredweight. Payments for reducing the minimum acreage were equal to 50 percent of the support price times the normal yield of the farm. Additional reductions of 20 to 40 percent of the base were paid at 60 percent of the county support rate. The program was voluntary, but the payments were so generous that farmers were strongly induced to enter the program. The government released Commodity Credit Corporation stocks to hold prices below loan levels and discourage farmers form staying out of the program. The program was expensive to the government, but it did restrain production and reduce the burdensome carryover. It was also popular with farmers. For these reasons it remained the basic form of feed grain program for the remainder of the 1960s.

THE FIGHT FOR MANDATORY CONTROLS

Willard Cochrane of the University of Minnesota was chosen to be Secretary of Agriculture Orville Freeman's chief economic adviser. Cochrane had argued forcefully and persuasively that burgeoning agriculture production induced by the technological revolution would bring huge commodity surpluses and chronically low farm prices and would ultimately destroy the family farm (cf. Cochrane, 1958). The only feasible solution according to Cochrane was mandatory supply management.

President Kennedy, Secretary Freeman, and Congressional leaders of the agricultural committees agreed with this approach and proceeded to sell it to Congress. The administration sent to Congress in late 1961 an omnibus farm program called the Cochrane Bill. One provision of the Cochrane Bill was that Congress would establish the broad guidelines for programs, but the decisions regarding allotment levels and price supports would be made by the Secretary of Agriculture. Congress could, however, veto a commodity program 60 days after it was submitted, but the delay likely would mean no program for that commodity for the coming year. The Cochrane Bill would also have strengthened the Department of Agriculture-affiliated Agriculture Stabilization and Conservation Service committee system which administered programs at the local level. Under the new plan the Service was to be given a strong voice in formation of policies.

The procedural aspects of the Cochrane Bill were designed to pave the way for mandatory supply-management programs. Although this satisfied many liberal spokesmen who felt, like Cochrane, that it was the only feasible way to get income protection to farmers, the 1961 procedural bill did not pass Congress. And the supply-management programs themselves had not yet been submitted (Hadwiger and Talbot, 1965, pp. 46-71).

The original version of the Food and Agriculture Act of 1962 was introduced into Congress in early 1962. The bill was designed to establish a comprehensive but flexible system of supply control for the major farm commodities, including feed grains and dairy products. The Secretary of Agriculture would establish allotment and quota levels. Producers of a particular commodity would vote in national referendum whether or not to approve the quotas and price supports. If not more than one-third disapproved, the quotas would become mandatory, and producers who violated the provisions would be penalized. The administration saw the Cochrane mandatory program as the only alternative to maintain farm income, stop the growth of stocks, and hold down government cost. President Kennedy, who was strongly urban oriented, felt pressure to cut the farm budget and release funds for other uses.

The bill was under the very able leadership of Congressmen Harold Cooley, chairman of the House Agricultural Committee, and Senator Allen Ellender, chairman of the Senate Agricultural Committee. It was supported by strong pressures from President Kennedy and Secretary Freeman, the National Farmers Union, and other farm organizations with the notable exception of the Farm Bureau. The bill was passed in the Senate 42 to 38 but was defeated in the House 205 to 215. A shift of six votes would have brought an extensive program of supply management. The vote was principally along party lines, although a sufficient number of Democrats defected to defeat the bill. The Farm Bureau and cattlemen's organizations effectively lobbied to defeat the bill. They were motivated not only by their ideological position but also by public opinion polls and wheat referendum votes which showed that farmers were increasingly unwilling to accept mandatory programs.

Following the Congressional defeat, the administration was not yet ready to settle for expensive voluntary programs and pressed for amendments making the mandatory program acceptable. The interesting result was that the Republicans pressed for continuation of the costly 1961 feed-grain program which they had opposed in 1961. The administration opposed its continuation though it had pressed for its enactment in 1961. The 1962 act, when finally enacted in amended form by Congress, contained an extension of the 1961 feed-grain program but made provision to submit to wheat-growers in 1963 the opportunity to accept in referendum essentially the mandatory wheat program that was included in the original Food and Agriculture Act of 1962 earlier defeated in Congress.

THE 1963 WHEAT REFERENDUM

Wheat-growers prior to 1963 had approved quotas annually in a national referendum. But surpluses accumulated because the minimum national acreage allotment, established by law, allowed production of more wheat than could be utilized. The 1962 act continued the mandatory 55 million acre allotment and referendum features. In addition, growers were paid to divert some of their allotment to soil-conserving uses. The law stipulated that in 1964 the 55 million acre minimum national allotment would be abolished, and the Secretary of Agriculture could set allotments as low as necessary to keep production in line with utilization. Farmers were to choose in referendum between two alternatives. One provided for the payment of penalties by farmers who overplanted acreage allotments and provided for issuance of marketing certificates based on the quantity of wheat to be used for domestic human consumption and a portion to be used for exports. The amount of wheat on which farmers received certificates was based on the quantity of wheat to be used for domestic human consumption and a portion to be used for exports. Certificate wheat would be supported between 65 and 90 percent of parity; the remaining portion would be priced at its value for feed. The 15-acre exemption from wheat controls that had been in effect and that benefited producers with small acreages was to be eliminated. The second alternative provided that the wheat of growers complying with allotments would be supported at 50 percent of parity, about $1.25 per bushel. There would be no penalty for overplanting, but noncomplying growers would not be eligible for price supports. The first alternative was defeated in a national referendum held in May 1963.

The administration and the Farm Bureau, which had highly conflicting views on farm policy, chose the 1963 wheat referendum as the battleground for deciding which direction the farmer would take. The Farm Bureau mounted an extensive campaign to defeat mandatory controls. The economic implications of the two alternatives proposed were clear: wheat would be priced at approximately $2.00 per bushel with the supply control "yes" alternative; at $1.10 to $1.25 per bushel with the "no" alternative. The second alternative would mean economic disaster for farmers who had to pay for land they had bought with the expectation of higher wheat prices. But economic issues were not all-important. The Farm Bureau said the issue was "freedom to farm" and "who would control agriculture, the bureaucrat or the farmers." Max Cooper interpreted the Farm Bureau information to mean "that a 'yes' vote [for mandatory controls] is a vote for slavery, bureaucrats, and big government. . . . If He [the farmer] votes 'yes' he'll start the whole country down the road to slavery. . . . One little 'yes' vote too many could throw the constitution out the window, overthrow the Supreme Court and make Congress useless" (Hadwiger and Talbot, 1965, pp. 300, 301). This is an exaggerated but colorful example of the tenor of the campaign.

The advocates of the mandatory program — the Democratic administration, the Agricultural Stabilization and Conservation Service, the National Farmers Union, and other groups — also mounted a strong, if less well-organized, campaign. Max Cooper summarized their views thus: "That a 'no' vote is a vote for depression, disaster, and big farmers. . . . If he [the farmer] votes 'no' he'll put us on a toboggan slide to a horrible depression. . . . And if we ever get into another depression a dictator will take over, they warn" (Hadwiger and Talbot, 1965, pp. 300, 301). Over a million farmers and their wives voted, five times the number who had participated in the referendum of the previous year. Only 48 percent voted for the mandatory program. It was a stunning defeat for the administration. The Farm Bureau had successfully captured the resentment that farmers had built up in past years toward the faulty administration of wheat programs.

The "no" alternative would indeed mean economic disaster to a great many farmers. The Farm Bureau had argued throughout the campaign that farmers would get a better program than the "no" alternative if they turned down the administration proposal. The campaigners for the administration program urged just as strongly that farmers would in fact be stuck with the "no" alternative if they voted for it.

The Farm Bureau did not have sufficient influence to get a better program after the referendum. Ironically, Democrats in Congress and the Administration realized that they could not live with the "no" alternative. They put forth a new program in 1964, which I will describe later, that was remarkably similar to the one wheat farmers had rejected. The new program was voluntary, however. The Farm Bureau, which had been so effective in defeating the program in referendum, could not stop the new program in Congress despite vigorous efforts.

Farmers had made a momentous decision, reversing the trend toward mandatory programs. Advocates of mandatory programs had argued that the public would not stand the high cost of voluntary programs needed to maintain farm income at satisfactory levels. They were wrong — the taxpayer was more tolerant of high Treasury costs than was the farmer of tight controls. And farmers continued to demonstrate their power to obtain favorable legislation even at high cost to taxpayers.

1964 Legislation

The Wheat-Cotton Act of 1964 established a certificate program for wheat. It, like the alternative rejected by farmers, was a two-price plan. But unlike the rejected alternative, it was voluntary. Complying farmers received $2.00 per bushel for 45 percent of their normal production, 70¢ of which was from purchase of certificates by processors. Another 45 percent was supported at a price of $1.55 per bushel. The remaining portion was supported at $1.30 per bushel. In subsequent years the domestic (40 to 45 percent) portion was supported at parity and the remainder at $1.25 per bushel. Processors paid 75¢ per bushel on the domestic or certificate portion.

Complying farmers were required to divert land from wheat to soil-conserving uses. In addition, land-diversion payments were used to reduce production below allotment levels. The program effectively reduced production and carryover. The 1964-type wheat program with modifications was used during the remainder of the 1960s.

Under the 1964 act, the Secretary of Agriculture was authorized to make subsidy payments to domestic handlers or textile mills to reduce the effective price of cotton consumed domestically to the export price level. Each farmer complying with his regular allotment was to have his cotton crop supported at 30¢ per pound. Farmers reducing acreage to a smaller domestic allotment were to receive a support price of 33.5¢ per pound (Rasmussen and Baker, 1966, p. 77). A payment-in-kind or PIK program was initiated for cotton whereby farmers were paid by the government with cotton from Commodity Credit Corporation stocks rather than in cash for participation in the cotton program. A similar program had been used on a small scale for grains in the early 1960s.

Food and Agriculture Act of 1965

The Food and Agriculture Act approved in November 1965 extended to 1969 the wheat and feed grain programs. The 1965 act was extended to 1970 by legislation enacted in 1968.

Cotton surpluses had continued to mount under the 1964 act, which called for a new policy. The market price of cotton under the 1965 act was to be supported at no more than 90 percent of the world price, thereby eliminating the necessity to subsidize cotton used in domestic mills and export. Participation in the program was voluntary although the monetary incentives of the program made participation overwhelmingly attractive. A minimum acreage diversion of 12.5 percent of effective allotment was necessary for a farmer to be eligible for supports. Provisions were also made for participation at higher levels. The program was effective in reducing cotton production and carryover.

The 1965 act also established a long-term, general land-retirement program called the Cropland Adjustment Program. The Secretary was authorized to enter into five- to ten-year contracts to retire cropland to conservation uses. Payments were to be not more than 40 percent of the value of probable crop production on the land, and new agreements were not to obligate more than $225 million per year.

COMMODITY PROGRAMS IN THE 1970S

The pattern of commodity programs established in the 1960s prevailed in the 1970s, although unique adaptations developed as well. For example, acreage "set aside" united the concepts of short-term acreage diversion and general land retirement to provide a greater market orientation.

The Agricultural Act of 1970

The Agricultural Act of 1970 was an act that no group advocated, no group supported, and no group greeted with enthusiasm. The act was opposed by all major farm organizations although for widely different reasons. Neither Congress nor Secretary of Agriculture Clifford Hardin had sufficient strength to enact a partisan bill. But Hardin and Congress wanted to help farmers and the result was a compromise that ultimately became widely accepted. The new three-year program eschewed allotments, acreage restriction, and marketing quotas for specific crops of wheat, upland cotton, and feed grains and substituted instead a short-term partial land-retirement program called *set aside*. To qualify for price support, the farmer was required to set aside a specific percentage of his cropland to soil-conserving practices. He could grow whatever he wished on remaining land, except for selected crops that remained under control. Payment limitations were established at $55,000 per crop (excluding commodity loans) for producers of upland cotton, wheat, and feed grains. This was the first real attempt to limit payments but the limit was too high to have much impact. A farm producing three covered crops could receive $165,000 of payments.

Wheat loans were made available to participants at not less than $1.25 per bushel for 1971 through 1973 and domestic marketing certificates covering a total of not less than 535 million bushels were established. The value of certificates paid to farmers was the difference between the wheat parity price and the average price received by farmers during the first five months of the marketing year. This feature was the forerunner of the target price used in later legislation.

To qualify for price supports, cotton planters were required to set aside 25 percent of the cotton allotment to soil-conservation uses. Loans were set at 90 percent of the average world price for the two previous years and deficiency payments were set equal to the difference between 65 percent of parity or 35¢ per pound, whichever was higher, and the average market price for the first five months of the marketing year. Deficiency payments were not to be less than 15¢ per pound, however. Because the problem in cotton was shifting from too much to too little production, the justification for restricting production by diverting acres became unclear.

Producers of wheat, feed grains, and upland cotton who failed to plant that commodity or an eligible substitute crop would incur a reduction of 20 percent in allotment the following year. If the allotment or substitute crop was not planted for three years, the entire allotment would be lost. This feature was criticized for encouraging farmers to plant their allotment when production needed to be reduced.

Price supports on corn were to be the higher of $1.35 per bushel or 70 percent of parity on October 1 and the loan not less than $1.00 nor more than 90 percent of parity as determined by the Secretary. A producer would receive a payment equal to the difference between the support price and market price in the first five months of the marketing year on half of his basic production (Rasmussen et al., 1976, p. 18).

Authorizations were continued for long-term land retirement at a pilot level of $10 million per year for each program.

THE AGRICULTURE AND CONSUMER PROTECTION ACT OF 1973

The Agriculture and Consumer Protection Act of 1973 was legislated in an entirely different atmosphere than farm programs since the 1930s — an atmosphere of excess demand. This emboldened Secretary of Agriculture Earl Butz to proclaim that the legislation represented "a historic turning point in the philosophy of farm programs in the United States." The statement may have reflected his personal philosophy, but the legislation and its implementation were the product of economic circumstances. The legislation continued past policies but the title served notice of an emerging element in farm commodity legislation — the consumer.

The 1973 legislation continued the set aside short-term partial land-retirement concept and introduced *target prices* which were to be used to set direct deficiency payments when market prices fell below the target price. *Deficiency payments* were direct transfers to farmers paid on the difference between target price and loan rate (or market price if higher) multiplied by normal production (permitted harvested acres times program yield). Total payments to any person under the wheat, feed grain, and upland cotton programs were limited to $20,000. This did not include loans or purchases.

In another departure from previous programs, the parity formula was dropped in setting future target prices. The 1974 and 1975 target prices were set at 38¢ per pound for upland cotton, $2.05 per bushel for wheat, and $1.38 per bushel for corn. Target prices remained at that level in 1974 and 1975 but 1976 and 1977 target prices were adjusted upward by the index of prices paid by farmers for items used in production and deflated downward by a higher most-recent national three-year average yield for each crop. The Secretary of Agriculture had considerable discretion in setting loan rates and raised loan rates for wheat from $1.37 per bushel (the minimum allowable) to $2.25 per bushel and for corn from $1.10 per bushel (the minimum allowable) to $1.50 per bushel in 1976.

Another feature of the 1973 act was disaster payments authorized for eligible producers prevented from planting any portion of allotment because of drought, flood, or natural disaster and other conditions beyond their control. Payments were made available when natural disaster prevented a farmer from harvesting two-thirds of his normal production of the crop allotment acreage. This created a "notch" problem: A farmer harvesting slightly over two-thirds of his normal production received no payments.

The support price for milk was to be at a level between 75 and 90 percent of parity but as determined by the Secretary of Agriculture "to insure an adequate supply of milk." The actual support rate was held at approximately 80 percent of parity from the beginnings of the 1973 act to 1977. President Carter raised the support rate after taking office in 1977 to fulfill a campaign pledge. Because the dairy program contained no production controls, high price supports generated costly surpluses sometimes dumped on foreign markets in violation of the General Agreement on Tariffs and Trade.

For cotton, the payment limitation for large producers was accompanied by provisions that cotton producers whose basis was 10 acres or less received deficiency payments at a rate 30 percent greater than other producers. This along with payment limitations was an attempt to benefit small producers relative to large producers.

Because of large foreign demand from 1973 to 1977 the act remained largely a backstop rather than an operating program. Except for long-term land contracts which had little effect on reducing output and except for restrictions on planting of winter wheat in 1973, production controls were not imposed under the 1973 act which expired in 1977. Additional details on the 1973 act and other postwar commodity legislation are given by Cochrane and Ryan (1976).

FOOD AND AGRICULTURE ACT OF 1977

The 1977 act was four-year legislation retaining the basic structure of the 1973 act but introducing several modifications. These included increases in loan and target prices, a departure from the outmoded allotment system in establishing set aside and disaster payments, adjustment of target prices over time according to changes in the variable cost of production, adjustment of loan rates for excess supplies, and a farmer-owned reserve of stocks.

For 1979 and subsequent years, all target prices escalated according to changes in variable costs of production. Deficiency payments were awarded based on the difference between the target and average market price in the first five months of the marketing year or loan rate, whichever was higher.

Statutory minimum loan levels remained constant for wheat and feed grains through the life of the act. However, a provision allowed the Secretary to adjust the loans for wheat and feed grains downward a maximum of 10 percent for the next crop year whenever the yearly national price of the commodity averaged below 105 percent of current loan rate. But the loan rate could not be lowered to less than $2.00 per bushel for wheat or $1.75 per bushel for corn. Loan levels for cotton and peanuts varied according to world and domestic market conditions but within statutory limits. Marketing quotas on cotton were eliminated so that the cotton program was similar to feed grain and wheat programs.

Total deficiency-payment limitations to wheat, feed grain, and upland cotton producers, $20,000 under the 1973 act, were raised to $40,000 in 1978 and $45,000 in 1979 under the 1977 act. In 1980 and 1981, a limitation of $50,000 applied to the total payments received on wheat, feed grains, upland cotton, and rice. In another change from the 1973 legislation, the 1977 act provided that, beginning in 1978, payments for disaster, certain resource adjustments, and public access for recreation were exempt from limitations.

The 1977 act stipulated that from enactment until March 31, 1979, the milk price support will be set between 80 and 90 percent of parity compared to 75 to 90 percent under the current, "permanent" law. After March 31, 1979, the support level reverted to permanent law — 75 to 90 percent of parity. In the absence of production controls milk prices as low as 75 percent of parity generated surpluses and high government costs for their removal from markets.

The 1977 act provided a new, *two-price peanut program* beginning with the 1978 crop. A minimum national allotment of 1,614,000 acres was assigned to allotment holders for total peanut production of quota and "additional" (nonquota) peanuts. For quota peanuts, the minimum support rate was $420 per ton for each year 1978-81, but the Secretary had discretion to increase the minimum on the basis of the index of prices paid by farmers.

For "additional" peanut production, the Secretary has discretionary authority to set lower support rates, taking into consideration projected crushing and edible export prices and such other guidelines as written into the legislation. Under the former program, production in excess of domestic food use had to be supported at high rates. The excess had to be disposed of in markets bringing prices considerably below the support rate, and the government cost of the program was large.

The legislation added a new provision mandating (previously, it had been discretionary) the Secretary to establish a soybean loan and purchase program for producers, but it provides flexibility for the Secretary to set the level of support in relation to competing commodities and foreign and domestic supply and demand factors. It did not authorize the Secretary to make deficiency payments; no target price nor set aside was authorized for soybeans.

The act stipulated that if the President of the United States or any other member of the executive branch of the federal government suspended commercial export sales of any major agricultural commodity, the Secretary of Agriculture must, on the day of the suspension, set the loan level for the commodity price at 90 percent of parity. However, this provision applied only in the case of embargoes because of "short domestic supplies," hence was easily circumvented.

The 1977 act established the *Farmer Owned Reserve* (FOR). Conventional storage loans on commodities under the 1973 and 1977 acts were for 11 months. The Secretary was authorized by the 1977 legislation to offer wheat and feed grain producers an extended loan program of three to five years' duration. The three-year feature was widely used. The quantity of wheat held under such a farmer-held reserve was not to be less than 300 million nor more than 700 million bushels. As an incentive for producers to participate in the program, the Secretary was authorized to pay storage costs (typically a payment of 26.5¢ per bushel per year) of the grain as well as to waive or adjust interest rates.

Whenever the national average price reached between 140 and 160 percent of the loan, the Secretary could discontinue the storage payments and charge interest to encourage the producer to redeem the loan and market the grain. If the producer redeemed his loan before the average price reached 140 percent, the producer would be subject to a penalty as determined by the Secretary.

Whenever the national average market price reached 175 percent of the loan, the Secretary could recall the loan. The Commodity Credit Corporation (CCC) resale price for government-owned stocks was set at 150 percent of the current loan whenever a similar kind of grain was held under the FOR. Otherwise the CCC resale price was 115 percent of the current loan.

OTHER 1970S LEGISLATION

In response to weak farm prices and incomes in 1977, Congress passed the Emergency Assistance Act in 1978. The act gave the Secretary of Agriculture discretionary authority to raise target prices. The wheat target price was increased from $3.05 to $3.40 per bushel. The act raised the loan rate on cotton, provided a $4 billion emergency loan program, and declared a moratorium on Farmers Home Administration foreclosures.

COMMODITY PROGRAMS IN THE 1980S

In January 1980 President Carter began a partial suspension of export sales to the Soviet Union. To protect farmers, nonrecourse loan rates were raised on wheat, corn, and soybeans; release and call rates were raised for the FOR which was now opened to all producers whether or not they participated in programs. More credit was made available to farmers through the Farmers Home Administration. Expansion of credit viewed as part of the solution to farm problems in the late 1970s and early 1980s soon was viewed as part of the problem. Excessive credit not only added to surplus resources and production, it also added to farm debt servicing problems.

The Federal Crop Insurance Act of 1980 expanded the federal crop insurance program, including subsidies to make it attractive to farmers. Congress at the same time terminated the very costly disaster payments which had been encouraging cropping in high-risk areas of the nation. Disaster aid was revived in drought year 1988, however.

AGRICULTURAL AND FOOD ACT OF 1981

The new Secretary of Agriculture John Block first proposed eliminating target prices and making other changes to reduce government costs but these were rejected by Congress. Principal new directions for the 1981 comprehensive farm bill included specific target prices for the four-year length of the bill, elimination of the rice allotment and marketing quotas to shift rice to the voluntary program used for other grains, and lower dairy supports. Although provision for set-aside was retained, the 1981 bill and subsequent legislation in the 1980s returned to restricting acreage of specific crops. Allotment base history was lost if soybeans were planted on base acres. In the later 1980s this did serious damage to soybean export markets.

The 1981 act was similar to the 1977 act but was designed by its framers to cost less because target prices were preset by Congress with only modest increases during the four-year life of the bill. Inflation turned out to be less than expected, however, and given the large worldwide supply relative to demand for farm output, the 1981 act set minimum loan rates at levels making the U.S. non-competitive in world markets.

The FOR was continued for grains but with the maximum level raised to 1 billion bushels for corn and a simple trigger price replaced separate call and release prices. Whereas the 1977 legislation provided dairy price supports at least 80 percent of parity, the

1981 legislation reduced supports to 70 percent of parity. Peanut and rice allotments were eliminated.

Anyone who desired could grow peanuts but would receive only the lower support price on "additional" peanuts. The mandatory marketing quota tobacco program which had been in place for some years was redesigned to ensure "zero net cost" defined as no cost to government in excess of administrative expenses.

To give greater protection to farmers under future embargoes, the 1981 act called for loan rates to be raised to 90 percent of parity if embargoes were extended for foreign policy or national security purposes. Total direct payments per person could not exceed $50,000 excepting disaster payments (available only in countries without all-risk federal crop insurance) which could not exceed $100,000 per person. Finally, the 1981 act expanded Public Law 480 and other programs to increase exports.

With worldwide recession, falling exports, growing surpluses, and real net farm income at the lowest level in 1982 since the Great Depression, the farm surplus mentality of the 1930s, 1950s, and 1960s returned. Secretary Block announced modest acreage reduction programs for 1982 and Congress passed the Omnibus Budget Reconciliation Act requiring a large land program for 1983.

The administration sought greater production cuts through a payment-in-kind (PIK) program. Congress balked so the administration proceeded anyway, reasoning that rather than pay huge storage costs it made sense to use burgeoning government stocks to pay farmers not to produce. Payments in kind were made equal to 80 percent of normal production per acre for corn, grain sorghum, cotton, and rice, and equal to 95 percent for wheat. The program differed from PIK programs of the 1960s by including more crops, making actual delivery to farmers from government stocks, and by making more generous payments. The result was the largest diversion program ever, 82 million acres, in 1983. Coupled with severe drouth, PIK brought excessive drawdown of stocks and a rise in commodity prices which exacerbated the already severe problems of overpricing and lack of world competitiveness caused by excessive price supports.

Tobacco and Dairy Programs

In response to groups attempting to terminate the tobacco program (the federal government allegedly was subsidizing a dangerous habit), the 1981 farm bill provided regulations to ensure that net program costs (except administrative expense) would be zero. The No Net Cost Tobacco Program Act of 1982 required tobacco farmers to contribute to a fund reimbursing the federal government for losses in tobacco price support operations. The Act retained the allotment and quota system but made it more likely that allotments would be owned by farmers. Corporate and institutional allotment owners had to sell their interest by the end of 1983 unless they were actively involved in farming.

The Dairy and Tobacco Adjustment Act of 1983 established a voluntary diversion program for dairy similar to that for crops. Farmers could cut their production 5 to 30 percent for payments of $10 per hundredweight of milk. Price supports were reduced and a "corresponsibility levy" of 50¢ per hundredweight was levied on milk to pay for the diversion program.

Dairy surpluses averaging nearly 10 percent of output continued and in 1986 the government instituted a massive *dairy cow buyout program.* The program was effective in reducing production but at a cost to the beef industry troubled by the glut of dairy cows being slaughtered. By 1987, dairy surpluses were beginning to mount again.

FOOD SECURITY ACT OF 1985

The Agricultural Programs Adjustment Act of 1984 made minor adjustment to stem the rising surplus in grains by freezing target prices at 1984 levels rather than continue the automatic upward adjustment provided under the 1981 act. The 1984 act also required acreage diversion in 1985.

The Food Security Act of 1985 provided a five-year comprehensive framework for the Secretary of Agriculture to administer various agricultural and food programs described in the act's 18 titles (Glaser, 1986). The 1985 legislation continued target prices from 1981 legislation (as amended by 1984 legislation) for two years and then reduced target prices by a total of 10 percent over the final three years. Subsequent legislation modified target prices, however.

The major new direction in the 1985 legislation was to sharply reduce loan rates and market prices. The basis loan rates were in line with those under the 1981 act for 1986 but for subsequent years were to be 75-85 percent of the simple average of the seasonal average prices received by producers during the preceding five marketing years, dropping the years with the high and low prices. The rate could not be lowered by more than 5 percent from the basic rate of the previous year. Under the *Findley Amendment* provision, however, the Secretary could reduce loan levels up to 20 percent if the average market price was 110 percent or less of the announced loan rate during the previous year or if the reduction was deemed necessary to retain markets. The Secretary elected to make use of the Findley Amendment provision to lower loan rates.

The Secretary of Agriculture had two tools available to drive down market prices to be competitive in world markets and to encourage participation in commodity programs. One tool, used extensively for wheat and feed grains, was generic, negotiable PIK. The generic PIK certificates were given to producers in lieu of cash. The recipient could use the certificates to redeem his own grain under CCC loan or could sell the certificates to the CCC at face value or could sell them to anyone else usually at a premium above face value. The buyer, say an exporter, could then use the certificates to obtain government grain stocks at a preferred location for export.

A second instrument, the *market loan,* was potentially available on all major crops but was used initially and with considerable success for rice and cotton. The market loan allowed the farmer to keep the commodity he had placed under nonrecourse loan and repay the government at the current market price (but not less than 70 percent of the base loan) rather than at the loan rate. This eliminated incentives for farmers to place stocks in government hands. The nonrecourse loan rate no longer placed a floor under market prices. Like PIK, the market loan reduced stocks and kept market prices competitive.

Extensive use of generic PIK and market loans not only kept U.S. market prices competitive, they also raised government costs of the program. Deficiency payments were enlarged because the difference between the market price (or loan rate in the case of wheat and feed grains) and the target price was large. The combination of low market prices and high target prices made participation so attractive some described the program as "mandatory." Participation rates were very high. The program designed to cost $35 billion the first three years actually cost approximately $50 billion its first two years.

Direct payments became a large part of net farm income with market price depressed by PIK and market loans. With market prices arbitrarily depressed, large farmers needed direct payments to survive. The payment limitation of $50,000 applied to deficiency payments; limitations per recipient allowed an additional $200,000 in payments under market loan and "Findley Amendment" (loans set by the Secretary of Agriculture at 20 percent below legislated levels).

Previous commodity programs typically set acreage bases (allotments) equal to crop acreage planted or considered planted in the previous two years. Higher program yields used to set deficiency payments could be "proven" from elevator receipts showing production and yields of the past three years. Hence program acreages and yields were expanded rather readily in response to higher loan and target prices. The ease of expanding bases and program yields in response to incentives made the government target price the supply price and caused overproduction and excess capacity. To reduce the tendency for farmers to expand resource use to produce for the program rather than for the market and to make the market price rather than the target price the supply price, the 1985 act established crop bases equal to the average of planted and considered planting of the crop for harvest on the farm during the five preceding crop years. The farm program yield for 1986-87 was the average program yield over the five years 1983-85 excluding the high and low year. The Secretary could continue the program yield as established above for subsequent years or could allow computing a new program yield based on the most recent past five years eliminating the high and low year. Actual rather than program yields for 1987 and later years would be using in calculating the moving average. The Secretary elected to freeze yields.

The 1985 act also attempted to reduce incentives to produce for programs with the so called *50/92 rule*. This rule provided deficiency payments on 92 percent of permitted acreage if at least 50 percent of the permitted acreage was planted to the program crop and remaining permitted acreage was in soil conserving uses. The provision was changed to a *0/92 rule* for 1988.

Another important feature was the Conservation Reserve Program provision which called for placing up to 45 million erosion prone cropland acres in soil conserving uses by 1990 under 10-year bid contracts. A producer who violated the *swampbuster* provision by cropping new wetlands or the *sodbuster* provision by cropping grazingland would lose eligibility for commodity program benefits. In addition, the *conservation compliance* feature would declare a producer ineligible for commodity program benefits if he failed to have an approved conservation plan for highly erodible cropland by 1990 and the plan implemented by 1995. The provisions making it difficult to expand bases, freezing of yields, and the 0/92 rule helped to *decouple* program benefits from incentives to expand production. Still, farmers were careful to maintain bases by not shifting such land to grass, trees, soybeans, or other higher priority uses. They reasoned that acreage and yields of the 1980s would be used to set program bases in the 1990s.

The 1985 act contained numerous provisions to expand demand. These included export credit, guarantee, and subsidy programs; expansion in Public Law 480 and exemption of non-PL 480 from U.S. vessel requirements but raising from 50 to 75 percent the proportion of PL 480 shipped in U.S. bottoms. The act established mandatory promotion programs for beef and pork. The Export Enhancement Program (export PIK) was aggressively used to expand exports of grains, mainly wheat. The Targeted Assistance Program subsidized exports of high-value commodities such as livestock, horticultural commodities, and processed foods.

The 1985 farm bill was the most comprehensive, expensive, and flexible farm bill ever enacted. The Secretary of Agriculture was given discretion to use virtually any type of diversion program whether unpaid voluntary (deficiency payments and loan benefits entice participation without specific diversion payments), paid voluntary (by PIK or cash), mandatory controls (if approved by voters in referendum), or long-term land retirement taking out whole or part farms. The Secretary could use direct payments, market or nonrecourse loans, FOR, or supply management to raise farm income. The most severe restraint on the Secretary's discretionary power was legislated target prices. The prices created a dilemma for the Secretary: Low loan rates raised deficiency payments and Treasury costs; high loan rates reduced deficiency payments but required large acreage diversion and priced the U.S. out of international markets. Either approach retained a large government presence in farming.

Lower loan rates and market prices forced down by PIK and market loans restrained resource use and spurred exports. By 1989, resource adjustments and drought sharply reduced reserve production capacity and stocks. Also in 1989, greater flexibility was introduced by allowing a nonallotment crop such as soybeans on allotment crop acreage without loss of base.

SUMMARY

Farm legislation has chronically been at a crossroads. In the 1920s the issue was direct intervention in the market versus improved credit, storage, and marketing efficiency to raise farm income. In the 1930s the issue was not whether, but to what degree, the government should be involved in supporting farm income. In the 1940s the issue was flexible versus fixed levels of price support. In the 1950s the issue was continued controls and supports or transition to a market orientation. In 1962 and 1963 one branch of the crossroads led to voluntary controls, the other to mandatory controls. In the 1970s a central issue was whether to phase out of commodity programs (Butz) or to focus them on farm structure (Bob Bergland's concern for family farms). In the 1980s, the issue was whether to remain competitive in export markets and rely on direct payments. The result of the directions taken in farm legislation has been a growing commitment of the federal government to support farm income. The direction has been to voluntary programs and direct payments; hence the Treasury costs are high. The Employment Act of 1946 committed the federal government to policies that would secure full national employment. The government since the 1930s has remained committed in a less formal though effective way to support farm income.

Some lessons are apparent from history. In the absence of effective controls, prices supported above free market equilibrium lead to a crisis of abundant stocks. Second, history has demonstrated that production can be controlled with voluntary programs, although at high Treasury cost. The problem becomes even more difficult as land, the focus of controls, is increasingly replaced in the production process by fertilizer and other nonfarm-produced inputs.

Third, the economic problems of agriculture are not solved by controlling the acreage or marketing of one crop. Cross-compliance is necessary to keep farmers from planting grain sorghum, for example, on land removed from corn production by allotments, thus depressing the price of crops not supported by the government. Farm production capacity behaves like a balloon filled with water. Compressing one part causes the volume elsewhere to expand. The volume can be reduced only by letting water out of the balloon. Controlling only one crop without cross compliance is like pushing in one part of the balloon; removing resources from production is like letting water out. Only the latter is effective in controlling production and raising income in the farming industry. Individual allotments without diversion of land to soil-conserving uses do little to raise the total income of farmers. Programs to improve the efficiency of markets, expand demand, and reduce marketing margins have been unable to cope with the more basic problem of too many resources in agriculture, low demand and supply elasticities, and unstable yields at home and abroad. Excess resources in farming can be temporarily diverted from production by voluntary or mandatory supply controls but can be permanently diverted only by ending price incentives for them to remain in farming.

Another lesson to be learned from history is that it is very difficult to change the direction of policy. The Brannan Plan, the Cochrane Bill, the proposals of Secretary Ezra Taft Benson, and the initial proposals of John Block in 1981 and 1985 represented considerable departures from previous directions and were rejected. The farm commodity programs since 1933 have been remarkably similar in structure, though different in style and emphasis. Major changes come mostly in times of crisis. This is not meant to imply that the trend in farm policy should not be changed but that a strong program of economic education must precede changes.

Current farm legislation is a remarkable potpourri of the many programs tried or proposed since the 1920s. Export subsidies reflect the two-price scheme contained in the McNary-Haugen bills of the 1920s. The direct-payment features characterized by deficiency payments and target prices bear traces of the Brannan Plan of 1949.

Widespread economic deprivation among farm people motivated programs in the 1930s. Since the 1930s, farmers as a class have not been economically deprived. The implication is that farm price and income supports would narrow from supports for a large number of commodities produced by a broad spectrum of farmers to a few programs benefiting those commodity groups having greatest political clout on making the best case that their programs hold reserve capacity, preserve the family farm, and induce price stability of benefit to consumers. In general, this trend predicted by theory has held. Fewer commodities receive price and income supports than in earlier years. Equity and economic efficiency conflicts remain, however. Some contend that programs for commercial farmers should be designed to reduce price instability and provide reserves at minimum public costs and that a separate negative income tax or other income-conditioned welfare programs for low-income farmers should be part of a nationwide income-maintenance policy.

The dynamic nature of agriculture requires flexibility in programs to respond to emerging realities. Long-term legislation is possible only if it allows change. Agricultural commodity legislation has provided increasing discretion and flexibility to the Secretary of Agriculture in setting loan rates and in adjusting production through controls on acreage of specific crops, short-term diversion, or long-term land-retirement contracts. Legislation provides considerable opportunity to be competitive in world markets, to establish a food reserve, and to allow freedom to farmers in production and marketing decisions. In recent years, however, Congress has seen fit to intervene with adjustments in farm legislation nearly every year.

The major criticism is that the presence of government in farming looms large and Treasury expenditures represent a potentially high claim on limited government receipts that have alternative, high-priority uses.

Congressional votes of the 1980s dispel concern that the declining proportion of farmers in the electorate spells the end of legislation favorable to farmers. Support has been stronger in Congress than from the executive branch for farm commodity programs. The President mainly has acted to constrain Congress in recent years. To the extent that Congress represents special interests and the President the public interest, that trend will continue. The public has reason to ask what it is receiving for outlays of $25 billion or more per year. Times have changed since major programs were initiated in 1933. Most farm output is produced by farmers who are wealthy by average taxpayer standards and most farmers have operations too small to be helped by commodity programs. The programs have not preserved the family farm and efforts to target benefits by payment limitations have motivated great ingenuity by farmers in circumventing restrictions. Farmers have also been ingenious in avoiding supply controls. Acquisition of commodity stocks raises farm prices in the short run but government have rarely found a propitious time to reduce stocks. Hence grain stocks well in excess of economically optimal levels have hung over markets to depress farm prices year after year. Food supplies would not be threatened by absense of commodity programs but farm and food prices would be somewhat more variable. The total cost of food and farm programs would be substantially reduced. Some existing family farms would be unable to compete without government subsidies but family farms are resilient and efficient and large numbers will be around for generations to come whether government intervention is large or nonexistent in farming. A transition program of perhaps 10 years duration would be necessary to remove government farm subsidies because many farmers have learned to depend on them. Net farm income would average as high or higher without government programs after the transition.

Large farms receive large payments per farm but much lower payments per dollar of output than family farms. Only 15 percent of output on large farms is from enterprises covered by commodity programs compared with over half on mid-size farms. For these and other reasons it is not possible to state that commodity programs have unfairly benefited large farms (see Tweeten 1984, pp. 31-33; 1986).

Finally, commodity programs have consistently distorted market prices. The 1981 farm bill supported farm prices above normal market levels, creating excess stocks and losing world markets. The 1985 farm bill held prices below normal market levels, angering export competitors and creating indecent federal exposure to deficiency payments. Neither policy can be defended for the long term.

REFERENCES

Benedict, Murray. 1953. *Farm Policies of the United States, 1790-1950*. New York: Twentieth Century Fund.

Cochrane, Willard W. 1958. *Farm Prices: Myth and Reality*. Minneapolis: University of Minnesota Press.

Dubov, Irving and E.L. Rawls. 1974. *American Farm Price and Income Policies: Main Lines of Development, 1920-73*. Tennessee Agricultural Experiment Station Bulletin 939. Knoxville: University of Tennessee.

Cochrane, Willard and Mary Ryan. 1976. *American Farm Policy, 1948-1973*. Minneapolis: University of Minnesota Press.

Glaser, Lewrene. 1986. Provisions of the Food Security Act of 1985. Agricultural Information Bulletin No. 498. Washington, D.C., ERS, USDA.

Myrick, Dana. 1967. Background of the Federal Crop Insurance Program. In *Crop Insurance in the Great Plains*. Bozeman: Agricultural Experiment Station, University of Montana, pp. 6-27.

Rasmussen, Wayne and Gladys Baker. 1966. A short history of price support and adjustment legislation and programs for agriculture. *Agricultural Economics Research* 18:68-79.

Teigen, Lloyd. 1987. Agricultural parity: Historical review and alternative calculations. Agricultural Economic Report No. 571. Washington, D.C.: ERS, USDA.

Tweeten, Luther 1984. Causes and consequences of structural change in the farming industry. NPA Report No. 207. Washington, D.C.: National Planning Association.

Tweeten, Luther. December 1985. Are current U.S. farm commodity programs outdated? *Western Journal of Agricultural Economics* 10:259-69.

Tweeten, Luther. 1986. Do large farms unfairly benefit from government commodity programs? In Gary Comstock, ed., proceedings of conference *Is There a Moral Obligation to Preserve the Family Farm?* Ames: Religious Studies Program, Iowa State University.

CHAPTER TWELVE

Commodity Programs: Analysis and Alternatives

The performance of commodity programs is judged by economic efficiency, equity, and other criteria in this chapter. Groups concerned with farm policy do not necessarily judge programs by these same criteria. Farmers want high income and minimum restriction on production and marketing decisions. Consumers want abundant and high-quality food and fiber at low prices. Taxpayers want low Treasury cost and low administrative burdens. Society wants economic efficiency, equity, and flexibility in programs to meet world food needs. These interests cannot be served simultaneously; they often conflict and lead to inconsistent, piecemeal legislation.

The first section in this chapter reviews analysis of aggregate performance, the second reviews individual commodity program performance, and the third section reviews advantages and disadvantages of numerous alternatives to current programs. The contribution of commodity programs to the level and distribution of farm income and to stability of farm and food prices is noted. Also at issue is whether commodity programs help alleviate farm problems of financial stress (cash flow and debt stress), instability, environment degradation, poverty, and loss of family farms.

AGGREGATE IMPACT OF PROGRAMS

The main conceptual framework for this chapter is classical welfare analysis presented in Chapter 2. Sources of economic inefficiency noted in that chapter include deadweight inefficiency (social cost defined as real national income lost) through output deviating from that of a well-functioning market.

A second social cost is from an inefficient resource mix. Farm commodity programs restrict land use to inefficient resource combinations — too little land relative to other inputs. Acreage-diversion programs idled up to 80 million cropland acres having little value in uses other than producing farm crops.

Farm labor has been in excess supply for several decades. As will be noted later, there is no evidence that farm programs slowed the exodus of farm labor over the long run nor slowed the trend toward fewer, larger farms. Programs may have helped to steady the flow of unneeded labor out of agriculture. But no substantive basis exists to conclude that withdrawal of land from production and other aspects of government programs markedly shifted the supply curve upward, raising real costs of producing a given output. Hence attention is focused on the social cost of too little or too much output in this chapter. Another social cost is national output foregone by resources devoted to obtain government transfers. Cost of administering programs is another social cost not given much attention herein but typically has averaged 10-15 percent of transfers to producers.

Costs and benefits of commodity programs can be measured in many ways. Most of the $20-$30 billion annually spent by the federal government on commodity programs in the mid-1980s was merely transfers from consumers or taxpayers to producers and was not an economic cost in terms of reduced national income. As a general rule, the more effective a program is in raising farm income the more it reduces national income. We sort such issues in this section.

PROGRAM SLIPPAGE

Slippage has increased as farmers have learned to divert inferior acres, expand their crop base, substitute fertilizer and other productive inputs for diverted land, and in other ways avoid production controls. Diversion of 80 million acres in some years of the 1980s or 21 percent of the nation's 380 million acre cropland base gives an inflated estimate of excess capacity. On average, each diverted acre removed only .66 harvested acres and each diverted acre potentially yielded only 80 percent as much as the average acre in production (Dvoskin, 1987). Thus one diverted acre had the effective productivity of .66(.8) = .53 acres in production. Alternatively stated, approximately two acres had to be diverted to reduce production by the equivalent of one acre in use. These results conform closely to earlier estimates by Tweeten (1979, p. 484).

Participation rates in programs vary widely but two-thirds participation is not unusual. Considering the three elements of slippage in a voluntary program (yield, acreage, and participation), it follows that a 30 percent diversion program typically is required to reduce production 10 percent.

IMPACT OF COMMODITY PROGRAM ON NATIONAL INCOME

Before analyzing the impact on specific commodities, the impact of commodity programs is illustrated with aggregate data and analysis. Excess capacity data will be used to measure net social cost. Aggregate analysis is advantageous in circumventing substitution problems not easily surmounted in individual commodity analysis. Excess capacity measured as diversion from the market by government programs of acreage diversion, stock accumulation, and subsidized exports averaged 15 percent for rice, 14 percent for wheat, 8 percent for feed grain, 5 percent for cotton and for soybeans, and 4 percent for dairy for the 1940-86 period (Dvoskin, 1987, p. 34). Excess capacity was 8-9 percent for the 1985-86 period. This was a record and compares with excess capacity averaging 5-6 percent in the 1960s.

Net social cost C of government commodity programs expressed as a percentage of gross farm receipts is approximated by the following formula:

$$C = 50\left(\frac{1}{\alpha} - \frac{1}{\beta}\right) \cdot \left(\frac{\Delta Q}{Q}\right)^2$$

where α is the supply elasticity, β is the demand elasticity, and $\Delta Q/Q$ is the proportion of farm output diverted from markets by government programs (Tweeten, 1979, p. 485). It is noted that social cost increases with the square of excess capacity. The following data show social cost as a proportion of receipts and in billion dollars for typical rates of excess capacity in the 1960s and mid-1980s.

Excess Capacity	**Social Cost**	
(Percent)	**(Percent of receipts)**	**(Bil. 1987 dollars)**
6 Percent		
Short Run	2.3	3.51
Long Run	.5	.81
9 Percent		
Short Run	5.27	7.90
Long Run	1.22	1.82

Short-run results are for $\alpha = .1$ and $\beta = -.3$; long-run results are for $\alpha = 1$ and $\beta = -.5$ and for farm gross receipts of $150 billion. According to results, programs which cost taxpayers $26 billion in 1986 and held excess capacity of approximately 9 percent reduced national income by nearly $8 billion in the short run and less than $2 billion per year in the long run. The greater opportunity for demand and supply quantity to adjust to changes in incentives reduces long-term social costs. Each Treasury dollar spent on farm programs reduces national income 31¢ in the short run and 8¢ in the long run from deadweight loss alone. If programs had been operated at levels of the 1960s and removed only 6 percent of farm output, net social cost would have been $3.51 billion in the short run and $.81 billion in the long run as noted above.

Cost-Effectiveness of Commodity Programs

Policymakers often appear to be more concerned with Treasury costs than social costs of programs to support farm income. Therefore it is well to examine *cost-effectiveness* of past programs, defined as the contribution to farm income per Treasury dollar spent. Mandatory controls are most cost-effective, but are unacceptable for most commodities. Mandatory controls entail the largest *transfer inefficiency* or social cost per dollar transferred to producers. The discussion below focuses on voluntary programs.

Programs that are most effective in removing farm production per Treasury dollar are also most cost-effective because demand for farm output is inelastic in the short run. The *effectiveness* E in removing value of production per program dollar G from an acre of cropland is given by the following formula (Sobering and Tweeten, 1964, p. 829):

$$E = \frac{PY}{G} = \frac{PY}{PY-C} = \frac{1}{1 - \frac{C}{PY}} \qquad C < PY \qquad 1 \leq E < \infty ,$$

where P is product price, C is variable cost of production per acre, and Y is yield per acre. In theory, the producer is equally well off by diverting acres or by producing a crop if the government pays rent equal to net returns G = PY - C, gross returns less variable costs of production. It is apparent for the most marginal cropland that PY = C and E = ∞, the upper limit of effectiveness. If variable costs are zero, E = 1, the lower limit of effectiveness. E becomes larger, other things equal as (1) product price P is reduced, (2) yield Y is reduced, and (3) variable cost of production C is increased.

It follows in theory that effectiveness becomes large as diversion is concentrated on marginal land and whole farms under long-term contracts so that more of the producer's costs become variable. Furthermore, E is enhanced by use of sealed bids that allow the government to discriminate among bids, paying only the amount necessary to retire land from production and selecting those bids that remove most production per program dollar. These considerations imply that long-term general land-retirement programs give the highest E, a conclusion supported by empirical evidence which indicated E for the early Conservation Reserve Program was approximately 3.0 (Christensen and Aines, 1961). But such programs have not been emphasized because they concentrate land retirement in selected geographic areas with attendant problems for nonfarm communities and because they are inflexible and difficult to adjust from year to year in response to emerging realities.

Based on results from a personal interview survey of wheat producers, Carr and Tweeten (1974) found that several voluntary programs potentially could remove two dollars of production per program dollar if care would be taken to avoid slippage in administering programs. The effectiveness of short-term diversion programs has been low in relation to reasonable estimates of potential effectiveness, however.

The ratio of market price to diversion payments per bushel is an approximate measure of E. Several past programs described in the previous chapter provided diversion payments per bushel approximately equal to half the price, hence in theory the effectiveness E of payments was 2.0. In view of these and other considerations, it is somewhat surprising to note that several estimates of actual effectiveness of programs were approximately 1.0 for feed grains and could be even lower for other commodities.

Iowa State University economists estimated that the 1961 feed grain program, a short-term acreage-diversion program, removed $1.14 of corn production per dollar of government cost (Shepherd et al., 1963, p. 22). Robinson (1966, p. 24) estimated that average annual production of feed grain was reduced 30-40 million tons by acreage withdrawal programs in the 1963-64 period. The cost to the government averaged $1.4 billion, and the value of production removed ranged from $0.86 to $1.14 per dollar spent on the program. Other estimates of the effectiveness of short-term acreage diversion and long-term land retirement range from 1.1 to 1.6 (cf. Tweeten et al., 1963). E has probably declined since these estimates.

Several factors explain the low effectiveness. One is that employees of the Agricultural Stabilization and Conservation Service administering programs are caught in the fallacy of composition at the local level. That is, producers in their country will receive greatest incomes if payments are generous and diversions lax, whereas for the nation E will be greatest under the opposite circumstances given a fixed national budget for programs. A second reason is that some programs entailed a large direct-payment component as opposed to supply control. Feed grain payments have had a high component of supply control relative to income supplement whereas wheat, cotton, and rice have had a low component.

Even if for acreage diversion programs E = 1, it can be shown that they are more cost-effective than direct-payment programs. Cost-effectiveness CE in raising net farm income per Treasury dollar is given by the formula (Tweeten, 1976, Appendix A):

$$CE = 1 - E(1 + F) + S$$

where E is effectiveness as defined above, F is the price flexibility of demand (inverse of aggregate output demand elasticity), and S is production cost savings per dollar of program cost. Each dollar increment in government diversion payments raises net income $1 from the payment itself, $2 through the macroeconomic impact E(1 + F) where E = 1 and F = -3, plus S = $.50 saved in production expenses. Thus each dollar spent on the diversion or set-aside program raises net farm income an estimated $1 + $2 + $.50 = $3.50. (This compares with CE = $1.00 for a direct-payment program requiring no diversion.) While in theory this is a conservative estimate, it is notable that on the average $100 spent on feed grain programs in the early 1960s actually increased farm income only $168 (Legislative Reference Service, 1964). Cost-effectiveness has tended to decline since the 1960s. In spite of the theoretically high cost-effectiveness of acreage diversion programs, the record is unimpressive.

Acreage diversion has advantages other than cost-effectiveness, including making program payments more acceptable to farmers who feel they are doing something to receive them by holding reserve capacity for future use and by conserving soil. But other program alternatives deserve consideration in view of the low cost-effectiveness of diversion programs, the high slippage factor, high social cost, inequities in program payments among producers, and concentration of diverted acres in selected geographic areas. If commodity programs are to focus benefits on farms in poverty, financial distress, or family farms, acreage diversion which distributes price benefits according to production would need to be replaced by payments targeted to farms "at risk."

Impact on Farm Income of Releasing Excess Capacity

Social cost calculations show impacts of programs on national income. We now turn to impact on farm income of releasing excess capacity. A measure of the short-run impact on farm prices and income can be made from the price elasticity of demand for farm output β. Given release of 1 percent more output on the market, the impact on prices received by farmers is given by the elasticity of price flexibility of demand $F = 1/\beta$, the impact on farm revenue is given by the elasticity of gross receipts with respect to output $1 + F$, and impact on net returns is given by the elasticity of net income with respect to output $(1 + F)R$ where

R is the ratio of gross receipts to net farm income. The assumption is that resources do not adjust to cut costs in the short run. Assuming β = -.33, the implication is that 9 percent additional output placed on the market would in the short run reduce prices received by 27 percent, gross receipts by 18 percent, and net receipts by 54 percent assuming a ratio R = 3.0 of receipts to net income. The drop in net farm income would be even greater if government deficiency payments would be terminated.

In the intermediate and long run, release of excess capacity on the market induces both a demand response (greater utilization) and a supply response (less resource use and output). The elasticity of excess capacity with respect to price is $\gamma = \alpha - \beta$. If the elasticity of supply α is .3 in the intermediate run and 1.0 in the long run and the elasticity of demand β is -.4 in the intermediate run and -.5 in the long run, the drop in price needed to eliminate excess capacity is the excess capacity divided by the elasticity of excess capacity with respect to price γ as follows:

		Excess Capacity	
Time	γ	**6 percent**	**9 percent**
Intermediate run	.7	9	13
Long run	1.5	4	6

Excess capacity of 1986 dimensions, 9 percent, could be eliminated by allowing price to fall 13 percent for 5 years (intermediate run) or 6 percent for 10 years or more (long run).

The impact of commodity programs on long-term farm income is quite different. Long-term gains are eroded by (1) an increasing elasticity of supply as length of run is extended, (2) an increasing elasticity of demand as the length of run is extended, (3) capitalization of program benefits into land values, and (4) security and capital provided by farm programs encouraging additional output. Accumulation of government commodity stocks well in excess of economically optimal levels overhangs markets and depresses price year after year.

The elasticity of supply may be only .1 in one year but is 1.0 in the long run. Output expands not only because of additional inputs brought into agriculture by prices supported above equilibrium levels by farm programs, but also because of the slippage elements as farmers learn to avoid production controls by diverting inferior cropland, bringing in new land, and substituting capital inputs for land. The elasticity of demand increases in the long run particularly in response to the impact of price on exports. The elasticity of demand in the long run for farm commodities is near -1, which implies that a change in output does not change the level of farm receipts.

Even if the supply and demand responses do not erode gains from farm programs, capitalization of program benefits into land value means that the sellers of farmland reap the gains and the additional income generated by programs is lost to buyers. The finding in Chapter 4 that land values adjust rapidly to changing conditions and that adequate-size, reasonably well-managed farms have tended to earn "parity" returns on resources most years indicates that benefits of commodity programs to new owners are lost quite rapidly. Renters, hired workers, and small farms also tend not to benefit from commodity programs given time for adjustments. In short, government programs raise farm income in the short run but not in the long run.

DISTRIBUTION OF COMMODITY PROGRAM BENEFITS

Table 12.1 shows the distribution of government payments among farms by economic class. Additional receipts generated by higher prices associated with less production under acreage diversion are nearly proportional to payments, hence the distributions of payments are indicative of the distribution of all commodity program benefits. Farms with sales of over $500,000 received payments of $37,499 per farm in 1985 while farms with sales of $5,000 or less received only $40 per farm. If equity requires larger payments to low-income producers than to high-income producers, commodity programs are not equitable. Commercial farmers who receive the lion's share of benefits have higher income and wealth than taxpayers on the average, hence commodity programs have not been equitable in redistributing income from taxpayers to producers.

TABLE 12.1. Distribution of Direct Payments by Economic Class of Farms, 1985

	Farm Size by Sales per Year								
Item	$500,000 and Over	$250,000 to $499,999	$100,000 to $249,999	$40,000 to $99,999	$20,000 to $39,999	$10,000 to $19,999	$5,000 to $9,999	Less than $5,000	All Farms
Number of farms (1,000)	27	66	221	323	230	243	268	896	2,275
Direct payments per farm ($)	37,499	21,783	12,845	5,193	2,040	678	233	40	3,387
Gross income per farm ($)	1,852,614	401,473	184,449	80,534	37,570	20,988	13,651	7,267	73,716
Direct payments per gross income ($)	.02	.05	.07	.07	.05	.03	.02	.01	.05

SOURCE: U.S. Department of Agriculture (November 1986).

A related issue attracting much attention is whether commodity programs have speeded the trend to fewer, larger farms. Conventional wisdom holds that larger payments to big farmers than to little farmers causes large farms to grow in size. Such reasoning is flawed. Economies of size are measured *per unit of output*. If large farms received $100,000 of receipts from wheat but were paid $2.00 per bushel while small farms received $20,000 of receipts from wheat but were paid $4.00 per bushel, we would say the pricing system favors small farms over large farms. Similar reasoning applies to commodity programs. Large farms received only 2¢ of government payments per dollar of output while medium-size farms received 7¢ per dollar of output (Table 12.1). Commodity programs have benefitted middle-size farms relative to small and large farms. In any given program, middle-size to large family farms have highest participation rates. That large farms receive only one-third as much payment per dollar of output as medium-size farms arises because large farms emphasize production of enterprises not covered by commodity programs such as cattle, broilers, fruits, and vegetables. Payment limitations reduce participation in programs on very large farms. Small farmers frequently do not participate in commodity programs because it isn't worth the bother. They also produce cattle, fruits, and vegetables not covered by commodity programs. In short, a number of features of commodity programs favor mid-size family farms.

On the other hand, some features of commodity programs hasten the trend to fewer, larger farms. One is the security provided by programs which enables a given equity to be leveraged to acquire more land. Commodity programs divert cropland so that an operator needs to obtain more acres to operate for efficiency. The most careful review to date indicates that commodity programs on the whole have been neutral, neither markedly encouraging nor constraining the trend to fewer, larger farms (Spitze et al., 1980).

More stringent payment limitations could direct an even larger proportion of benefits to family size farms. Two problems are apparent, however. One is that payment limitations are very difficult to enforce. Farmers are very clever in devising strategies to break up farms on paper to remain eligible for benefits while the farm is operated as before. Payment limitations reward the devious. Second, tight payment limitations have severe shortcomings if production is to be controlled with paid diversion programs. Larger farms which produce most of the output must be included if production is to be controlled. To exclude large farms would require that smaller farms be entirely removed from production — hardly a fitting way to preserve them.

A final difficulty is that payment limitations may unfairly discriminate against large farms. Commodity programs may be designed to bypass large farms but they are not generally intended to destroy them. Under the Food Security Act of 1985, payment-in-kind placed huge government stocks on the market to bring prices well below free market equilibrium. The strategy was to reduce stocks and prices so that prices *eventually* would rise but to provide farmers with direct payments through the difficult transition period. To arbitrarily reduce prices to well below normal free market levels while denying large farmers direct payments arbitrarily disadvantages such farmers.

PRICE VARIABILITY

On the whole farmers and consumers are risk averse, hence their utility is raised by greater price stability. Empirical evidence indicates that commodity programs have stabilized prices (Tweeten, 1981, pp. 145, 146). The principal benefit of commodity programs has been to help stabilize variation in food supplies and farm prices, albeit at a level many farmers consider to be too low. Much of the period of most intense government involvement from the mid-1950s to 1972 and from 1982 to 1987 would have been, in the absence of commodity programs, one of generally depressed farm prices. Release of reserves probably held down prices in the short-supply years of 1972 to 1977. Thus it appears that the programs dampened economic instability. The unwieldy surpluses generated by programs proved of considerable value in World War II, the Korean War, the 1966-67 food crisis, the 1970 corn blight, and at other times. Benefits were often achieved at considerable short-run social cost and with large transfer payments, however.

Summary Comments

Commodity programs have reduced instability, helping to keep some financially weak family farmers in business who would have failed during the cyclical economic downturn of the 1980s and at other times in previous decades. The record of commodity programs in the long run has been less favorable. There is no evidence that commodity programs have improved the long-term level or distribution of farm income, have reduced environmental degradation and poverty, or have slowed the exodus of family farms.

At the same time commodity programs have entailed sizable social and Treasury cost. After impacts of specific programs are reviewed in the next section, the final section examines alternatives that might better serve needs of farmers and society.

ECONOMICS OF MAJOR COMMODITY PROGRAMS

This section illustrates the economics of commodity programs for grains, dairy, sugar, and tobacco. Most of the "cost" of commodity programs to the U.S. Treasury is a transfer rather than a deadweight loss in foregone real income in the nation. It is useful to detail the redistribution of income among consumers, producers, and taxpayers as well as the impact on national income as measured by net social cost. The redistributions and losses are measured as deviations from a competitive market equilibrium for the following programs. Graphic presentations are followed by empirical estimates for 1978 and 1986.

The Grain Program

Figure 12.1 shows the short-run and the long-term impact of grain programs. Compared to market equilibrium at price p_e and quantity q_e, government programs raise the market price of grain to the loan rate p_L which would cause quantity q_L to be supplied in the short term (Figure 12.1A). A paid diversion program costing taxpayers a value equal to area 6 to 10 reduces quantity from q_L to q. In practice the cost to the government is equal to the value diverted but in theory the government cost is less than the value diverted. In addition, producers receive a direct deficiency payment equal to the difference between the target price p_T and the loan rate on quantity q, a value equal to area 1 + 2. In the short run the grain program increases the price to consumers from p_e to the market price p_L, causing a loss in consumer surplus of area 3 + 7. The loss to taxpayers is the deficiency payment plus the diversion payment. Producers gain in the short run. One of the losses to them is area 5 which is defined broadly as an X-inefficiency loss caused partly by reduced

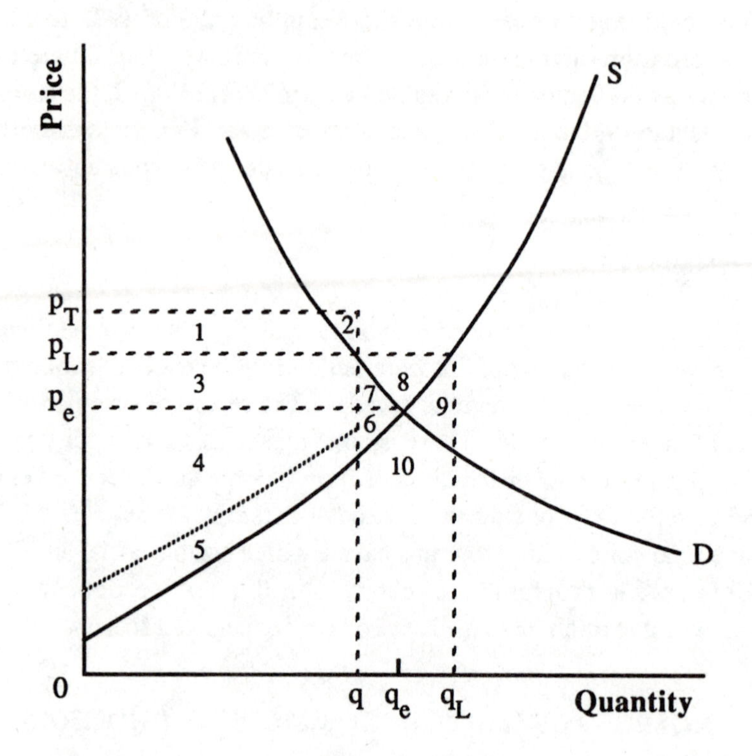

FIGURE 12.1A. Short-Run Impact of Grain Programs

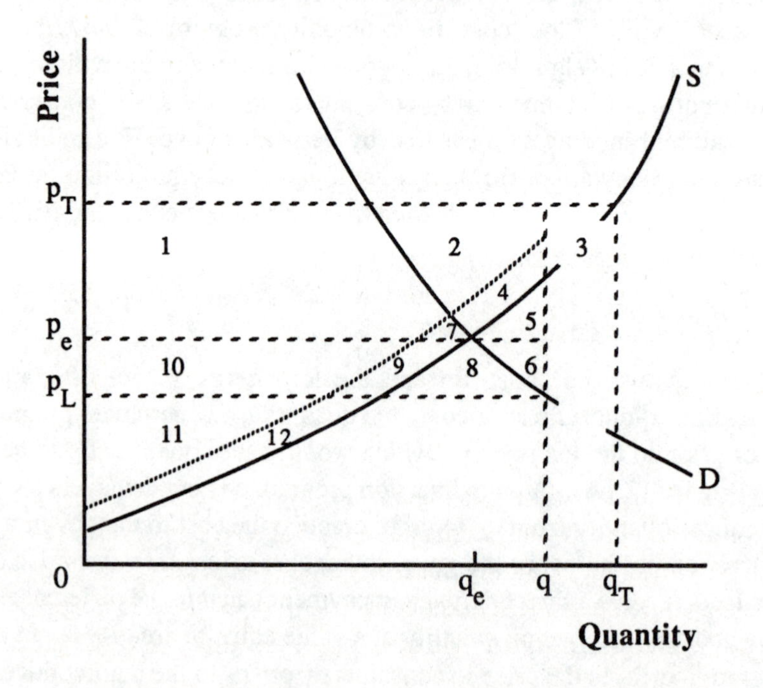

FIGURE 12.1B. Long-Run Impact of Grain Programs

incentives to combine resources efficiently and partly by land diversion bringing an inefficient mix of land to other resources. The latter could be shown to be a deadweight loss apparent in supply and demand curves for resources. A net social cost or deadweight loss in foregone national output is area 5 + 6 + 7 in the short run. In summary, the results are as follows:

Loss to consumers	3 + 7
Gain to producers	1 + 2 + 3 - 5 - 6 + 6 to 10
Loss to taxpayers	1 + 2 (deficiency payment) 6 to 10 (diversion payment)
Net social cost	5 + 6 + 7.

The impact of the government grain program in the long run is difficult to illustrate from an economic standpoint because it is not certain whether market price will be above or below the long-term equilibrium price. But in 1987 a strong case could be made that the grain loan rate and market price were below the long-term equilibrium as shown in Figure 12.1B. The deficiency payment equal to the difference between target price p_T and the loan rate p_L (assumed to be equal to the market price) paid on quantity q entails a large cost to taxpayers. Compared to competitive market equilibrium, the gain to producers and consumers is less than the loss to taxpayers. Results for the long run are summarized as follows:

Gain to consumers	8 + 9 + 10
Gain to producers	1 + 2 + 3 - 9 - 12
Loss to taxpayers	1 + 2 + 4 to 10 (deficiency payment) 3 (diversion payment)
Net social cost	4 to 7 + 9 + 12.

Commodity programs typically entail administrative expenses equal to 10-15 percent of overall Treasury costs. In addition, resources with high opportunity cost (foregone worthwhile activity) devoted to lobbying for expanding or continuing commodity programs are an economic loss to society. These administrative and so-called PEST costs are not included in Figures 12.1-4 or subsequent Tables 12.1 and 12.2.

THE DAIRY PROGRAM

The dairy program illustrated in Figure 12.2 recognizes two segments of the milk market — the more inelastic fluid demand D_f and the less inelastic manufactured-use demand which is not shown explicitly but is added to D_f to form a total demand for milk D_T. Supply is S and competitive market equilibrium price is p_e in all markets and total quantity is q_e. The equilibrium quantity is q_f in the fluid market and $q_e - q_f$ in the manufactured milk market.

Under the two-price milk marketing order, p_f is charged in the fluid market and a lower price is charged in the manufactured milk product market. The new total demand

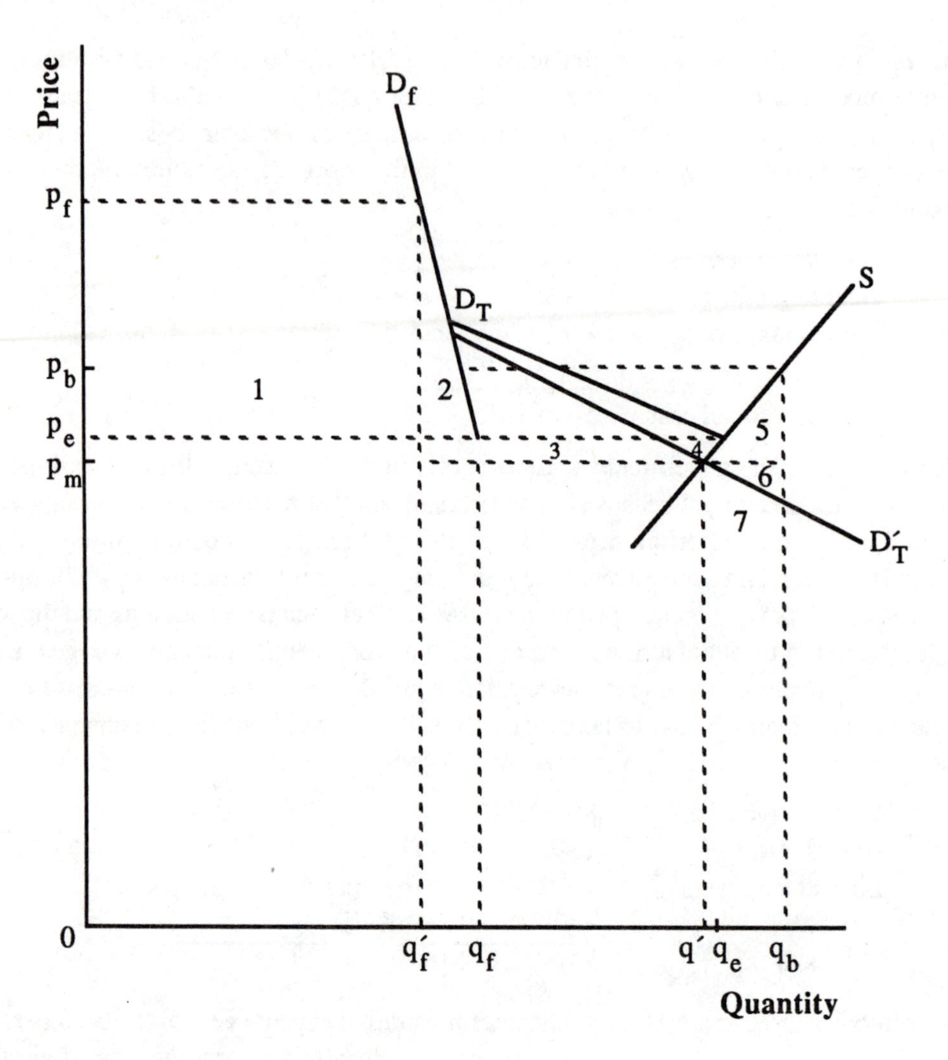

FIGURE 12.2. Illustration of Dairy Program with Milk Demand for Fluid Use D_f, for All Uses D_T, and Supply S

curve is D'_T for the dairy industry. The new curve is formed by adding horizontally the manufactured milk demand to the fluid demand, the latter with q'_f the origin rather than D_f. The market is cleared with quantity q'-q'_f and a price p_m in the manufactured milk market. Quantity q'_f is sold in the fluid milk market at price p_f.

Under the federal dairy program, a blend price p_b is paid producers which according to supply curve S brings production q_b. The government purchases quantity q_b - q' off the market to clear at a price for manufactured milk products of p_m. If the government disposes of the milk to increase consumer surplus by area 7, the deadweight loss is modest. However, if the government allows the excess purchased off the market to spoil in storage or in other ways to not enhance consumers' needs, the deadweight loss is increased by the area 7.

Compared to competitive equilibrium, the results in terms of redistributions and social loss are summarized as follows:

Loss to fluid milk consumers	1 + 2
Gain to mfg. milk consumers	3
Gain to other milk consumers	7
Gain to producers	1 - 3 - 4 - 5
Loss to taxpayers	6 + 7
Net social cost	2 + 4 + 5 + 6
Net social cost without gain to "other" milk consumers	2 + 4 + 5 + 6 + 7.

The loss in consumer surplus is area 1 + 2 but producers gain only area 1 so net social cost is 2. The gain to "other" milk consumers is actually only the portion of area 7 to the right of q_e, hence the welfare impacts in the manufactured milk market are not precisely as indicated in the above accounting. The net social cost is approximately area 2 + 4 + 5 + 6 if full value is obtained by the government from disposal of surpluses as shown by area 7. However, net social cost (reduction in real national income) is increased by area 7 if the surplus milk products have no value.

The milk marketing order decreases the quantity in the fluid market and increases the quantity in the manufactured market. Because the marginal utility of income may be higher for consumers of fluid milk than for consumers of manufactured milk and for milk producers, the redistribution of income from fluid milk consumers to others under the dairy program may reduce utility by greater amounts than revealed by the net social cost.

Figure 12.2 also serves as a conceptual framework to analyze a more nearly true two-price program such as the peanut program which does not control production or pay a blend price. Total demand is D_fD_T as before and market price is p_e and quantity is q_e without market distortion. Under the peanut program, quota peanuts bring price p_f; peanuts for oil and other less inelastic uses receive the lower price p_m in the two-price scheme.

Redistributions and loss to society from the two-price peanut program versus free market equilibrium are summarized as follows from Figure 12.2:

Loss to "quota" peanut consumers	1 + 2
Gain to producers	1 - 3 - 4
Gain to "other" market consumers	3
Net social cost	2 + 4.

Thus the milk blend price raised social cost by 5 + 6 compared to the peanut program.

THE SUGAR PROGRAM

The domestic demand for sugar D_d and domestic supply of sugar S_d intersect at the competitive equilibrium price p and quantity q in Figure 12.3 in isolation if foreign markets

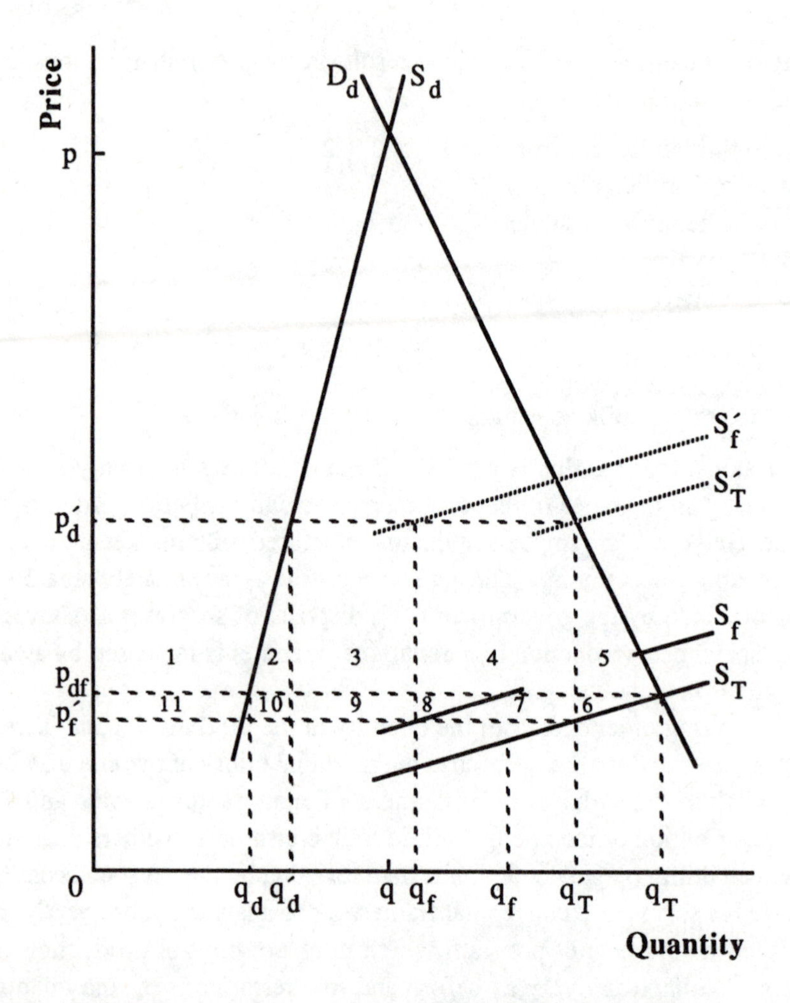

FIGURE 12.3. Illustration of Sugar Program with Domestic Supply S_d and Demand D_d, Foreign Supply S_f and Demand D_f, and with Tariffs and Duties $S_f' - S_f$

are omitted. Production costs are lower in the rest of the world and the United States lacks both absolute and comparative advantage in production of sugar. The import supply of sugar is shown by the somewhat elastic foreign supply curve S_f. Horizontal summation of S_f and the domestic supply curve S_d results in the total supply curve S_T in Figure 12.3. The resulting competitive equilibrium price and quantity are respectively p_{df} and q_T.

Imposition of import duties and tariffs causes the U.S. import supply curve to be shifted upward to S_f'. When added to domestic supply S_d, the total supply curve is S_T'. The equilibrium price in the domestic market is now p_d' and total quantity demanded is q_T'. Quantity q_d' supplied by the domestic market with tariffs is greater than the quantity q_d supplied without tariffs. On the other hand, quantity supplied by the foreign market q_f' with tariffs is less than quantity q_f provided under a free market.

The welfare redistribution and net social cost of the sugar program is illustrated as follows:

Loss to consumers	1 + 2 + 3 + 4 + 5
Gain to domestic producers	1
Loss to foreign producers	8 + 9 + 10 + 11
Gain to taxpayers	3 + 4 + 9 + 10 + 11
Net social cost	2 + 5 + 8.

The loss to consumers is the value of areas 1 through 5. Domestic producers gain area 1 and foreign producers lose areas 9 through 11. The gain to taxpayers is area 3 + 4 + 9 + 10 + 11 (note that $q_d' = q_T' - q_f'$). The net social cost is area 2 + 5 + 8. (Figure 12.3 and others herein are not drawn to scale.)

THE TOBACCO PROGRAM

The tobacco program has been attacked for subsidizing consumption of a substance entailing greater social than private cost because individuals do not bear the entire cost of health maintenance caused by use of tobacco. In competitive private market equilibrium with private demand D and private supply S, equilibrium is at price p_e and quantity q_e (Figure 12.4). Tobacco quotas restrict quantity to q', raising price to p'. If S is the social supply curve, the welfare impact is as follows:

Loss to consumers	1 + 2 + 3
Gain to producers	1 + 2 - 5 - 7
Net social cost	3 + 5 + 7.

By this accounting, the real national income foregone by the tobacco program is measured by area 3 + 5 + 7 where 7 is the increase in cost of production induced by excessive use of other inputs relative to land.

In a more realistic scenario it is well to consider the social supply curve as being S'. The welfare-maximizing equilibrium then is at p' and q', and the costs of the free market must be measured by deviation from that efficient equilibrium. The welfare impact of failing to intervene when social cost exceeds private cost is:

Gain to consumers	1 + 2 + 3
Loss to producers and others	1 + 2 + 3 + 4 - 7
Net social cost	4 - 7.

Consumers gain areas 1 + 2 + 3 from a free market but the loss to producers and others is the same area plus area 4 - 7. Given equilibrium at price p_e and quantity q_e the net social cost of a market outcome is area 4 - 7 which of course will be positive if area 4 is greater than area 7 as seems likely. In this case the net social cost is zero for the tobacco program when the social supply curve S' is accounted for and price is p' and quantity q'.

The tobacco situation may be treated as a divergence between private and social costs or between private and social benefits. Benefits to society are much less than benefits to consumers of tobacco, and that divergence may be defined as equal to the divergence in

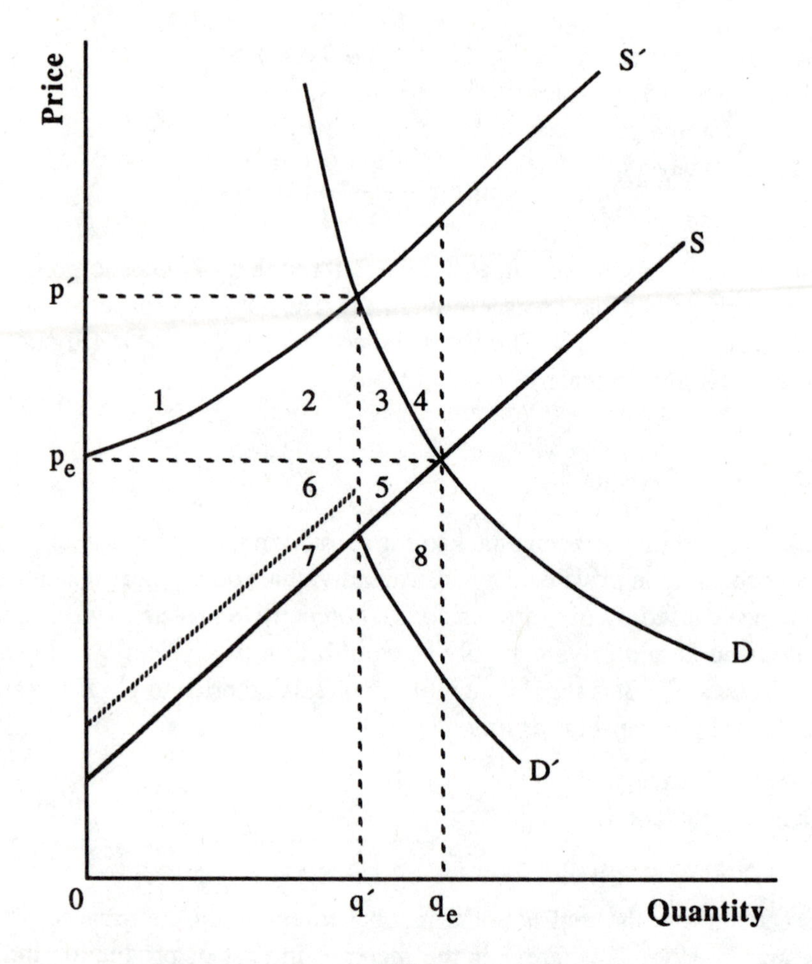

FIGURE 12.4. Illustration of Tobacco Program with Demand D, Private Supply S, and Social Supply S′

cost. That is, S' - S is equal to D - D' at any given quantity. Under such circumstances whether the externality is arbitrarily associated with producers (supply) or consumers, the welfare implications remain unchanged. The net social cost is area 8 (equal to area 4) of overconsumption for failure to control production and raise prices when the social demand curve is D' and the private demand curve is D.

In the case of tobacco, the amount by which the social supply and demand curves differ from the private curves is unknown. In view of the fact that the tobacco program provides benefits to small and low income farmers to a greater degree than other programs and in view of the divergence between private and social costs (benefits), it is possible that the "no-net-cost-to-government" tobacco program entails lower social cost than other commodity programs despite its bad press.

SUMMARY

Table 12.2 from Gardner (1981, p. 73) and Table 12.3 from The Council of Economic Advisors (1987, p. 159) summarize measures of redistributions and net social loss from commodity programs. Estimates in Table 12.2 indicate that in 1978/79 commodity programs were relatively modest intrusions into farm markets. Losses to taxpayers and consumers exceeded gains to producers by $.9 billion, which was only .4 percent of total farm receipts. With a net social cost of 5 percent of receipts, the sugar program entailed the greatest relative inefficiency. The milk program entailed the greatest absolute inefficiency as measured by lost national income.

With much larger commodity programs in 1986 under the Food Security Act of 1985, transfers and net social loss were greater (Table 12.2). The wheat and milk programs entailed the greatest absolute net losses but the cotton, wheat, rice, and tobacco programs had the greatest net losses relative to receipts. No correction was made for divergence between private and social cost for tobacco, hence social cost in Table 12.3 is overestimated. Net social cost averaged 8.1 percent of receipts of covered commodities but averaged 3.7 percent of all receipts for the farming industry. The aggregate social cost of $4-6 billion in Table 12.3 is between the aggregate short-run social cost of $8 billion and long-run social cost of $2 billion estimated earlier in this chapter from macro parameters.

COMMODITY POLICY ALTERNATIVES[1]

Each of the commodity policy alternatives and options examined in this section has advantages as well as disadvantages. No alternative or option is likely to elicit early approval simultaneously from farmers, taxpayers, and consumers. But the study of these strengths and weaknesses provides background for the public to make necessary hard choices and compromises.

AN AUGMENTED PRIVATE SECTOR

A free market for agriculture is not feasible but a case can be made that the nation would benefit as a whole by relying more upon markets to determine farm prices and incomes (CAST, 1983, p. 11). Despite shortcomings of past programs, there is reluctance to rely solely upon markets because of periodic instability in commodity prices and consequent periods of low incomes. A limited governmental role in price or income stabilization is an alternative.

The first option considered here is the general step of moving away from federal intervention in commodity markets that either raises or lowers commodity prices or farmers' incomes averaged over a period of years. The specific options are alternative approaches to achieving the general goal of less government involvement in farm

[1]Parts of this section are from the Council for Agriculture Science and Technology (1983).

TABLE 12.2. Summary Estimates of the Effects of Farm Programs in 1978/79[a]

Commodity	Percentage of U.S. Total Farm Output	Ratio of Price With Programs to Price Without Programs	Taxpayers Cost	Consumers Cost	Producers Gain	Net Loss	
	(%)		---(Million)---				(%of receipts)
Wheat	6.0	1.175	$1,100	$400	$1,300	$200	3.2
Feed Grains	12.0	1.060	1,020	700	1,500	220	2.0
Rice	1.0	1.020	10	0	10	-----	----
Cotton	4.0	1.020	90	0	90	-----	----
Tobacco	2.0	1.330	0	425	400	25	1.0
Sugar	1.0	1.880	-770	1,440	600	70	5.0
Peanuts	1.0	1.400	0	200	200	-----	----
Cattle	21.0	1.035	-40	1,200	1,100	60	0.2
Milk	12.0	1.110	0	1,500	1,200	300	2.2
Wool	0.2	1.300	30	-15	15	-----	----
All other	40.0	1.000	0	0	0	-----	----
Total (or mean)	100.0	1.060	$1,440	$5,850	$6,415	$875	0.4

SOURCE: Gardner (1981, p. 73).

[a]See original for footnotes.

TABLE 12.3. Annual Gains and Losses from Income-Support Programs Under the 1985 Food Security Act and Trade Restrictions

Commodity	Percent of Total Farm Output	Consumer Loss	Taxpayer Cost[a]	Producer Gain	Net Loss	
	(%)	---(Billion)---				(%of receipts)[c]
Corn	11	$0.5-1.1	$10.5	$10.4-10.9	$0.6-0.7	4.1
Sugar I[b]	2	1.8-2.5	0.0	1.5-1.7	0.3-0.7	18.1
Sugar II[b]	2	1.1-1.8	0.0	1.0-1.4	0.1-0.4	9.1
Milk	20	1.6-3.1	1.0	1.5-2.4	1.1-1.7	4.9
Cotton	3	small	2.1	1.2-1.6	0.5-0.9	18.4
Wheat	6	0.1-0.3	4.7	3.3-3.6	1.4-1.5	18.4
Rice	1	.02-.06	1.1	0.8-1.1	.06-.32	20.1
Peanuts	1	0.2-0.4	0.0	.15-.40	0.0-.05	2.4
Tobacco	2	0.4-0.7	0.1	0.1-0.2	0.4-0.6	18.4
Total	36	$5.7- 10.0	$19.5	$18.5- 21.6	$4.2-6.2	3.7

SOURCE: Complied by the Council of Economic Advisers (1987, p.159).

[a]Includes CCC expenses after cost recovery.

[b]Case I assumes U.S. policies do not affect world sugar prices. Case II takes into account the fact that U.S. policies reduce world sugar prices.

[c]Calculated at midpoint of estimates.

commodity markets. The first specific option is to supplement forward-pricing markets. The second specific option is to provide insurance, which farmers would buy and which would make them eligible for indemnity payments when receipts fall below an insured level. The third specific option is price stabilization through a buffer-stock program that returns all commodities taken off the market back to the market at a later time, and so does not constitute a price-support program in a long-term context.

The General Alternative of Greater Reliance upon Markets. As noted in Chapter 4, there is ample evidence that adequate-size, well-managed farms have earned and will continue to earn adequate rates of return on resources. After a period of adjustment they would do so without government programs. The farming industry is capable of adjusting to emerging disequilibrium if the current excess capacity is worked off with help from a government transition program.

This general policy approach recognizes that hardly anyone defends past governmental management of farming and is probably unable to improve on the situation generated by unrestricted markets for agricultural commodities. It assumes that there is no chronic tendency toward overproduction or shortage in an unregulated market context. It assumes that in such an environment commercial farmers can be expected, over the long term, to earn returns to their labor, management, and investment comparable to returns earned in the nonfarm sector. However, this alternative views short-term instability as partly a market failure; hence, some supplementation of markets by governmental action is necessary to deal with market fluctuations.

The wisdom of accepting the general approach involves the truth or falsity of the assumptions just listed. Against them it may be argued that there are inflexibilities in resource adjustment, a lack of free international markets due to policies of our competitors abroad, and imperfections in the markets to which farmers sell and from which they buy. Commodity programs as traditionally established, however, are poor remedies for these imperfections, and our historical experience with legislation of agricultural prices has been disheartening.

The transition to minimum government support would not be easy in a time of excess capacity and stocks. Hence, a transition program seems essential, scaling down direct payments and supply controls until phased out completely after, say, five or ten years.

Advantages of the general alternative of greater reliance upon markets include the following:

— Long-run prices different from unregulated commodity-market-clearing prices result in surpluses or shortages, either of which is socially costly. Combined food and Treasury costs will be minimized.

— Even well-intentioned efforts to correct real inadequacies of markets tend to run afoul of a political environment dominated by narrow-interest-group politics. Social cost of government inventions exceeds market failures the interventions were designed to correct.

— Historical experience with price-support policies is chastening in that low-income and inefficient farmers have not been saved, nor have bankruptcies been prevented. The option abandons the attempt to prop up inefficient producers, which past programs have proved unable to accomplish anyway. Benefits of programs have been capitalized into land value and lost to renters, hired workers, and new landowners.

— The federal budget would be reduced by nonintervention, freeing scarce federal funds for other purposes, reducing taxes, or reducing deficits. The vast bureaucracy administering programs would be freed to do something more useful.

— With less intervention, U.S. commercial farmers would be in a better position to produce efficiently for the international market, increasing the productivity and competitiveness of U.S. agriculture and the nation as a whole. The trade deficit would be reduced.

Disadvantages of the general alternative of greater reliance upon markets include the following:

— Unregulated markets may give rise to a series of years of short supplies and high food prices or excessive supplies and low farm income which only reserves held as diverted acres can adequately dampen. There is little danger of domestic food shortages, but higher prices could burden some low-income food-importing countries.

— Our government intervenes to generate inflation, recession, and export embargoes; therefore, it needs to intervene with farm-commodity price supports to alleviate the consequences.

— Current programs are a response to democratic political forces that should be respected.

— Some family farmers would not survive in a competitive market, especially during the transition period.

Other arguments used to justify continued government intervention to raise farm prices and incomes lack substance. Unsupported contentions are that: (1) farmers will earn low returns in a world of imperfect competition in agribusiness and international trade, (2) the family farm will not survive without government intervention, and (3) giant conglomerates will take over agriculture, creating a shared monopoly to raise food prices.

Current programs are indeed the result of a democratic process that should be respected. But at issue is the role of special pleadings, lack of knowledge by the public of farm policy economics, and the inability of current programs to solve farm problems.

Specific Alternatives to Bring Greater Reliance upon Markets. Farmers traditionally have combined the role of production and risk-taking, so that their personal income is a return to their labor, management, and risk. The most unstable portion of their personal income is the return to risk, which is often negative. Production and risk roles can be separated, however, with the risk shifted to professional speculators and insurance agencies. On the whole, those who assume risk require compensation. Thus farmers' personal income would be somewhat lower but more stable over a period of years with risk shifted to others.

Futures markets are currently available to hedge price risk. Some shortcomings of the futures markets are: (1) only 7 percent of farmers use them; (2) contracts are available only a little over one year into the future and, hence, with tricky rollover strategies cannot hedge risks arising from a farm economic slump exceeding that duration; and (3) the speculator side of the market may be too thin to avoid substantial price discounting (low forward prices), especially for more distant contracts, if a large portion of farm output were hedged.

Some farmers complain that they do not like to hedge into a loss. The future price is likely to be the best predictor of future price, however, and the only thing worse than hedging into a loss is to "speculate" by not hedging and then take a larger loss. A superior strategy is to use the price information of the futures market and adjust production to minimize loss or to profit on production of another commodity.

One alternative would be for the government to prespecify a price band in the futures market for commodities, intervening to maintain prices within that band. Commodity stocks and direct payments are two tools that could be used on occasion in this effort. Futures contracts would be made available for markets more than one year ahead, perhaps up to three years. An extensive educational program would be mounted to inform producers in the use of futures markets and perhaps assist them directly until they become familiar with procedures.

A supplemental approach is greater use of *put* options for the main agricultural commodities. A producer purchases an option to sell a commodity as a specified price. If market prices exceed the specified price, the option is not exercised. One advantage over hedging in the futures market is that put options do not require margin calls. The put option provides security against an unexpected price decline. But unlike the hedge, which insures against either a price gain or loss, the put option allows the farmer to reap the benefit of a price increase. By buying both crop insurance and put options (or selling futures contracts), a producer could essentially insure against either low yield or low prices. The cost to the producer would depend upon the level of coverage.

Another alternative for stabilization of farm prices and incomes, the major if not the only real benefit of past commodity programs, is with a modest commodity stock reserve policy. Some economists contend that private costs exceed social costs of storage, hence markets fail in that the private sector acting alone will not hold optimal buffer reserves to stabilize the farm and food economy. Futures markets and income-stabilization measures for farmers would do little, if anything, to stabilize food supplies and market prices. If serious market failure is apparent in price and food-supply instability in unregulated markets, additional supplementation of private-sector activity would be necessary. To avoid the pitfalls of current commodity programs, subsidies to production or controls on output could not be a part of this governmental intervention.

According to the price-stabilization concept in this option, the government would acquire stocks when commodities are cheap and would sell when commodities are expensive, hence stabilizing markets. Direct payment and supply control would be phased out. A policy tool for accomplishing the desired stabilization of market prices would be governmental management of commodity buffer stocks. Because acquisitions would equal releases over time, the effect on long-term average prices would be small. Although buffer stock stabilization schemes have been used historically, they went beyond stabilization objectives and evolved into long-term price-support policies.

A preferred mechanism to public management of stocks would be to have the government simply pay a buffer stock storage subsidy of say 3¢ per bushel per month to private firms and individuals, essentially a farmer-owned-reserve program without acquisition or trigger prices. Any private storer, not just the producer, would be eligible for the buffer stock storage payment on grains. Such a policy would decrease the tendency to depart from stabilization objectives but would raise concerns that firms and individuals would not acquire and release stocks at the proper times. Although the concern may be valid, government guidelines often have served no better than private decisions in the past (Gardner, 1981).

A third alternative is to provide producers with revenue insurance. The central thrust of pre-New Deal programs under the Federal Farm Board was commodity market stabilization through governmental purchases when prices were low for resale during periods of stronger markets. This basic idea has been continued, with refinements, through the "Ever-Normal Granary" to the "Farmer-Owned Reserve" and finally to international buffer-stock approaches. Past attempts at such stabilization tended to develop into more or less permanent price-support programs, with consequent buildup of unmanageable surpluses. These in turn have led to production controls, restraints on international trade, subsidized sales of stocks, and depressed farm prices.

Consequently, there has been experimentation with programs to support farm income during low-price periods without direct intervention in commodity markets. These programs, such as the current system of deficiency payments, have provided cash payments to farmers. These payments have tended to encourage overproduction. Proposals have been made to move away from these payments while still providing some insurance against low farm incomes (see, for example, Swerling, 1961; Schuh, 1981).

Stabilization of income could be accomplished without direct effects on market prices. The mechanism would be a program making direct payments to producers when their incomes are low. The most straightforward scheme of this type is income insurance. Insurance of gross farm income appears to be more feasible than insurance of net income. An insurance policy on gross receipts would pay an indemnity when a producer's revenue falls below, say, 80 percent of a five-year average of market prices times the producer's normal output. Or the payment could be made to the insured when county price times yield falls below 75 percent of the five-year average of county price times yield. This insurance could be sold by the federal government or by private insurance companies. Premiums would be adjusted as necessary to cover costs over a long-term period. The government could pay part of the premium and administrative costs while substantially reducing Treasury outlays for commodity programs. An insurance program of this type is used for grains in the Prairie Provinces of Canada.

Advantages of this option include:

— This alternative would alleviate unacceptable short-term instability while allowing the market to allocate resources efficiently over the long term.

— Because premiums would be charged for the insurance, induced output effects of unwarranted redistribution to wealthy farmers could be minimized.

— Resources would be allocated more efficiently if farmers' risks were reduced.

— Producers would be able to choose the degree of income protection most suitable to their individual situations and their willingness to pay. They would do so without the governmental costs and potential resource misallocation generated by government price supports and supply control.

— Political pressures for inefficient forms of intervention would be defused by this program.

Disadvantages of the farm-income-stabilization option include the following:

— The insurance probably would not be purchased by producers unless the premiums were heavily subsidized. If the program did not contain subsidies, it would have no chance politically.

— If farmers wanted income insurance enough to pay for it, the market would already be providing it. Crop yield insurance and price hedging in the futures market are strategies now available but not used by most farmers.

— Income insurance is not actuarially or managerially feasible in the near term, and may never be. A series of years of low income would deplete fund reserves.

— The program could be abused by farmers who fail to make wise production and marketing decisions. This "moral hazard" could lead to high insurance premiums and payments, inequitable treatment among participants, and low participation. In time, however, such problems could be reduced by adjusting the premiums of individual producers according to the individuals' historic record of claims or by basing payments on county averages.

DIRECT OR COMPENSATORY PAYMENTS

Government commodity programs featuring direct or compensatory payments would stabilize incomes of agricultural producers while letting market prices and production adjust to economic conditions. Cash payments could be made to producers on the shortfall of market prices below some guaranteed minimum price or on the shortfall of income below some minimum income specified in advance. Limits on payments per recipient could be established. In principle, a direct-payment program would allow market forces the freedom to establish market-clearing prices without supply control. Unsupported commodity market prices would benefit consumers and would keep exports competitive in world markets. Government ownership of stocks would be avoided, along with handling costs and disposal problems for surplus commodities. Payments potentially could be targeted to special groups, such as low-income farmers, financially stressed farmers, or families operating small farms. Government costs of these programs could be monitored readily. Costs would be allocated to taxpayers on the basis of their tax contributions.

Large farms do not need commodity programs — they are efficient and wealthy enough to not warrant subsidies. Part-time smaller farms depend on off-farm income and also do not warrant subsidies. A relatively few mid-size family farms are at risk and the public may wish to intervene to preserve them but at a fraction of the cost of current commodity programs by targeting direct payments to them. Direct payments minimize social costs of transfers if they are *decoupled* from production incentives or supply management.

Direct or compensatory payments have been part of farm legislation for many years. They were introduced for wool in the 1950s. The Agriculture and Consumer Protection Act of 1973 and subsequent legislation have provided for deficiency payments when market prices fall below target prices for wheat, feed grains, rice, and cotton. The *market loan* was mandated for rice and cotton and was permitted for other crops under the Food Security Act of 1985. The market loan allows the producer to pay off the nonrecourse loan at the market price and keep his commodity, hence converts the traditional loan into a direct payment. Extension of the market loan to wheat, feed grains, and soybeans coupled with scaling down of target prices and phasing out production controls (perhaps with the exception of the Conservation Reserve) would be a convenient transition to direct payments.

In theory, paying farmers to reduce output is more cost-effective than direct payments to raise farm income. To be effective, diversion programs must include large farms and not simply concentrate output reductions on small and medium-sized farms. Thus it is more feasible to focus compensatory payments than acreage-diversion payments on smaller farms. A principal argument against payment limitations is administrative problems. It is difficult to establish a system of payment limitations which cannot be circumvented by division of land ownership, reallocation of management responsibility, and other mechanisms.

Other Modifications of Payment Programs. Shortcomings of commodity programs, such as capitalization of benefits into production bases, encouragement of overproduction, and benefits tied to cropland rather than people, are not unique to direct payments. But the flexibility of payments offers some unique remedies, only some of which offer promise.

Benefits of direct payments tied to a production base are capitalized into the base; hence, benefits are lost to the new owner when the base is sold. As a remedy, some have proposed that payments be tied to the producer rather than the production base. Implementation of this proposal would enhance human resource mobility but would raise claims of unfairness if payments were to continue to those who leave farming. Other nonfarmers, of course, would receive no payment.

Payments adjusted to changing yields and cropland acreage become a "supply price" and encourage overproduction. A remedy would be to *decouple* payments by never changing the acreage and yield base. A problem with this remedy would be that payments could continue for apparently unjustifiable reasons when the base is shifted to urban development, forest, or other uses unrelated to the original intent of the payments. The only true decoupling of payments from production incentives is to phase out payments at a prespecified rate over (say) a ten-year period.

Direct payments in the form of a negative income tax or related personal-income-tested program are difficult to administer and raise issues of equity between sectors if only agriculture is included. Also income-tested programs may not serve social objectives such as maintaining the family farm or stabilizing food and fiber prices and quantities in a national food policy.

Summary of Advantages and Disadvantages. Any effort to consider alternative programs to reduce instability in a market-oriented agricultural economy must recognize trade-offs. Some will gain while others lose. No program ensures that all will be better off.

The potential advantages associated with a direct-payments program include the following:

— Lower market prices for farm commodities would increase the competitive position of U.S. farm products in international trade.

— Price-depressing stocks of commodities and associated storage costs would be reduced. Unit production costs would be reduced because cropland would not be diverted but could be combined with other production resources in the least-cost combination to produce output.

— Consumer prices for food would be reduced relative to programs designed to restrict production and raise commodity prices.

— Costs of the program would be allocated to U.S. citizens on the basis of progressive income taxes rather than "regressive" food costs. Because low-income consumers pay a higher proportion of their income for food than do other consumers, the relative burden of higher food costs from supply control fall disproportionately on those with low income.

— The program could be adapted to perishable fruits, vegetables, and livestock products as well as for storable commodities if necessary or appropriate.

— Direct payments limited to each day-to-day farm operator would be an incentive to create more family farms.

— Direct payments could be targeted more effectively than supply-control programs to serve farm-structure objectives such as preserving small and medium-size family farms. Upper limits on payments to individual businesses could be established.

— Direct payments without production controls would give farmers more freedom to make production and marketing decisions while eliminating government administrative costs of supply control.

— Direct public costs of the program would be readily apparent to the public. Effective monitoring of expenditures thus would be encouraged.

— Much of the county, state, and federal structure, as well as the data base, necessary to carry out such a program is already in place or could be adapted readily.

— Livestock feeders would benefit from low feed prices in the absence of supply control.

The disadvantages of a direct-payments program include the following:

— Continuity of direct payments might be threatened in times of budget stringency. Payments to farmers would be highly visible in terms of costs, while consumer benefits could not easily be documented or identified. (Social costs to consumers of tight production controls might be much greater than under direct payments, but costs would be less visible.)

— The market-clearing mechanism, so crucial to making a direct-payments program successful, might not be fully effective in stabilizing or moving supplies in the short run. Thus, direct payments, compared with existing programs, might add to consumers' price instability while buffering farm income.

— The political acceptability of a direct-payments program to farmers would be uncertain. Other things equal, farmers prefer income from the market to income from the government.

— Payment limitations, an attractive feature of direct payments, would be difficult to administer. Limits for each commodity would allow large total payments for a diversified producer. A limit per recipient farmer would require aggregating commodities for the business as a whole. Operators could be expected to divide their businesses "on paper" to avoid limitations. Payment limitations reward farmers who engage in deceit.

— International market repercussions, particularly with competitors, would be likely if our farm export prices were to become more unstable and were viewed as subsidized by payments in the absence of production controls.

— Budgeting the costs of compensatory payments would be difficult because the difference between an established target price and the future market-clearing price would be unknown. Potential government costs would be high.

— Establishing "equitable" target prices for commodities or products would be difficult if the aim were to provide a measure of income insurance in times of surplus but not to provide incentives for production beyond expected effective demand.

— Payments would be difficult to divorce from production incentives. Payments could lead to some kind of production controls.

— Federal outlays go further to raise farm incomes when used to reduce farm output and, hence, raise prices.

— Direct payments might delay resource adjustment out of commercial agriculture in areas or situations where continued production would be uneconomic without subsidies.

— The only way to fully decouple payments from incentives is to continue payments whether farmers produce or not and indeed whether they remain on the farm or not. But if farmers receive payments for doing nothing and while living and working in the city, nonfarmers wonder why they cannot receive similar generosity from government.

— A negative income tax or related income-maintenance program not tied to production might be a more cost-effective method to target income protection to the needy.

DEMAND AND REVENUE EXPANSION

Demand- and revenue-expansion programs are preferred by agriculture producer groups in the United States because such programs promise to increase commodity receipts and solve farm problems without use of government payments or supply control. In this section, alternatives are separated into general categories: (1) options that increase demand

and (2) options such as multiple price plans to obtain more revenue from a given demand without supply control.

Efforts to Increase Demand. Demand expansion results in higher prices for a given supply. Options to increase the demand for food and fiber must take into consideration the factors that cause the demand to increase: (1) tastes and preferences of consumers, (2) disposable income per capita, (3) population size, and (4) price and availability of substitutes.

Little can be done by producer groups to enhance demand by increasing disposable income and population or by influencing prices and quantities of substitutes. Options do exist, however, for changing the tastes and preferences of consumers.

Advertising and promotion campaigns may increase demand for particular food items by changing consumers' tastes and preferences. U.S. agricultural producers have used checkoff programs to fund product-promotion campaigns domestically over the past 30 years. These promotion campaigns emphasize generic advertising for a particular type of food or fiber such as pork, milk, oranges, beef, or cotton. A commodity group organized and funded by producers sometimes cooperates with the government to increase demand through advertising.

Calories and pounds of food consumed by each person is affected little by advertising. More domestic demand for a particular food group, say, beef, usually results in less demand for other food groups due to substitution by consumers. Advertising and other promotion, however, can raise total receipts for agriculture when consumers substitute higher-priced food items for lower-priced items or where a new use of a product, say for fuel, does not interfere with other uses.

Development of alternative uses for a raw product can increase the demand for some agricultural commodities. The use of corn to produce fructose is an example of demand expansion through alternative uses for the crop. Alternative uses for a commodity, however numerous, are of little benefit to producers if uneconomic at typical price relationships. For example, production of ethanol from grains for motor fuel has been economically feasible only with substantial subsidies.

Foreign market development efforts of various groups, including the Foreign Agricultural Service (FAS) of the U.S. Department of Agriculture, have increased export demand for U.S. agricultural commodities. FAS manages market-development activities that are planned, implemented, evaluated, and financed jointly by FAS and a cooperating producer organization. Activities of FAS emphasize market information, technical assistance to importers, buyer awareness, and consumer education. The producers' share of the costs generally is financed by a checkoff program made feasible by enabling legislation.

Well-financed market-development programs to increase export demand work best if the U.S. commodity is available for export at competitive prices. In comparison with commodity programs, the cost to the government and producer groups is small for demand expansion. Other activities to influence demand for exports include reduction of trade barriers and export-subsidy programs.

The government has subsidized exports under credit guarantee, credit subsidy, export payment-in-kind, and other programs. These programs have been especially useful to counter subsidies used by other exporters. The programs have angered competing, nonsubsidizing exporters and ineligible importers.

PL 480 (Food for Peace) exports tend to increase demand through their market-development actions. The program may have been designed originally for humanitarian and supply-disposal purposes, but it helped build commercial markets in Japan, South Korea, Taiwan, Brazil, Spain, and elsewhere.

Government food programs to expand domestic food demand have included commodity donation, food stamps, the Women-Infant-Children Supplemental Food Program, and the school-lunch program. Some surplus foodstuffs acquired by the Commodity Credit Corporation in its price-support activities have been given to institutions and needy families under the commodity-donation program.

Under the school-lunch program, USDA donates food plus a small cash subsidy per lunch to reduce the cost of school lunches and increase the nutritional intake of school children.

Additional federal spending for food programs tends to displace commercial sales. However, Wetmore et al. estimated in 1959 that food programs could feasibly expand domestic demand up to 4 percent. The percentage is probably less today.

Efforts to Increase Revenue with a Given Demand. Revenue-enhancing approaches discussed here are of two general types: (1) obtaining greater access to markets by reducing trade barriers, by trade agreements, and by other means and (2) price discrimination by export subsidies or administered lower prices in export markets than in domestic markets. The distinction between demand enhancement discussed in the preceding section and revenue enhancement discussed here is often blurred, however.

Even without supply control or direct payments, industry revenue potentially can be increased through *multiple-price plans* by shifting sales among markets of different elasticities. For this approach to work, markets for a commodity (1) must be separable and (2) must have different price elasticities. Administrative procedures must be suitable to allocate quantities and charge different prices in each market.

Revenue can be expanded by charging a higher price in inelastic domestic markets and a lower price in elastic export markets. Export demands faced by U.S. producers for grains and for cotton are considered to be relatively elastic in the long run. The peanut program and the milk program are examples of division of markets and prices to increase revenue.

Various options, some actually used and others merely proposed, for increasing revenue are discussed in this section. Two-price and other means to raise revenue from a given demand without supply control also are discussed in other sections.

Long-term bilateral trade agreements can also be used to increase quantities exported. Long-term bilateral trade agreements specify annual minimum and maximum quantities of agricultural commodities that will be traded between two countries over a given time period. Prices for these exports are not negotiated, but depend upon prevailing market conditions. The long-term grain agreements between the United States and the Soviet Union negotiated in 1975 and again in later years are examples of bilateral trade agreements. Such agreements offer a way to expand existing markets in foreign countries at low cost. A disadvantage to U.S. agricultural producers and world markets is that bilateral trade agreements may increase price instability and impede trade among countries not included in the agreements.

Marketing boards have been proposed by some producer groups interested in increasing export prices. A central government authority or marketing board (similar to Canada's Wheat Marketing Board) would direct foreign grain and fiber sales to secure higher prices for exports. Price enhancement rather than increased exports would be the primary goal of the board. At the extreme, a central marketing board could make major production and marketing decisions for producers to stabilize domestic prices and control output. If the marketing board reduced marketing to obtain a higher price, production would have to be restrained.

Another approach patterned after the McNary-Haugen proposals of the 1920s would be for a government corporation to accept any commodities that would not sell in domestic markets at say 75 percent of parity.[2] The corporation would sell the "surplus" in export markets for the world price and return the proceeds to producers. Marginal output would receive the world price, discouraging overproduction. This approach would be very difficult to administer. It would raise domestic food prices, and could bring charges by competitors that we are dumping surpluses in international markets and by domestic consumers that they are being forced to pay more for food than are foreigners.

Multinational cartels for food and fiber producers have also been proposed as a means of increasing prices. Demand-quantity expansion in foreign markets is a secondary objective for export cartels. The main objectives are supply control, price enhancement, and price stability on an international scale. To be effective, cartels require member discipline, close coordination of domestic farm programs, and provisions for allocating market shares to participants. Because demand is more inelastic for several countries in a cartel, a cartel is more successful than action by one country to raise revenue by withholding exports.

A cartel for agricultural exports could increase and stabilize domestic prices in the short run. However, there would be considerable pressure to undercut the cartel's established export price. Grains and fiber can be produced in a large number of countries. A cartel's efforts to raise the world price would encourage production in other countries to erode the cartel's export market and reduce receipts below unrestricted market levels in the long run (Hillman, Peck, and Schmitz, 1978, pp. 69-70). Other disadvantages of a cartel are apparent in the section on mandatory controls.

Full-cost-of-production pricing for exports has been proposed for grains and fibers. such a policy would require setting minimum export prices for grains and fibers equal to

---

[2]One method to operate such a program would be to issue certificates to producers. The certificates would be required to attend sales of the commodity for domestic use at a specified price which would be above the world price level. Tariffs would be required to protect the domestic market from imports at lower prices.

their full cost of production, while letting domestic prices be set by the "free" trade forces of demand and supply. Such action would be the opposite of the traditional price-discrimination solution for maximizing revenues when faced with two markets. Charging the higher price in the elastic export market and the lower price in the inelastic domestic market would reduce total revenue. This option would make U.S. agricultural commodities less competitive, reducing quantity more than increasing price in the long run. Full-cost-of-production for exports would likely reduce domestic prices of affected food and fiber in the short run due to the price-depressing effect of reduced export quantities shifted to domestic markets. Our producers rather than the foreign buyer would bear the burden of full-cost export pricing. Full-cost export pricing would minimize rather than maximize long-term revenue.

The advantages and limitations of demand- or revenue-enhancing proposals are summarized as follows:

The advantages of demand-expansion programs may be summarized as follows:

— Demand expansion does not interfere with individual producers' production and marketing decisions.

— Advertising funded by a producer checkoff increases demand without large government costs.

— Developing alternative uses for farm commodities makes the total demand more elastic and thus less price-sensitive to random changes in supply.

— Government food programs not only enhance farm income but also are a basic income supplement to low-income consumers, improving their nutritional intake.

The disadvantages and limitations of expansion program include the following:

— Unexploited opportunities to expand demand above current levels at favorable benefit-cost ratios are few indeed and would have little impact on excess capacity.

— Using advertising to expand consumption is detrimental to society because the greatest nutritional problem in the U.S. is chronic overeating and obesity.

— Demand-expansion efforts cannot easily be turned on and off to cope with economic problems of instability, cash-flow, and commodity surpluses in agriculture.

— Demand expansion by advertising and promotion, by finding new uses for outputs, and by domestic and export subsidies may be worth pursuing where benefits exceed costs, but the payoff is primarily of longer-term benefit to producers.

— Many alternative uses of major farm commodities are not economical without subsidies. Encouraging resource adjustment out of agriculture is often a lower-cost alternative.

— Advertising that increases the demand for one food item reduces the demand for substitute food items. Beef, pork, and poultry advertisers merely offset each other and preserve market shares with no benefit to each commodity. Promotion may expand the demand for food processing rather than for raw food ingredients.

— Foreign-market-development activities are unproductive if the commodity for which demand is created cannot be provided by U.S. farmers at competitive prices.

— PL 480 exports often replace commercial exports, discourage agricultural development in recipient countries, and may be controlled for diplomatic purposes.

— Supply responses to higher prices reduce some of the long-run benefits from demand expansion.

— Once initiated, advertising campaigns must be continued to maintain demand. Increases in demand are not necessarily permanent.

— Attempts to increase domestic food programs substantially would result in wasted food in school-lunch programs, food stamps spent for nonfood items, and transfers of government funds to those who can afford to pay.

— Export subsidies anger competing exporters and non-eligible importers, contributing to trade wars.

— The U.S. has much to gain from more open international markets; unfair exploitation of those markets undermines our moral position in calling for freer trade.

— Generating revenues by utilizing demand more fully is a long-term remedy which does not deal with farm-price instability and cash-flow problems.

— Bilateral trade agreements increase price instability, impede trade, and may interrupt supplies for countries excluded from the agreement, many of which are developing nations.

— Export subsidies not only burden U.S. taxpayers but also discourage agricultural progress in developing countries by providing food imports at "unfair" subsidized prices.

— Direct dollar export subsidies are prohibited under the General Agreement on Tariffs and Trade if they increase market share. Export subsidies raise claims abroad that the United States is "dumping;" this weakens our moral bargaining position to stop dumping by foreign countries in the U.S. market.

— A marketing board, an export cartel, or administered full-cost export pricing could make U.S. exports less competitive, perhaps increase costs for supply control or carrying surpluses, increase domestic food prices, and reduce total commodity revenues in the long run.

Mandatory Controls

The referendum-quota approach, here called mandatory controls, refers to programs requiring producers to reduce output or marketings without diversion payment. The lower quantity raises commodity price and, if demand is inelastic, receipts. This option contrasts with direct payments in that higher farm income comes from consumers in the marketplace rather than from federal taxes. Such programs can be directed by organized farmers or by government. The latter is emphasized here, but the principles would remain the same for whomever administers the program.

Marketing quotas and acreage allotments are the major instruments of control. A marketing quota limits the quantity of a specific commodity that each farmer is allowed to place on the market; an acreage allotment limits the number of acres that each farmer can plant or harvest. Once the program has been authorized by Congress, it can be implemented only after being submitted in a referendum to producers of the crop. If a specified proportion (usually two-thirds but one-half in the Harkin-Gephardt proposal) of these producers vote in favor of the program, the provisions become binding for all producers.

Marketing quotas and/or acreage allotments have been used at various times over the past 40 years for several commodities including cotton, wheat, peanuts, tobacco, and milk. Marketing quotas are still in operation for tobacco, peanuts, and hops. Marketing orders are used to control the marketing of some fruits and vegetables.

The overall purpose of this section is to analyze the feasibility of extending mandatory controls to a broad range of crops. After examining current mandatory control programs for tobacco, the discussion focuses on acceptance by farmers and on the economic impact of extending mandatory controls to other crops. The final portion summarizes advantages and disadvantages of mandatory controls.

Current Mandatory Control Programs. The tobacco program is an example of a mandatory-control approach that has survived five decades, through evolutionary processes. This program gives insight into the potential structure of mandatory controls applied to a broad range of crops.

Tobacco production has been controlled through acreage allotments and/or poundage quotas since the Agriculture Adjustment Act of 1938. Currently, production of barley and flue-cured tobacco is restricted by poundage quotas. Some of the lesser-volume types are still produced under acreage allotments. The tobacco price-support program works through farmer-owned cooperatives which buy surplus tobacco with government loans. Under the "no-net-cost" tobacco program enacted by Congress, producers contribute fees to a fund covering any loans the cooperative cannot repay. Another feature of the no-net-cost tobacco program provides for the lease and transfer of allocated quotas within a county. Owners of allotments can lease or sell these rights separately from the farm to which the quotas are attached. The national marketing quota for tobacco has steadily declined through time. The declining quota is associated with the declining share of U.S. tobacco in world market and is one outgrowth of high, rigid price supports.

Potential of Mandatory Controls for Other Crops. The relative success of mandatory controls for tobacco, as measured by producers' acceptance and monetary costs, raises the issue of extending this program to other crops such as wheat, feed grains, cotton, and soybeans. The 1985 farm bill required a nonbinding referendum on mandatory controls among wheat producers and allowed the Secretary of Agriculture to impose such controls. The 54 percent vote for mandatory controls in 1986 was viewed as inconclusive, and mandatory controls were not imposed.

Economic problems of agriculture are not confined to specific crops. Surplus aggregate production capacity cannot be eliminated by extending mandatory production controls or bushel or poundage quotas to one or a few crops. Resources diverted from the production of the controlled crop would be switched to other corps which were not controlled. Hence, in the absence of associated acreage-division programs, mandatory production controls would have to be implemented for virtually all crops to address the overall problem of surplus crop-production capacity. If tobacco producers were required to divert acres to soil conserving uses as would be required in a comprehensive multi-commodity control program, they would find mandatory controls less financially rewarding.

In the absence of large export subsidies, receipts to export markets would sharply decline with a one-price system of mandatory controls supporting prices at 70-80 percent of 1910-14 parity as called for in the Harkin-Gephardt bill. One option would be a two-price plan. Producers would be provided a domestic allotment accompanied by certificates which would attend any sales in the domestic market at a required price of, say, 75 percent of parity. Producers would market additional output in export markets at the lower world price. To avoid charges of dumping, producers would be required to divert land to soil conserving uses for no payment. Such a program would provide less income to producers than under 1987 programs but more income than under a free market in the short run.

Raising crop prices would have impacts upon other sectors of the farm and nonfarm economies, including consumers. Higher farm prices would not be absorbed by marketing firms and would be passed on to consumers. Higher crop prices would have a direct impact on retail prices of bread, cereals, and other foods made from these products. In addition, higher prices on crops fed to livestock and poultry would impact on producers and consumers of beef, pork, lamb, poultry, eggs, and milk.

Evaluation of Mandatory Controls. Advantages of mandatory control programs include the following:

— Mandatory controls raise and stabilize farm commodity prices without the government costs required for voluntary paid-diversion programs or direct payments.

— Mandatory controls potentially are more effective in supply control than voluntary programs because all producers are required to participate in the programs. hence, the "free-rider" problems associated with voluntary programs are lessened.

— If poundage or bushel marketing quotas rather than acreage allotments are used to reduce supplies, the tendency for farmers to raise yields on the allotted acreage is less and producers can combine inputs in a least-cost manner.

Although mandatory-control programs have been used to achieve such policy objectives as increasing farm commodity prices and reducing payment costs, they have also been subject to considerable criticism. Disadvantages of the mandatory supply-control program include the following:

— Reduced output achieved through quota restrictions would raise food prices for consumers. The higher food prices would constitute a regressive, cost-of-living drain on low-income consumers since they spend a higher proportion of their income for food than do high-income consumers.

— Producers would gain less than consumers and taxpayers would lose because mandatory controls bring inefficiency — deadweight loss as discussed earlier.

— Many fewer jobs would be saved in farming than would be lost in the agribusiness sector which constitutes nine-tenths of the food and agriculture industry.

— Higher commodity prices resulting from mandatory controls would severely restrict U.S. exports of farm products. Exports of grains, soybeans, and cotton could be at least cut in half in the long run under mandatory controls without export subsidies. Demand might be so responsive to price in international markets that revenue not only from exports but also from combined domestic and export sales of these commodities would fall after a few years.

— Allotments and quotas prevent efficient producers from increasing production and tend to keep inefficient producers in production.

— Mandatory controls freeze production patterns which may become increasingly inefficient over time. Producers would be unable to produce at least cost. Allotment rigidities would slow the adoption of technological and market innovations.

— Anticipated income benefits from mandatory controls would be capitalized into marketing quotas or acreage allotments. When allotments were sold, the original owner would receive the expected future benefits through the capitalized value of the control instrument, while benefits would be lost to subsequent buyers or renters. Hired farm laborers and renters would lose opportunities in farming and would be worse off with mandatory controls.

— Producers failing to participate in mandatory programs would face civil penalties, a severe problem if noncompliance were widespread. Administrative monitoring to avoid cheating would become increasingly intensive, burdensome, and costly. Inducements to circumvent the program would rise with the program's success in raising prices and returns. Forced compliance would intensify efforts to circumvent the program. Administering mandatory controls to grain producers who feed the grain on their own farms would be especially difficult. And if such producers are exempt from controls, they eventually would drive out commercial producers who must purchase grain at high prices or would force commercial feeders to vertically integrate into producing feed.

— Because mandatory controls increase income through market prices, it would not be possible to target benefits to needy farmers.

— Beneficiaries would spend large amounts lobbying to retain or expand government-bestowed economic rents. Resources used for lobbying would be socially wasteful.

— Livestock producers would be hurt by high feed costs induced by mandatory controls.

— Mandatory controls by themselves do not necessarily reduce price instability or assure stable food supplies to consumers. Hence, price supports and commodity stock-reserve programs might need to be linked with mandatory controls. Also tendencies to increase the output of commodities not covered by controls using resources released from controlled commodities would bring pressure for control of all commodities, for a government resource-diversion program, or for both.

— Mandatory controls could speed consolidation into larger farms. The economic inefficiency of mandatory controls could be reduced by making allotments negotiable in an open national market so that production would move to the least-cost areas and producers. Negotiable allotments in a mandatory dairy program in Canada speeded the movement toward fewer, larger dairy farms — an outcome opposite to what the program was supposed to achieve (Dvoskin, 1987, p. 31).

SHORT-TERM VOLUNTARY LAND DIVERSION

Acreage diversion and related programs in recent years have been effective in curtailing production of specific crops such as corn and wheat. These were the "bread-and-butter" programs of recent decades. Like the land-retirement program, they have a double-barrelled effect on farm income and can make government dollars go farther than would direct payments to raise farm earnings. Farmers increase their income, first, by receiving a cash payment to remove land from production. Second, by reducing national production, they reduce the aggregate supply quantity and get better prices in the market. As noted earlier, cost-effectiveness of voluntary acreage diversion programs has been disappointing compared to the potential.

Advantages of short-term diversion programs are summarized as follows:

— They are proven programs with which producers and policymakers feel comfortable.

— Annual diversion programs are flexible from year to year. They can be general set-aside of cropland or restrictions on acreage of one crop.

— Voluntary programs give farmers freedom to participate or not and to make their own production and marketing decisions.

— Acreage diversion programs work well with other programs such as direct payments and long-term land retirement.

— Voluntary programs are easier to administer and police than mandatory programs.

— Because acreage diversion is oriented to better land than long-term land retirement, it reduces output with a smaller acreage withdrawal.

Disadvantages of voluntary acreage diversion include:

— The programs are costly to the Treasury; slippage is a serious shortcoming. Producers learn to substitute fertilizers and pesticides for land.

— Acreage diversion programs are difficult to target to needy farmers if production is to be controlled. Large farms benefit most.

— Any diversion program tends to price the nation out of international markets and to accumulate stocks.

— Acreage diversion is not as effective as the whole-farm land-retirement program in inducing movement of farm workers, operating capital, and marginal land to uses in the public interest.

DAIRY POLICY

Most policy options in this chapter relate directly to crops. Economic problems in the dairy industry are unique and receive separate treatment in this section.

Various approaches have attempted with mixed success to stabilize dairy production and prices. As noted earlier, the blend-pricing scheme inherently causes overproduction. Measures to remedy the situation have included (1) lower support prices, (2) coresponsibility levies whereby producers were taxed on their output and the tax proceeds in turn were used to pay them not to produce, and (3) a dairy cow buyout program, whereby the government paid farmers to send whole herds to slaughter. None of these measures succeeded in controlling production except on a transitory basis.

The list of existing and proposed dairy policy instruments is long and varied. To avoid becoming bogged down in the morass of specific policy mechanisms and losing sight of the important policy options, only a short list of selected options is discussed. Some options are to continue the current support program, a milk-tax plan, a sales-expansion plan, and a quota plan.

The main features of the current support program could be left intact with changes in the price mechanism. The program support price has been moved by the indexes of prices paid and received in the farm sector and by prices administered by Congress. A parity formula could be developed for the dairy sector which would be more responsive to productivity, feed costs, and other factors influencing milk production and consumption. Various mechanisms have been proposed to trigger higher or lower support prices based upon USDA dairy purchases.

There are advantages of modifying the current support program. Many of the options for dairy such as a greater market orientation and expansion of demand are similar to those discussed earlier and are not repeated here. A central issue of dairy is whether to move to a market quota on either all or some portion of production. Various types of quota plans have been proposed as a means of balancing milk production and consumption at predetermined prices. The quota could be fixed on the basis of production history or could be modified over time by changes in future production and consumption. Most quota plans would involve two or more prices — a higher, administered price for milk produced within the quota and a lower, market-clearing price for milk in excess of the quota. Milk produced in excess of the quota would receive the market price rather than a blend price. Milk quota plans now are administered by several state milk-control agencies and dairy cooperatives.

One option would be to provide for higher administered prices on an assigned proportion of milk production (base) and a free market on additional milk. Actually receiving the market price on additional output would stop excess production. The government would not purchase excess supplies but would help to administer the program. The issue of whether to have a base is separate from the issue of what constitutes the base. One approach would be to include only Class I milk in the base. Because manufactured milk products are stored and transported more easily than fluid milk, they might be allocated by the market alone. The higher price on the base milk would provide stability, cover additional resource costs, and assure supplies. The base would provide some income security to producers. Reliance would be placed upon the private trade to manage storage and stock acquisition and release functions. This approach would cause capitalization of the value of the milk quota whether the quota is negotiable or not. Unless the quotas could be sold or rented, they would tend to freeze production patterns with resulting built-in inefficiency. Quotas might be given a specified life, after which they would either be renegotiated or terminated. An alternative to reduce the unfair "milk tax" on fluid milk consumers would be for the government to replace the premium charged consumers with a direct payment. The payment might be financed in part by resale of bases each five years by the government.

Some of the advantages of quota plans are:

— Tax-based costs could be nil.
— Production and consumption could be kept in balance.
— Price signals reflecting the marginal value of milk would be clearer than at present.
— Management decisions might be improved.

Some disadvantages are:

— Entry barriers would be raised.
— The value of quotas would become capitalized into control instruments.
— Resource allocation would be distorted compared to a more market-oriented program.
— Administrative cost could be high.
— Milk consumers would pay a premium for quota milk.

Other advantages and disadvantages were discussed in the mandatory control section.

SUMMARY AND CONCLUSIONS

The advantages and disadvantages of the farm commodity programs have been well documented. Proponents argue that the programs have reduced instability in farm prices and incomes, have provided a strategic reserve of production capacity to meet unpredictable emergencies such as wars and droughts, have provided an orderly outmovement of surplus farm labor, have conserved farm resources for future generations, and have lessened the trauma of low farm income.

Opponents argue that the programs have cost taxpayers too much money, have benefited only large producers, have regressively distributed income from taxpayers of modest means to wealthy farmers, have diverted public attention and support from real problems of rural poverty, have interfered with freedom of farmers to produce and market as they please, have lost their effectiveness in raising income through capitalization of benefits into land or through slippage (bringing in new cropland, using more fertilizer, etc.), have interfered with commercial exports of farm products, and have caused inefficiency through freezing of production patterns and idling of land resources which have little value for anything but agricultural uses.

Suggestions to improve farm programs have been offered. It has been suggested that allotments be made negotiable, that acreage allotments be shifted to bushel or poundage quotas, that "normal" yields be set once and for all so farmers are not encouraged to expand yields to get more payments, that a farmer not be allowed to move allotments from a poor farm which he purchases to the good land on his "home" farm, that the farmer actually receive the market (rather than a blend) price for his marginal production so as to constrain output expansion in a two-price program, that long-term land retirement be expanded so as to remove marginal land from production and to reduce government costs, that program administration be streamlined at the local level, that programs be announced before farmers plant, that payments be cut off or graduated for large farmers, and that program formulation be placed in the hands of an Agricultural Board patterned after the Federal Reserve Board. Many of these changes in programs have merit, but changes in commercial farm policies come slowly.

Some have proposed a tax or limits on fertilizer use. Higher fertilizer and pesticide prices can raise farm income because input demand is more elastic than product demand (see Tweeten, 1979, pp. 253-258). Such measures might be justified for environmental reasons, but are unlikely to be accepted by farmers.

Another proposal is to restrict agricultural research and extension. The long-term ability of agriculture to provide adequate, safe food supplies at low cost to consumers while protecting the environment and competing internationally will require more rather than less application of science.

With each new farm bill, the search goes out for a magic program that will simultaneously resolve agriculture's economic problems. Laws of economics operate like laws of physics, conditioning what is feasible. Increased farm income must come from consumers, taxpayers, or greater efficiency, for example. The major program alternatives were reviewed in this chapter — new programs are variants of these basic programs.

In formulating realistic policies, it is well to recognize that commodity programs do not raise the net income of farm people over the long run. Farmers have demonstrated that they are capable of adjusting to changing conditions. Adequate-size, reasonably well-managed family farms have and likely will continue to earn a favorable return on resources. A greater market-orientation in farming threatens neither the family farm nor food supplies. There is little point in holding farm prices above the free-market equilibrium level over extended periods. The benefits of higher prices will be capitalized into land or other control instruments and will eventually be lost to farmers anyway.

In concluding this chapter, it is well to review the problems of agriculture within the context of concern for the level, distribution, and variability of farm economic outcomes. The intent is to note government programs that will address the problem consistent with social welfare objectives. A precise match between the objectives listed below of alleviating five farm problems and of policy instruments is achieved by having an equal member of each. However, it is well to watch for complementaries among instruments which can reduce the cost of addressing problems. It may be noted that interventions suggested to address farm problems are far less intensive and costly than are current programs.

Financial Stress: The principal instrument is monetary-fiscal policy to achieve national economic growth with price stability — including interest and exchange rates. In addition, programs of personal and financial counseling, improved general and vocational-technical education, job information, and mobility assistance can ease the pain of adjustment for those who have difficulty coping with the managerial and economic size challenges of a viable agriculture.

Instability: Annual and cyclical instability remains the number one problem of commercial agriculture. The suggested main policy instrument is publicly subsidized grain stocks. The instrument might be as simple as a 3¢ per bushel per month subsidy to anyone who agrees to store buffer stocks of grain from one year to the next. In addition, the government might hold an emergency wheat reserve as it has done from time to time.

It is difficult to justify continuing massive transfers from lower income taxpayers to more wealthy farmers when such transfers lose their intended benefits for hired farm workers, renters, and new owners. There is little point in holding farm price above long-term market clearing levels. By use of futures options, storage, insurance, and drawing on equity capital reserves, well-managed farms can cope with instability. However, if some intervention is deemed necessary, the simplest approach is extension of the market loan to all grains and soybeans as well as cotton. The target price and production controls would be phased out over a ten-year period, leaving only a market loan set at some proportion of the five-year moving average of past market prices. Decoupling would be attained by limiting the market loan to some proportion of the historic base production.

Environment: The principal commodity policy instrument to deal with problems of the environment and natural resource depletion would be the long-term Conservation Reserve Program. The whole- or part-farm program might be expanded to 50 or 55 million acres on a bid basis using perpetual cropping easements. Under the easement, the government would provide a loan on bid basis to the landowner in return for cropping rights on cropland prone to erosion or irrigated by nonrenewable water supplies. The farmer would graze or hay the land under an approved conservation plan, but could not crop it. The easement loan could be temporarily suspended by declaration of government in times of food needs or permanently by the owner repaying the loan plus interest. Because easement would allow noncrop dryland uses of land, it would have less impact on rural communities that other diversion programs (which would be terminated). Up to 40 percent of cropland could be removed in a county — even more if deemed necessary to conserve groundwater. The conservation easement programs not only would conserve soil and water but would remove excess capacity, provide reserve production capacity if needed, and would help to stabilize the farm economy. Many specific measures to protect food and water from chemical residues and in general assure high quality, safe food supplies would accompany the above effort.

Poverty: A principal problem of non-commercial farms is poverty. The main policy instrument to alleviate poverty is human resource development programs and improved welfare programs such as outlined in Chapter 10. It is difficult to build a case for special programs to address poverty and underemployment on farms alone. Programs of welfare reform and human resource development would be expected to simultaneously address problems of poverty on farms, rural areas, suburbs, and ghettoes. However, the special needs of farm people such as the high incidence of underemployment, isolation, and the working poor in intact families need specific emphasis.

Family Farm: The family farm is a remarkably resilient institution which will be around for generations to come. The part-time family farm is especially durable and may be increasing in numbers. Market and production economies now extend beyond the family farm size and the middle-size family farm of $40,000 to $150,000 sales is increasingly marginal. It will feel pressures to either become larger or obtain more off-farm earnings. The latter is difficult because such farms demand much time.

Land reform is unwarranted because it would sacrifice low-cost food supplies from larger farms and because the government would find it difficult to make better decisions than the market regarding which operators should be free to grow and which should be pruned. Relatively few rural people and communities today rely on farming for their economic base and those who do would be better served by policies of human resource development and part-time farming rather than by land reform.

The principal commodity policy instrument to assist family farms, if the public deems such policy is warranted, would be a premium provided family-size farms under the market loan direct payment discussed above to provide greater economic stability. The premium would be provided only to the person responsible for the day-to-day operation of the farm. No operator could receive more than a specified payment. The maximum size of the payment is arbitrary and is a political decision. However, it is difficult to make a case for a larger payment to farmers than the value of food stamps and AFDC payments received by a family on welfare.

REFERENCES

Carr, A. Barry and Luther Tweeten. 1974. *Comparative Efficiency of Selected Voluntary Acreage Control Programs in the Use of Government Funds*. Oklahoma Agricultural Experiment Station Research Report P-696. Stillwater: Oklahoma State University.

Christensen, Raymond P. and Ronald O. Aines. 1962. *Economic Effects of Acreage Control Programs in the 1950s*. Economic Report No. 18. Washington, D.C.: U.S. Department of Agriculture.

Council for Agricultural Science and Technology (CAST). 1983. *The Emerging Economics of Agriculture*. Report No. 98. Ames, Iowa: CAST.

Council of Economic Advisors. January 1987. Economic report of the President. Washington, D.C.: U.S. Government Printing Office.

Dvoskin, Dan. 1987. *Excess Capacity in U.S. Agriculture*. Staff Report No. AGES870618. Washington, D.C.: Resources and Technology Division, ERS, USDA.

Dvoskin, Dan. May 1987. Some international experiences with mandatory supply controls. *Agricultural Outlook* AO-130:29-33.

Gardner, Bruce. 1987. *The Economics of Agricultural Policies*. New York: Macmillan.

Gardner, Bruce. 1981. *The Governing of Agriculture*. Lawrence: Regents Press of Kansas.

Gardner, Bruce, ed. 1985. *U.S. Agricultural Policy: The 1985 Farm Legislation*. Washington, D.C.: American Enterprise Institute.

Legislative Reference Service, Library of Congress. 1965. *Farm Programs and Dynamic Forces in Agriculture*. U.S. Senate Committee on Agriculture and Forestry, 89th Congress, 2nd Session. Washington, D.C.: Government Printing Office.

Robinson, K.L. 1960. Possible effects of eliminating direct price support and acreage control programs. Pp. 5813-20 in *Farm Economics*. Ithaca, New York: Cornell University.

Shephard, Geoffrey et al. June 1963. *Controlling Inputs*. Bulletin B-798. Columbia, Missouri: Agricultural Experiment Station.

Sobering, Fred D. and Luther G. Tweeten. 1964. A simplified approach to adjustment analysis. *Journal of Farm Economics* 47:820-34.

Spitze, Robert, Daryll Ray, Allen Walter, and Jerry West. 1980. Public agricultural food policies and small farms. Paper I of NRC Small Farm Project. Washington, D.C.: National Rural Center.

Tweeten, Luther. 1976. Objectives of U.S. food and agricultural policy and implications for commodity legislation. Pp. 41-63 in *Farm and Food Policy 1977*. Print 75-040 0 of the Committee on Agriculture and Forestry, U.S. Senate, 94th Congress, 2nd Session. Washington, D.C.: Government Printing Office.

Tweeten, Luther. 1981. Prospective changes in U.S. agricultural structure. Pp. 113-146 in D. Gale, ed., *Food and Agricultural Policy for the 1980s*. Washington, D.C.: American Enterprise Institute.

U.S. Department of Agriculture. November 1986. *Economic Indicators of the Farm Sector: National Financial Summary, 1985*. ECIFS5-2. Washington, D.C.: ERS, USDA.

Glossary of Agricultural Policy Terms

ACREAGE ALLOTMENT. An individual farm's share, based on its previous production, of the national acreage needed to produce sufficient supplies of a particular crop. Also used to refer to *base acreage* defined as acres planted for harvest or diverted of a specific commodity on a farm in a specific period, say the last two-year average or the last five-year average dropping the high and low years.

ACREAGE-REDUCTION PROGRAM (ARP). A voluntary short-term land retirement system in which farmers reduce their planted acreage of a specific crop from their base acreage. Farmers are usually not paid for ARP participation (although it can be required for participation in other agricultural programs), but they are made eligible for nonrecourse loans and deficiency payments defined below.

AGRICULTURAL STABILIZATION AND CONSERVATION SERVICE (ASCS). A USDA agency responsible for administering farm price- and income-support programs as well as some conservation and forestry cost-sharing programs; local offices are maintained in nearly all farming county seats.

COMMODITY CREDIT CORPORATION (CCC). A wholly owned federal corporation within USDA. CCC functions as the financial institution through which all money transactions are handled for farm price and income support.

CONSERVATION RESERVE PROGRAM (CRP). A long-term general (not crop specific) land retirement program under which farmers voluntarily contract to take cropland out of production typically for ten years and devote it to conserving uses. In return, farmers may receive an annual rental payment for the contract period and payment assistance either in cash or in kind (PIK) for carrying out approved conservation practices on the conservation acreage.

DEFICIENCY PAYMENT. Government payment made to farmers who participate in feed grain, wheat, rice, or cotton programs; payment rate is per bushel, pound, or hundredweight, based on the difference between a target price and the market price or the loan rate, whichever difference is less. Payment base is base acreage (less diversions of a crop) times *program yield.* See *Target price.*

FARM. Defined by the Bureau of the Census in 1978 as any place that has or normally would have had $1,000 or more in gross sales of farm products.

FARMER-OWNED RESERVE (FOR). Farmers place their grain in storage and receive extended nonrecourse loans for three years, with extensions waived, and farmers may receive annual storage payments from the government. Farmers cannot take grain out of storage without penalty unless the market price reaches a specified "release price." When the release price is reached, farmers may elect to remove their grain from the reserve but are not required to do so. However, at that point the storage and interest incentives may be reduced or eliminated.

FEDERAL MARKETING ORDERS AND AGREEMENTS. To promote orderly marketing, a means authorized by legislation for agricultural producers to collectively influence the supply, demand, or price of particular commodities. If approved by a required number of commodity's producers, the marketing order is binding on handlers of the commodity. It may limit total marketings, prorate the movement of a commodity to market, or impose site and grade standards.

LOAN RATE. The price per unit (bushel, bale, or pound) at which the government will provide loans to farmers to enable them to hold their crops for later sales.

MARKETING LOAN. Authorizes producers to repay their commodity loan at a lower "market" price level.

NONRECOURSE LOANS. Price-support loans to farmers to enable them to hold their crops for later sale, usually for nine months or within the marketing year. The loans are nonrecourse in that farmers can forfeit without penalty the loan collateral (the commodity) to the government as full settlement of the loan. If the market price rises above the loan rate, the producer can repay the loan plus interest and sell the commodity for the market price. Hence, the loan rate tends to place a floor under market prices except when PIK is released on the market to drive prices down. See *Loan rate*.

NORMAL CROP ACREAGE. The acreage on a farm normally devoted to a group of designated crops. When a set-aside program is in effect, a farm's total planted acreage of such designated crops plus set-aside acreage cannot exceed the normal crop acreage, if the farmer wants to participate in the commodity loan program or receive deficiency payments.

PAID DIVERSION PROGRAM (PDP). A voluntary short-term land retirement program whereby farmers are paid a prescribed rate (say $2 per bushel of wheat) times program yield to divert up to a prescribed percentage of base acreage.

PARITY PRICE. Originally, the price per bushel, pound, or bale that would be necessary for a bushel today to buy the same quantity of goods (from a standard list) that a bushel would have bought in the 1910 - 14 base period at the prices then prevailing. In 1948, the parity price formula was revised to make parity prices dependent on the relationship of farm and nonfarm prices during the most recent ten-year period for nonbasic commodities. Basic commodities, including wheat, corn, rice, peanuts, and cotton use the higher of the historical formula or the new formula.

The *parity index* is the index of prices paid by farmers for inputs and including interest, taxes, and wage rates. The *parity ratio* is the ratio of the index of prices received by farmers for all crops and livestock divided by the parity index, and the ratio is often expressed as a percent of the 1910 - 14 average.

PAYMENT-IN-KIND (PIK). Used by CCC in both export and domestic commodity programs, PIK certificates, expressed as a dollar value, may be redeemed either for commodities, or in some cases, for cash.

PROGRAM YIELD. The farm commodity yield of record determined by averaging the yield for the past five years, dropping the high and low years.

SET-ASIDE. An annual general land retirement program to limit production by restricting the use of land. Restricts the amount of farmer's total cropland base used for production rather than on the acres used to produce a specific crop. See *Normal crop acreage*.

TARGET PRICE. A price level established by law for wheat, feed grains, rice, and cotton. If the market price falls below the target price, an amount equal to the difference (but not more than the difference between the target price and the price-support loan levels) is paid in cash or in kind to farmers who participate in commodity programs. See *Deficiency payment*.

Index